T0180579

Studies in Computational Intelligence

Volume 551

Series editor

Janusz Kacprzyk, Polish Academy of Sciences, Warsaw, Poland
e-mail: kacprzyk@ibspan.waw.pl

For further volumes:
http://www.springer.com/series/7092

About this Series

The series "Studies in Computational Intelligence" (SCI) publishes new developments and advances in the various areas of computational intelligence—quickly and with a high quality. The intent is to cover the theory, applications, and design methods of computational intelligence, as embedded in the fields of engineering, computer science, physics and life sciences, as well as the methodologies behind them. The series contains monographs, lecture notes and edited volumes in computational intelligence spanning the areas of neural networks, connectionist systems, genetic algorithms, evolutionary computation, artificial intelligence, cellular automata, self-organizing systems, soft computing, fuzzy systems, and hybrid intelligent systems. Of particular value to both the contributors and the readership are the short publication timeframe and the worldwide distribution, which enable both wide and rapid dissemination of research output.

Janusz Sobecki · Veera Boonjing
Suphamit Chittayasothorn
Editors

Advanced Approaches to Intelligent Information and Database Systems

 Springer

Editors
Janusz Sobecki
Institute of Informatics
Wroclaw University of Technology
Wroclaw
Poland

Veera Boonjing
King Mongkut's Institute of Technology
Ladkrabang (KMITL)
Faculty of Science
Department of Computer Science
Bangkok
Thailand

Suphamit Chittayasothorn
King Mongkut's Institute of Technology
Ladkrabang (KMITL)
Faculty of Engineering
Department of Computer Engineering
Bangkok
Thailand

ISSN 1860-949X   ISSN 1860-9503   (electronic)
ISBN 978-3-319-38237-1   ISBN 978-3-319-05503-9   (eBook)
DOI 10.1007/978-3-319-05503-9
Springer Cham Heidelberg New York Dordrecht London

Printed on acid-free paper

Springer is part of Springer Science+Business Media (www.springer.com)

Preface

Intelligent Information and Database Systems have been known for over thirty years. They are applied in most of the areas of human activities, however they are also the subject of the scientific research. In this book quite deal of interesting issues related to the advanced methods in intelligent information models and methods as well as their advanced applications, database systems applications, data models and their analysis, and digital multimedia methods and applications are presented and discussed both from the practical and theoretical points of view. All the presented chapters address directly or indirectly to the concepts of intelligent information or database systems. These chapters are extended versions of the poster presentations at the 6th Asian Conference on Intelligent Information and Database Systems – ACCIDS 2014 (7–9 April 2014, Bangkok Thailand).

The content of the book has been divided into four parts: Intelligent systems models and methods, Intelligent systems advanced applications, Database systems methods and applications, Multimedia systems methods and applications. In the first part filtering and recommendation and classification methods as well as estimation and personalization models are presented. In the second part we can find several applications of intelligent systems in the following areas: scheduling, stock market and revenue forecasting, crime classification and investigation, e-learning and software requirement prioritization. The third part presents data models and their analysis together with advanced database systems applications, such as: ontology applied to information sharing and social network recommendation, opinion mining, simulation of supply management, agent-based systems modelling, usability verification, workflow scheduling and specialized chemical databases. The last part contains chapters that are dealing with different aspects of digital multimedia systems methods and their applications, i.e. positioning systems, image classification, ophthalmology measurements, vehicular systems modelling, edge detection in chromosome images, texture analysis, image based tracking and ciphering on mobile devices.

April 2014

Janusz Sobecki
Veera Boonjing
Suphamit Chittayasothorn

Contents

Part I: Intelligent Systems Models and Methods

Part II: Intelligent Systems Advanced Applications

Part III: Database Systems Applications, Data Models and Analysis

Part IV: Digital Multimedia Systems Methods and Applications

Part I
Intelligent Systems Models and Methods

Towards Intention, Contextual and Social Based Recommender System

Romain Picot-Clémente, Cécile Bothorel, and Philippe Lenca

Institut Mines-Telecom, Telecom Bretagne, UMR CNRS 6285 Lab-STICC, France
{romain.picotclemente,cecile.bothorel,
philippe.lenca}@telecom-bretagne.eu

Abstract. This article proposes a recommender system of shopping places for users in mobility with given intentions. It considers the user intention, her context and her location-based social network for providing recommendations using a method based on association rules mining. The context is considered at multiple levels: first in selecting the interesting rules following to the current session of visits of the user; then into a relevance measure containing a geographic measure that considers her current geographic position. Besides the ge-ographic measure, the relevance measure combines a social measure and an in-terest measure. The social measure takes into account the habits of the user's friends and the interest measure depends on the current intention of the user among predefined intention scenarios. An important aspect of this work is that we propose a framework allowing the personalization of the recommendations by acting only on the relevance measure. This proposition is tested on a real da-taset and we show that the use of social and geographic features beyond usage information can improve contextual recommendations.

Keywords: Location-based social networks, contextual recommender system, association rules, shopping recommendation.

1 Introduction

With the development of social networks and the widespread use of smartphones, users are sharing more and more contents in mobile situations. In Location-Based Social Networks (LBSNs) like Foursquare[1], sharing places with friends is even the finality. In this paper, we are interested in the recommendation of shopping places to users in mobility conditions, considering information from a LBSN. We consider the case where the information available on users is limited to their history of visits, their social network, their intention and their current context. This latter includes the current geographic position of the user and her current session of visits. We limit the intention of a user to one of the following intent scenarios: - "discovery" scenario where the user is not in a purchasing perspective; - "efficiency" scenario where the

---

[1] https://fr.foursquare.com/

J. Sobecki, V. Boonjing, and S. Chittayasothorn (eds.), *Advanced Approaches to Intelligent Information and Database Systems*, Studies in Computational Intelligence 551,
DOI: 10.1007/978-3-319-05503-9_1, © Springer International Publishing Switzerland 2014

user is visiting stores with a buying goal. These intent scenarios are targeted by a social shopping mobile application[2].

Many studies have focused on the recommendation of places. Recent works [1], [2] have shown that taking into account the social and geographical data could improve significantly the recommendations. However, despite the obvious influence of the user context on her everyday choices, none has focused (to our knowledge) on the use of contextual data to provide different kinds of recommendations of places that depend directly on the user's intention. Thus, we propose in this paper a recommender system based on association rules mining that can adapt to different user intentions and contexts, and that takes into account the visits of their friends (from the social network) and visits history to recommend personalized shopping places to users.

The remainder of this paper is as follows. The first part will address the related works of the literature, tackling place recommendation, contextual recommendation and association rules for making recommendation. The next section will present our method for making recommendations to a user, based on her context and intention, and her history of visit and friends' habits. Finally, we will present some experiments to show the benefits of using our solution for making recommendations.

2   Related Works from Literature

2.1   Recommendation of Places

Recommendation of places is a very active research field. Among the reference works, we can cite [3], [4] proposing recommender systems using collaborative filtering models to provide personalized recommendations of places to users. For the similarity computation, [3] proposes to calculate it only on users who are close in profile, for limiting the computation time. [4] shows that considering only friends for this computation gives equivalent results than methods relying on all users. The papers [1], [2] show that using both geographic information and social information beyond the user preferences can improve significantly the recommendations of places. In [5], the authors show users have temporal patterns of daily activities and usually visit places close to their home. They also show that the recommendation of categories of places gives better results than the recommendation of exact locations.

Although some place recommender systems have been proposed for LBSNs, none of them considers the context of the user (including her current position, session of visits, mood, etc…) and her intention in order to make contextual recommendation of places in mobility conditions. Then, the next section will discuss works of the literature that consider context to provide recommendations.

2.2   Contextual Recommendation

The main objective of a contextual recommender system is to provide relevant information to a user according to her current context. In the literature, many contextual

---

[2]   This work has been supported by the French FUI Project Pay2You Places, "labelled" by three Pôles de Compétitivité (French clusters).

recommender systems have been proposed. [6], [7] show the interest of including the user context into the collaborative filtering approach to improve the quality of recommendations. Although they take into account the context for the similarity computation between users, they are not designed to make recommendations to users according to their current context and intention, as we focus in this paper. The field of music recommendation is very active in considering the user current context to propose adapted music. In [8], the authors use the Case-Based Reasoning algorithm to provide music recommendations based on the user current context (season, weather, temperature, etc.). They show that the precision of recommendations can be improved by considering contextual data. More recently, [9] has proposed a music recommender system considering the current activity of the user (working, walking, shopping, etc.). The article shows that the use of context can provide good recommendations even in the absence of user ratings or annotations, a situation in which traditional systems only can recommend songs randomly.

All these works show a real relevance in considering the context into recommender systems. Nevertheless, to our knowledge, the problem of finding personalized geolocated items to users according to their current context and intention, and taking into account the social network and visits history has not been tackled in the literature.

2.3   Association Rules and Recommendation

Association rules were introduced by Agrawal et al. [10] for analyzing the food basket in order to discover relationships between the purchased items. The principle of the extraction of association rules is as follows. First, the system considers a set of transactions consisting of elements (called items). To extract association rules based on these transactions, the first step usually consists in extracting groups of items (itemsets) that are frequently present in transactions. A threshold on the occurrence frequency in transactions called "support threshold" is set and allows eliminating itemsets that are infrequent in the transactions (whose support is low). Given these itemsets, the second step consists in generating, for each itemset, every possible combination of association rules between the items from the itemset. An association rule r_i is of the form $r_i: ant(r_i) \rightarrow cons(r_i)$ where $ant(r_i)$ and $cons(r_i)$ are sets of items, respectively called antecedent and consequent of the rule. Generally, a measure is applied to each rule to rank them and / or restrict their number using a threshold. The confidence measure is the most used measure, it highlights the rules that are often realized when the antecedent of the rule is present in the transaction. However, many other measures exist (surprise, lift, conviction, etc.) and their choice is highly dependent on the rule extraction objective [11]. In our context, controlling these measures allows us to take into account the visiting intention of shops (discovery, efficiency, rarity, popularity, friendly, etc.).

The methods for extracting association rules have often been successfully used in real-time recommender systems of web pages [12], [13], [14]. In [15], the authors propose a recommender system based on geo-located pictures for recommending itineraries of places to users according to the current picture location they just took.

In these works, the general principle is as follows. An offline component generates the association rules based on usage histories and an online component is in charge of providing real-time recommendations to users based on matching association rules with the current usages of users. In [16], the authors propose a dating recommender system by providing a list of potentially interesting men or women. They show their recommender system provides significantly better recommendations than traditional collaborative filtering systems.

These works show that association rules mining is well suited for real-time recommendation applications, which is one of our needed features for the system. Moreover, another feature that guided us to the use of association rules mining is that it is easily scalable on large datasets [17], a frequent case when dealing with LBSNs. Finally, the last feature is that the principle of selecting and classifying rules according to a relevance measure depending on the extraction objective is particularly suited to our application case since the adaptation to different intentions could be done by acting directly and in a real-time fashion on this measure.

3   Intention-Oriented, Social-Based and Contextual Recommender System of Places

3.1   Constraints and Assumptions

We detail here the different constraints and assumptions of the application case that have brought the proposition described in the following part. First, as explained in the introduction, the user will have the choice between these two intention scenarios:

— "discovery" scenario: While the user is walking, proposals for shops with offers are made to "feed her curiosity". In this scenario, the idea is to lead the user to make impulsive purchases in shops that would not have been visited without these recommendations, but that are related to the visits she made during her walk.
— "efficiency" scenario: The user is walking with the perspective of buying something. While she is visiting some shops, recommendations of other shops are done according to her previous visits with the goal to bring her to an "efficient" shop. We try to guide the user to the most relevant shop for her future purchase.

Then, relying on [18], [19], [20], [21] showing that the check-in probability of users according to the distance from their other check-ins follows generally a power law function and relying on [22], [23] showing that users tend to share common interests, we admit the two following assumptions are true for our shopping application case:

— Assumption 1: The shops that have been visited by friends of a given user are most likely to interest him than other shops.
— Assumption 2: The probability that a user would go to a shop at a given distance from her current location follows a power law function. Farther the shop is from the current position of the user, less he would tend to get there.

3.2   Proposition

Our approach in this article is based on the use of association rules to offer contextual and intention oriented recommendations of shops to users. It consists in two steps: the generation of the association rules model, and the generation of the recommendations for a given user. Fig. 1 shows an overview of these steps.

Step 1: Generation of the association rules model.
This first step consists in generating the association rules model by extracting association rules from the history of visits of all users according to the method presented in section 2.3, except that no measure for ordering or selecting rules is applied at this level. This step is performed periodically, every day or week, to take into account the evolution of users behavior. Several methods can be used to extract association rules. We have chosen the FP-Growth algorithm [24] which is one of the most efficient in terms of computation time. The association rules model is denoted R and a rule $r_i \in R$ is composed of an antecedent $ant(r_i)$ and a consequent $cons(r_i)$ in the next step.

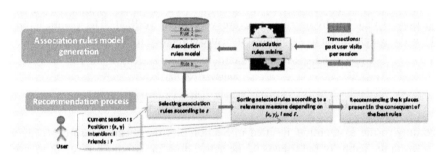

Fig. 1. Overview of the association rules model generation and the recommendation process

Step 2: Generation of recommendations according to the user context.
This second step focuses on the recommendation process which is triggered regularly, or if the user asked for it, to take into account changing contexts and intentions.

Let u_a be the active user and $m_j \in M$ be a shop where $M = \{m_1, m_2, ..., m_l\}$ is the set of all known shops in the system. The active session of u_a is $sa_t = (m_{u_a1}, m_{u_a2}, ..., m_{u_at})$ and consists in her last t visited shops.

First, the system seeks all the rules from R that have a subset of sa_t as antecedent. To do this, all the subsets of sa_t denoted ss_{sa_t} are generated, then for each subset ss_k, the system seeks if it is equal to the antecedent of a rule from R. The resulting set of rules, corresponding to the active session sa_t of the user is denoted R_t. A reduction space phase is then launched on R_t. In this phase, the system removes the rules $r_i \in R_t$ that verify the following conditions:

— $cons(r_i) \in sa_t$.
— $cons(r_i) \in hist_{u_a}$ AND $scenario = $ "discovery"

$hist_{u_a} \subset M$ is the history of shops' visits of the user u_a (the whole set of previous visits). The first condition avoids providing places that the user has visited in her current session, in which case the recommendation would be completely useless. The second condition avoids providing places that the user already knows, if he is in the "discovery" scenario because her intention is precisely to discover new places.

The objective is to evaluate the score of each m_j that is consequent to a rule of R_t. In the literature, the score assigned to a shop is either the sum of the confidence of each rule associated to this shop, either the maximum of these confidences [25]. In our approach, the given score is not based on the confidence but on a personalized measure, called relevance measure, depending on the intention, the geographic position, and the friends of the user. The relevance measure denoted M_p is applied to each rule from R_t and combines three measures: an interest measure M_i, a social measure M_s and a geographic measure M_g, as follows:

$$M_p(r_i) = M_g(r_i)\big(\alpha M_i(r_i) + (1 - \alpha)\, M_s(r_i)\big) \tag{1}$$

The interest measure M_i depends on the intention scenario of the user. If the scenario is "discovery", then M_i is the surprise measure (Equation 2) that highlights rules that are not obvious. If the scenario is "efficiency", then M_i is the confidence measure (Equation 3) that highlights rules that are very often verified. The parameter $\alpha \in [0,1]$ is the weight given to the interest measure compared to the social measure.

$$suprise(r_i) = \frac{support(ant(r_i) \wedge cons(r_i)) - support(ant(r_i) \wedge \neg cons(r_i))}{support(cons(r_i))} \tag{2}$$

$$confidence(r_i) = \frac{support(ant(r_i) \wedge cons(r_i))}{support(ant(r_i))} \tag{3}$$

According to the assumption 1 stated in Section 3.1, the social measure M_s allows considering the habits of the friends of the active user for assigning a weight to a rule depending on whether this rule is often or rarely verified for the friends:

$$M_s(r_i) = confidence_{friends}(r_i) \tag{4}$$

According to the assumption 2, the geographic measure M_g allows considering the position of the user and assigns a weight to a rule r_i depending on the distance between the consequent shop $cons(r_i)$ and the user position pos_{u_a}. This measure follows a power law function and is defined such that:

$$M_g(r_i) = a \times d\big(cons(r_i), pos_{u_a}\big)^b \tag{5}$$

a and b are the parameters of the power law function that fits the density of probability that the user will visit a shop according to its distance to her previous visits $m_i \in hist_{u_a}$. These parameters are computed periodically for each user together with the calculation of the model to not slow down the recommendation process.

The relevance measure is calculated on every rule from R_t. Each shop m_j that is consequent of these rules is assigned a score $S_t(m_j)$ which is the maximum of the relevance values of the rules whose it is the consequent. We denote R_{t,m_j} the set of rules whose m_j is the consequent.

$$S_t(m_j) = \max\left(M_p(r_i)\middle|\forall r_i \in R_{t,m_j}\right) \tag{6}$$

Finally, the shops are sorted in ascending order of relevance and the k shops with the highest scores are recommended to the user.

4   Evaluation of the Method

This section proposes some experiments to show the interest of our method. In the literature, there are several measures to evaluate a recommender system [26] [27]. As many studies, we choose to evaluate it on a Gowalla[3] dataset by measuring how the system predicts a future usage of the user, using the recall and precision measures. However, the act of measuring the quality of recommendation by comparing the best predictions with real usages in an existing dataset implies that we can only measure the quality of recommending methods that have a predictive ability. Thus, we will be able to only evaluate the method on the "efficiency" intention scenario and not the method for the "discovery" scenario, because this latter is supposed to produce unusual recommendations. Gowalla dataset actually does not reflect visits of users exposed to this kind of recommendations. In other words, the recommendations for the "discovery" scenario do not have a predictive ability unlike to the recommendations for the "efficiency" scenario. Measuring the quality of the method in the "discovery" intention scenario will be the subject of future works when user feedbacks in real condition of use will be possible on our project. We will then process user-centric methods [28].

The Gowalla dataset for the evaluating the method in the "efficiency" intention scenario consists in a dataset of 196591 users, 950327 friendship relations and 6442890 check-ins of these users in 1279228 different places. The period of check-ins goes from Februray 2009 to October 2010. Gowalla was a LBSN which closed in March 2012 after being acquired by Facebook in late 2011. This dataset has been chosen because it is publicly available and often used in research papers. To our knowledge, no LBSN dataset dedicated to shopping is available.

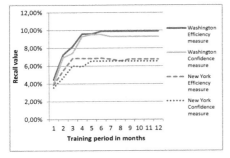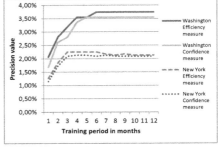

Fig. 2. Recall and precision values for New York and Washington based on the efficiency scenario and the confidence measure

---

[3]   Available on http://snap.stanford.edu/data/#locnet

The evaluation method we are proposing is as follows. First we limit the tests to the cities of New York and Washington. For each city, an active period is arbitrary chosen (2010-05-01 to 2010-05-10 for Washington, and 2010-08-01 to 2010-08-10 for New York). Then, the association rules model is constructed for each city according to the visits that occurred before the active period. Support thresholds, eliminating rare items, have been defined by hand to limit the computation time of the association rules due to time constraints for these tests (support threshold of 4 for Washington and 7 for New York). The training period for mining association rules will vary in order to see how the number of rules (and the age) influences the quality of recommendations. Then, each session of one day of visits for each user is taken in the active period (107 sessions for Washington and 489 sessions for New York). The average number of visits in the sessions is about 5 visits for Washington, and 6 visits for New York. Each day of each user is split into two subset of visits: an *active session* and a *test session* with a 2/3 ratio. The active session is used to compute the recommendations and the test session is used to measure the quality of recommendations. The last visited place of the active session defines the current position of the user. According to the method described in Section 3, 5 recommendations with the "efficiency" intention scenario are generated. The parameters a and b of Equation 5 are calculated using the gradient method, once for each considered user [29]. The parameter α in the relevance measure (see Equation 1) is first set to 0.5. For comparison, five recommendations are provided by replacing our proposed relevance measure M_p with the conventional confidence measure. The comparisons are done based on the *recall* and *precision* measures according to the real visits in each test session.

Fig. 2 presents the recall and precision values for the "efficiency" intention scenario (called efficiency measure on the figure) and the confidence measure, for each city. These values are presented for each training period of the rules model varying from 1 to 12 months before to the active period. These results show that the method using our relevance measure based on the "efficiency" scenario gives at least as good results (it is generally better) as the method using only the confidence measure. This shows that taking into account the social and geographic data for contextual recommendation (with a predictive objective) is relevant for improving the quality of recommendation based on the recall and precision measures. In addition, they show that more the considered period for training the rules is important (i.e. more the number of rules is important given the fixed support threshold), better the recommendations based on our method are, with a ceiling of about five months (that can vary according to the data).

Beyond these experiments, we looked at what would be a measure of relevance taking into account only the interest and the geographic measures ($\alpha = 1$). Results show that even if we take only into account the geographic criteria beyond the use of the confidence measure, it gives better results than using the simple confidence measure and gives worse results than taking into account social and geographic criteria. Thus, this shows clearly that each criterion (geographic and social) improves the recommendations compared to a simple confidence measure.

The reader may notice the low quality of both precision and recall, but one has to keep in mind that only 5 recommendations are triggered by only considering 2 or 3 visits (the context is really short, which is not the case in musical listening sessions). When we consider the history (no context, or a context that would last several months), 5 recommendations lead to a recall of 30% for Paris Gowalla subset but is only of 6% in a place like San Francisco [2].

Despite these interesting results, we cannot deduce that it would be identical for the final application based on the recommendation of shops. It will be necessary to later evaluate our method on a real use case of the system, especially for testing the "discovery" intention scenario. Nevertheless, we are proposing in this paper a new methodology for providing contextual and intention-based recommendations to a user considering habits of her friends. Future works will consists in showing that contextual data are really relevant for this recommendation type, especially by comparing with classic recommendation methods that do not consider the user position, her session of visits and her intention.

5   Conclusion

This article presents a contextual recommender system for visiting shops based on a location-based social network. We are proposing a methodology based on extracting association rules for generating recommendations that take into account the user context, her shopping visit intention and the habits of her friends. The context consists in the current user location and her current session of visits. It is first considered in the selection of rules which is done according to the visits in the active session of the user. Then, the context together with the user intention and friends habits is considered in the computation of a relevance measure which combines a geographic measure, an interest measure and a social measure. Thus, the relevance measure allows assigning a relevance score to an association rule depending on the current position of the user, the habits of her friends and an intention scenario. An important aspect of this work is that our proposition allows personalizing the recommendations on acting only on the relevance measure by changing the interest measure according to the user intention and by modifying the weights given to the social and the interest measures.

Our few benchmarks show that taking into account social and geographic criteria for making contextual recommendations can improve the quality of recommendation. Nevertheless, we tested our method on a Gowalla dataset that is not focused on visiting shopping places. This induces that we cannot conclude it will be the same for the future application that will be based on check-ins in shopping places. The context-based recommendation problem remains a difficult problem since it is necessary to generate few recommendations in real-time and on the basis of a context of reduced effective things.

References

1. Ye, M., Yin, P., Lee, W.-C., Lee, D.-L.: Exploiting geographical influence for collaborative point-of-interest recommendation. In: ACM SIGIR Conference on Research and Development in Information Retrieval, NY, USA (2011)

2. Picot-Clemente, R., Bothorel, C.: Recommendation of shopping places based on social and geographical influences. In: 5th ACM RecSys Workshop on Recommender Systems and the Social Web, Hong Kong (2013)
3. Horozov, T., Narasimhan, N., Vasudevan, V.: Using location for personalized poi recommendations in mobile environments. In: SAINT, pp. 124–129 (2006)
4. Ye, M., Yin, P., Lee, W.: Location recommendation for location-based social networks. In: SIGSPATAIL, pp. 458–461 (2010)
5. Rahimi, S.M., Wang, X.: Location Recommendation Based on Periodicity of Human Activities and Location Categories. In: Pei, J., Tseng, V.S., Cao, L., Motoda, H., Xu, G. (eds.) PAKDD 2013, Part II. LNCS, vol. 7819, pp. 377–389. Springer, Heidelberg (2013)
6. Oku, K., Nakajima, S., Miyazaki, J., Uemura, S.: Context-Aware SVM for Context-Dependent Information Recommendation. In: Mobile Data Management (2006)
7. Ahn, H., Kim, K., Han, I.: Mobile Advertisement Recommender System using Collaborative Filtering: MAR-CF. In: Proceeding of the Korea Society of Management Information Systems (2006)
8. Lee, J.S., Lee, J.C.: Context Awareness by Case-Based Reasoning in a Music Recommendation System. In: Ichikawa, H., Cho, W.-D., Satoh, I., Youn, H.Y. (eds.) UCS 2007. LNCS, vol. 4836, pp. 45–58. Springer, Heidelberg (2007)
9. Wang, X., Rosenblum, D., Wang, Y.: Context-aware mobile music recommendation for daily activities. In: MM 2012 Proceedings of the 20th ACM International Conference on Multimedia (2012)
10. Agrawal, R., Imielinski, T., Swami, A.: Mining Association Rules Between sets of Items in Large Databases. In: Proceedings of the ACM SIGMOD Conference on Management od Data, pp. 207–216 (1993)
11. Lenca, P., Meyer, P., Vaillant, B., Lallich, S.: On selecting interestingness measures for association rules: User oriented description and multiple criteria decision aid. European Journal of Operational Research, 610–626 (2008)
12. Mobasher, B., Dai, H., Luo, T., Nakagawa, M.: Effective personalization based on association rule discovery from web usage data. In: Proceedings of the 3rd ACM Workshop on Web Information and Data Management (WIDM 2001), Atlanta, Georgia (2001)
13. Mulvenna, M., Anand, S., Buchner, A.: Personalization on the net using Web mining. Communications of the ACM 43(8), 123–125 (2000)
14. Forsati, R., Meybodi, M.R., Ghari Neiat, A.: Web Page Personalization based on Weighted Association Rules. In: International Conference on Electronics Computer Technology, pp. 130–135 (2009)
15. Tai, C., Lin, L., Chen, M.: Recommending personalized scenic itinerary with geo-tagged photos. In: Multimedia and Expo, pp. 1209–1212 (2008)
16. Özseyhan, C., Badur, B., Darcan, O.: An Association Rule-Based Recommendation Engine for an Online Dating Site. In: Communications of the IBIMA, pp. 1–15 (2012)
17. Yang, X., Liu, Z., Fu, Y.: MapReduce as a programming model for association rules algorithm on Hadoop. In: IEEE (ed.) 2010 3rd International Conference on Information Sciences and Interaction Sciences (ICIS), pp. 99–102 (2010)
18. Brockmann, D., Hufnagel, L., Geisel, T.: The scaling laws of human travel. Nature 439, 462–465 (2006)
19. Couldry, N.: Mediaspace: Place, scale and culture in a media age. Routledge (2004)
20. Jiang, B., Yin, J., Zhao, S.: Characterizing human mobility patterns in a large street network. arXiv preprint arXiv:0809.5001 (2008)
21. Noulas, A., Scellato, S., Mascolo, C., Pontil, M.: An empirical study of geographic user activity patterns in foursquare. In: ICWSM (2011)

22. Jamali, M., Ester, M.: A matrix factorization technique with trust propagation for recommendation in social networks. In: 4th ACM RecSys Conf., pp. 135–142 (2010)
23. Konstas, I., Stathopoulos, V., Jose, J.: On social networks and collaborative recommendation. In: ACM SIGIR Conference on Research and Development in Information Retrieval (2009)
24. Han, J., Pei, H., Yin, Y.: Mining Frequent Patterns without Candidate Generation. In: Press, A. (ed.) Conf. on the Management of Data (SIGMOD 2000), Dallas, TX, New York (2000)
25. Brun, A., Boyer, A.: Towards Privacy Compliant and Anytime Recommender Systems. In: Di Noia, T., Buccafurri, F. (eds.) EC-Web 2009. LNCS, vol. 5692, pp. 276–287. Springer, Heidelberg (2009)
26. Herlocker, J., Konstan, J., Terveen, L., Riedl, J.: Evaluating collaborative filtering recommender systems. ACM Transactions on Information Systems (TOIS) 22, 5–53 (2004)
27. Shani, G., Gunawardana, A.: Evaluating Recommendation Systems. Technical Report, Microsoft Research (2009)
28. Pu, P., Chen, L., Hu, R.: A user-centric evaluation framework for recommender systems. In: Proceedings of the Fifth ACM Conference on Recommender Systems, pp. 157–164 (2011)
29. Bottou, L.: Large-Scale Machine Learning with Stochastic Gradient Descent. In: Lechevallier, Y., Saporta, G. (eds.) International Conference on Computational Statistics (COMPSTAT 2010), pp. 177–187 (2010)

Improved Negative Selection Algorithm with Application to Email Spam Detection

Ismaila Idris[1] and Ali Selamat[2]

[1] Software Engineering Department, Faculty of Computing,
Universiti Tecknologi Malaysia
Ismi_idris@yahoo.co.uk
[2] Software Engineering Department, Faculty of Computing,
Universiti Tecknologi Malaysia
aselamat@utm.my

Abstract. Email spam is an unsolicited message sent into the mail box; it requires the need of an adaptive spam detection model to eliminate these unwanted messages. The random detector generation in negative selection algorithm (NSA) is improved upon by the implementation of particle swarm optimization (PSO) to generate detectors. In the process of detector generation, a fitness function that calculates the reach-ability distance between the non-spam space and the candidate detector, and use the result of the reach-ability distance to calculate the local density factor among the candidate detector was also implemented. The result of the experiment shows an enhancement of the improved NSA over the standard NSA.

Keywords: Negative selection algorithm, Particle swarm optimization, Local outlier factor, Spam, Non-spam.

1 Introduction

Different techniques as been proposed in dealing with unsolicited email spam; the very first step in tackling spam is to detect spam email, this brought about the constant development of spam detection models. The quest for an adaptive algorithm is the reason negative selection algorithm is proposed. Most work on negative selection algorithm (NSA) and particle swarm optimization (PSO) solves problems of anomaly detection and intrusion detection. The implementation of particle swarm optimization with negative selection algorithm to maximize the coverage of the non-self space was proposed by [1] to solve problem in anomaly detection. The research of [2] focuses on non-overlapping detectors with fixed sizes to achieve maximal coverage of non-self space; this is initiated after the random generation of detectors by negative selection algorithm. No research as implement particle swarm optimization to generate detector in negative selection algorithm. This paper will replace the self in the system as non-spam while the non-self in the system will be referred to as spam.

In this paper, we propose a new algorithm for detector generation with the aim of upgrading the existing random detector generation in negative selection algorithm.

J. Sobecki, V. Boonjing, and S. Chittayasothorn (eds.), *Advanced Approaches to Intelligent Information and Database Systems*, Studies in Computational Intelligence 551,
DOI: 10.1007/978-3-319-05503-9_2, © Springer International Publishing Switzerland 2014

The proposed algorithm yields two advantages namely, increasing the accuracy and secondly helping to provide a viable detector that can be applied in the learning of negative selection algorithm. The local outlier factor is also implemented as fitness function to calculate both the reach-ability distance and the local density factor of the candidate detector. The rest of the paper is structured as follows: Section 2 discusses the related work in negative selection algorithm and spam detection, the methodology is presented in Section 3 while section 4 explains the experimental setup and section 5 presents results and conclusion.

2   Related Work

Many researchers have done useful work on detector generation in negative selection algorithm, though, not with respect to spam detection [3]. The work on the biological immune system with regard to spam detection is proposed by [4], it uses pattern matching in representing detectors as regular expression in the analysis of message. A weight is assigned to detector which was decremented or Incremented when observing expression in spam message with the classification of the message based on threshold sum of the weight of matching detectors. The system is meant to be corrected by either increasing or decreasing of all matching detector weight with 1000 detector generated from spam-assassin heuristic and personal corpus. The results were acceptable base on few number of detectors used. A comparison of two techniques to determine message classification using spam-assassin corpus with 100 detectors was also proposed by [5]. This approach is like the previous techniques but the difference is the increment of weight where there is recognition of pattern in spam messages. Random generation of detector those not help in solving problem of best selected features; though, feature weights are updated during and after the matching process of the generated detectors. The weighting of features complicates the performance of the matching process. In conclusion, the present techniques are better than the previous due to its classification accuracy and slightly improved false positive rate. More experiment was performed by [6] with the use of spam-assassin corpus and Bayesian combination of the detector weight. Massages were scored by simple sum of the message matched by each non-spam in the detector space and also the use of Bayes scores.

A genetic optimized spam detection using AIS algorithm was proposed by [7, 8]. The genetic algorithm optimized AIS to cull old lymphocytes (Replacing the old lymphocyte with new ones) and also check for new interest for users in a way that is similar. In updating intervals such as the number of received messages, the interval is updated with respect to time, user request and so on, many choices were used in selecting the update intervals which was the aim of using the genetic algorithm. The experiment was implemented with spam-assassin corpus with 4147 non-spam message and 1764 spam message. A swam intelligent optimization algorithm was proposed by [9]. The work optimized parameter vector with swarm intelligent algorithm which is made up of parameter of classifier and that of feature extraction approach which considers the spam detection problem as a optimization process. The experiment was conducted with the use of support vector machine as its classifier, fireworks algorithm as its swarm intelligence algorithm and the local concentration method as its feature extraction approach.

3   Methodology

3.1   The Original Negative Selection algorithm

Negative selection algorithm (NSA) has been used successfully for a broad range of applications in the construction of artificial immune system [10]. The standard algorithm was proposed by Forest et al [11]. The algorithm comprises of the data representation, the training phase and the testing phase. In the data representation phase, data are represented in a binary or in a real valued representation. The training phase of the algorithm or the detector generation phase randomly generate detector with binary or real valued data and then consequently use the generated detector to train the algorithm [12]; while the testing phase evaluate the trained algorithm. The random generation of detectors by a negative selection algorithm makes it impossible to analyze the type of data needed for the training algorithm. Figures 1 and 2 show the training and testing phase of NSA.

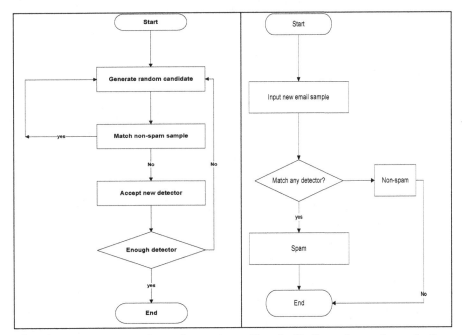

Fig. 1. Detector generation of negative selection algorithm

Fig. 2. Testing of negative selection algorithm

The main concept of the NSA as developed by Forrest et al. [11] was meant to generate a set of candidate detectors, C, such that $\forall\, x_i \in C$ and $\forall\, z_p \in S$, $f_{MATCH}(x_i, z_p) < r$, where x_i is a detector, z_p is a pattern and $f_{MATCH}(x_i, z_p)$ is the affinity matching function. The algorithm of NSA as given by Forrest et al [11] is presented in figure 3.

3.2   Generation of Detector with Particle Swarm Optimization

Particle are made up of 57 features $\{f_{57}\}$ while the accelerated constant C value is 0.5.

The Position and velocity of particle swarm optimization are represented in the N-dimensional space as:

$$P_i\ (p_{i,}^1\ p_{i,...}^2 p_i^n) \tag{1}$$

$$V_i\ (v_{i,}^1\ v_{i,...}^2 v_i^n) \tag{2}$$

Where p_{id} is the binary bits, $i = 1,2\ldots m$ (m is set to be the total number of particles), $d = 1,2\ldots n$ (n is the dimensionality of the data). Each particle in the generation updates its own position and velocity base on equation (19) and (20).

In this scenario we generate random candidate detector as follows.

$$r = (r_1 \ldots r_n) = \begin{bmatrix} r_{11} & \cdots & r_{1m} \\ \vdots & \ddots & \vdots \\ r_{k1} & \cdots & r_{km} \end{bmatrix} \tag{3}$$

$r_{ij} \in (0,1)^m,\ i = 1\ldots,k;\ j = 1,\ldots,m$

r_i is said to be the ith detector and r_{ij} is the jth feature of the ith detector.

In generating a random initial velocity matrix for random candidate detector we have $v\,(0)$.

$$v\,(0) = \begin{bmatrix} v_{1\,1}(0) & \cdots & v_{1_m}(0) \\ \vdots & \ddots & \vdots \\ v_{j\,1}(0) & \cdots & v_{j_m}(0) \end{bmatrix} \tag{4}$$

Equation (16) and (17) calculate the new velocity and particle position

$$v_{id}\,(t+1) = v_{id}(t) + c(Pbest_{id}(t) - x_{id}(t)) \tag{5}$$

$$x_{id}(t+1) = x_{id}(t) + v_{id}(t+1) \tag{6}$$

//*pb* is the local best
//*gb* is the global best
//*LOF* is the local outlier factor
//*p* is the population size
 Initialize all particles
 Initialize
 Repeat
 For each particle *i in p do*
 Compute the fitness function as shown in   equation (11) and (12)
   with   LOF
 If fitness value better than best fitness x_i
 //Update each particle best position
   If $f(x_i) < f(pb_i)$ *then*
 $pb_i = x_i$
End if
 //Since we need pbest as the optimal   solution
 //Update local best position (pb_i)
   If $f(pb_i) = f(gb)$ *then*
 $pb_i = gb$
  End if
  End for
 //Update particle velocity and position for each particle *i in p do*
  For each dimensional *d in D do*
 $v_{i,d} = v_{i,d} + C_1 * Rnd(0,1) * [pb_{i,d} - x_{i,d}] + C_2 * Rnd(0,1) *$
 $[pb_d - x_{i,d}]$
 $x_{i,d} = x_{i,d} + v_{i,d}$
  End for
  End for
 While maximum iteration or stop criteria reached.

Fig. 3. Particle swarm optimization algorithm

3.3   Implementation Model

Each of the particles was initialized at a random position in the search space. The position of particle *i* is given by the vector:

$$x_i = (x_{i1}, x_{i2}, ... , x_{iD})$$ (7)

Where *D* is the problem dimensionality with velocity given by the vector:

$$v_i = (v_{i1}, v_{i2}, ... , v_{iD})$$ 8)

The movement of the particles was influenced by an implemented memory: In the cognitive memory.

$$p_i = (p_{i1}, p_{i2}, ..., p_{iD})$$ (9)

The best previous position visited by each individual particle i is stored.

$$p_{best} = (p_{best1}, p_{best2}, \dots \dots \dots \dots p_{bestD}) \tag{10}$$

The vector in equation (24) contains the position of the best point in the search space visited by all the particles. At each iteration, each pbest is use to compute the density of the local neighborhood.

$$lrd\ (i) = 1 / \left(\frac{\sum_{s \in N_k(i)} reachability - distance_k\ (i,s)}{|N_K(i)|} \right) \tag{11}$$

After which the local reach-ability density is compared with those of its neighbour reach-ability distance.

$$LOF_K\ (i) = \frac{\sum_{s \in N_k(i)} \frac{lrd(s)}{lrd(i)}}{|N_k(i)|} = \frac{\sum_{s \in N_k(i)} lrd\ (s)}{|N_K(i)|} / lrd\ (i) \tag{12}$$

Given each particle a degree of being an outlier each iteration of pbest velocity were updated according to equation (27)

$$v_i(t+1) = w.v_i(t) + n_1 r_1 (p_i - x_i(t)) + n_2 r_2 (p_{best} - x_i(t)) \tag{13}$$

Where w is the local outlier factor for each particle of the velocity, n_1 and n_2 are positive constants called "cognitive" and "social" parameters that implements the local outlier factor of two different swarm memories and r_1 and r_2 are random number between 0 and 1. The proposed procedures those not require the swarm to perform a more global search with large movement, it only require a small movement and fine tuning in the end of the optimization process. After calculating the velocity vector, the position of the particles was updated based on equation (28).

$$x_i(t+1) = x_i(t) + v_i(t+1) \tag{14}$$

We then employ a maximum number of iteration as termination condition for the algorithm based on the task.

3.4   Fitness Function Computation

The technique will model the data point with the use of a stochastic distribution [13] and the point is determined to be an outlier base on its relationship with the model.

- Let's assume $k$distance (i) be the distance of the candidate detector or particle (i) to the nearest neighborhood (non-spam).
- Set of k nearest neighbor (non-spam element) includes all particles at this distance.
- Set S of k nearest neighbor is denoted as $N_k(i)$.
- This distance defines the reach-ability distance.
- Reach-ability-distance$_k$ $(i, s) = \max\{k - distance(s), d(i, s)\}$

The local reach-ability distance is then defined as:

$$\text{lrd}(i) = 1/\left(\frac{\sum s \in N_k(i) reachability-distance_k\ (i,s)}{|N_K(i)|}\right) \tag{15}$$

Equation (15) is the quotient of the average reach-ability distance of the candidate detector i from the non-spam element. It is not the average reach-ability of the neighbour from i but the distance from which it can be reached from its neighbor. We then compare the local reach-ability density with those of its neighbor using the equation below:

$$\text{LOF}_K\ (i) = \frac{\sum s \in N_k(i)\frac{lrd(s)}{lrd(i)}}{|N_k(i)|} = \frac{\sum s \in N_k(i)lrd\ (s)}{|N_K(i)|}/lrd(i) \tag{16}$$

Equation (16) shows the average local reach-ability density of the neighbour divided by the particle own local reach-ability density. In this scenario, values of particle approximately 1 indicates that the particle is comparable to its neighbour (not an outlier), value below 1 indicates a dense region (which will be an inlier) while value larger than 1 indicates an outlier. The major idea of this technique is to assign to each particle degree of being an outlier. The degree is called the local outlier factor (LOF) of the particle. The methodology for the computation of LOF's for all particles is explained in the following steps:

Let's assume G as the population of particles, S is the non-spam space and i is the ith particle in G

$$\text{For each particle } i \text{ we have } i \in G.\ \text{Max (k-dist.}(s) \tag{17}$$
$$/\ \text{Reach-dist } G/^*\ \max\ (dist(s,i)) \tag{18}$$
$$|G|^*\ (Minpt\ (s,i)) \tag{19}$$
$$|G|^*\ (similarity\ (i,G) \tag{20}$$

3.5   Computing the Generated Detector in the Spam Space

Let's assume the non-spam space to be S
S is defined as follows:

$$s = (s_1 ... s_n) = \begin{bmatrix} s_{11} & \cdots & s_{1m} \\ \vdots & \ddots & \vdots \\ s_{n1} & \cdots & s_{nm} \end{bmatrix}$$

$S_{ij} \in K^m,\ i = 1, \cdots, n;\ j = 1, \cdots, m$
S is normalized as follows:

$$S_i = \frac{S_i}{\|S_i\|} \tag{21}$$

From equation (21) of the normalized non-spam space, the non-spam space is represented in equation (22) with radius Rs as:

$$S = \{X_i | i = 1, 2, \cdots m; Rs = r\} \tag{22}$$

X_i is some point in the normalized N-dimensional space of both velocity and position as defined in equation (1) and (2)

$$X_i = \{x_{i1}, x_{i2}, x_{i3} \cdots x_{iN}\}, i = 1, 2, 3 \cdots m \tag{23}$$

All the normalized sample $space^{I \subset [0,1]^N}$, the spam space can then be represented as S = I-NS. Where S is spam and NS is non-spam.

$$d_j = (C_j, R^d j) \tag{24}$$

Equation (24) denote one detector where $C_j = \{C_{j1}, C_{j2}, C_{j3} \cdots C_{jN}\}$ is the detector center respectively, R_j is the detector radius. The Euclidean distance was used as the matching measurement. The distance between non-spam sample X_i and the detector d_j can be defined as:

$$L(X_i, d_j) = \sqrt{(x_{i1} - C_{j1})^2 + \cdots + (x_{iN} - C_{jN})^2} \tag{25}$$

$L(X_i, d_j)$ is compared with the non-spam space threshold Rs, obtaining the match value of $\ltimes$

$$\ltimes = L(X_i, d_j) - Rs \tag{26}$$

The detector d_j fails to match the non-spam sample X_i if $\ltimes > 0$, therefore if d_j does not match any non-spam sample, it will be retained in the detector set. The detector threshold R^d, j of detector d_j can be defined as:

$$R^d, j = \min (\ltimes), \text{ if } \ltimes \le 0 \tag{27}$$

If detector d_j match the non-spam sample, it will be discarded. This will not stop the generation of detector until the required detector set is reached and the required spam space coverage attained. After the generation of detectors in the spam space, the generated detectors can then monitor the status of the system. If some other new email (test) samples matches at least one of the detectors in the system, it is assume to be spam which is abnormal to the system but if the new email (test) sample does not match any of the generated detectors in the spam space, it is assume to be a non-spam email.

4   Experimental Setup

The corpus bench mark is obtained from spam base dataset which is an acquisition from email spam message. In acquiring this email spam message, it is made up of 4601 messages and 1813 (39%) of the message are marked to be spam messages and 2788 (61%) are identified as non-spam and was acquired by [14]. The evaluation of the NSA model and the proposed improved model was implemented by the division of the dataset using a stratified sample approach with 70% training set and 30% testing set to investigate the performance of the new model on an unseen data. The training set was used in the construction of the model by training the dataset on both models while evaluating the capability of the model with the testing set. The process of implementation did not use any ready-made code and all functions needed are coded using Delphi 5 platform. The evaluation of both NSA model and its improved model are implemented with a threshold value of between 0.1 and 1 while the number of generated detector is between 100 and 8000. The different threshold value and number of detector generated as tremendous impact on the final output measure. The criteria used to measure the performance evaluation are Sensitivity (SN), Specificity (SP), Positive prediction value (PPV), Accuracy (ACC), Negative prediction value (NPV), Correlation coefficient (CC) and F-measure (F1).

5   Results Discussion and Conclusion

The testing results for 4601 dataset with 8000 generated detectors and threshold value of 0.1 to 1; gives summary and comparison of results in percentage for NSA and NSA-PSO model.

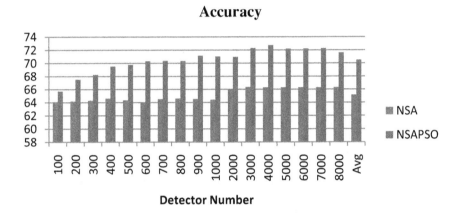

Fig. 4. Accuracy of NSA and NSA-PSO

From the plot presented for accuracy, it could be easily observed that accuracy with threshold value of 0.4 at 5000 detector generation is best with negative selection algorithm (NSA) model at 68.86% and the proposed improved negative selection algorithm-particle swarm optimization (NSA-PSO) model at 83.20%. Other evaluation criteria are as thus: with threshold value of 0.4 at 5000 generated detectors, correlation coefficient (MCC) of NSA model is 36.07% while NSA-PSO model is 64.40%. F-measure (F1) is also best at 5000 generated detector and threshold value of 0.4 with NSA at 36.01% and the proposed hybrid NSA-PSO at 76.85%. Sensitivity (SN) is best at 7000 generated detector and threshold value of 0.3 with NSA model at 37.87% and the proposed hybrid NSA-PSO model at 75.37%.

5.1   Conclusion

The detector generation phase of NSA determines how robust and effective the algorithm will perform. PSO implementation with local outlier factor (LOF) as fitness function no doubt improved the detector generation phase of NSA. The proposed improved model serves as a better replacement to NSA model. Spam-base dataset was used to investigate the performance of NSA and the improved NSA-PSO model.

Performance and accuracy investigation has shown that the proposed improved model is able to detect email spam better than the NSA model.

Acknowledgements. The Universiti Teknologi Malaysia (UTM) under research grant 03H02 and Ministry of Science, Technology & Innovations Malaysia, under research grant 4S062 are hereby acknowledged for some of the facilities utilized during the course of this research work.

References

1. Wang, H., Gao, X., Huang, X., Song, Z.: PSO-Optimized Negative Selection Algorithm for Anomaly Detection. In: Applications of Soft Computing, vol. 52, pp. 13–21. Springer, Heidelberg (2009)
2. Gao, X.Z., Ovaska, S.J., Wang, X.: Particle Swarm Optimization of detectors in Negative Selection Algorithm. In: 2007 ISIC IEEE International Conference on Systems, Man and Cybernetics, October 7-10, pp. 1236–1242 (2007)
3. Sotiropoulos, D.: Artificial Immune System-based Machine Learning Methodologies. PhD thesis, University of Piraeus, Piraeus, Greece (2010)
4. Oda, T., White, T.: Developing an Immunity to Spam. In: Cantú-Paz, E., et al. (eds.) GECCO 2003. LNCS, vol. 2723, pp. 231–242. Springer, Heidelberg (2003)
5. Oda, T., White, T.: Increasing the accuracy of a spam-detecting artificial immune system. In: The 2003 Congress on Evolutionary Computation, CEC 2003, December 8-12, vol. 391, pp. 390–396 (2003)
6. Oda, T., White, T.: Immunity from Spam: An Analysis of an Artificial Immune System for Junk Email Detection. In: Jacob, C., Pilat, M.L., Bentley, P.J., Timmis, J.I. (eds.) ICARIS 2005. LNCS, vol. 3627, pp. 276–289. Springer, Heidelberg (2005)
7. Mohammad, A.H., Zitar, R.A.: Application of genetic optimized artificial immune system and neural networks in spam detection. Applied Soft Computing 11(4), 3827–3845 (2011)
8. Yevseyeva, I., Basto-Fernandes, V., Ruano-Ordás, D., Méndez, J.R.: Optimising anti-spam filters with evolutionary algorithms. Expert Systems with Applications 40(10), 4010–4021 (2013)
9. He, W., Mi, G., Tan, Y.: Parameter Optimization of Local-Concentration Model for Spam Detection by Using Fireworks Algorithm. In: Tan, Y., Shi, Y., Mo, H. (eds.) ICSI 2013, Part I. LNCS, vol. 7928, pp. 439–450. Springer, Heidelberg (2013)
10. Balthrop, J., Forrest, S., Glickman, M.R.: Revisiting LISYS: parameters and normal behavior. In: Proceedings of the 2002 Congress on Evolutionary Computing, pp. 1045–1050 (2002)
11. Forrest, S., Perelson, A.S.: Self nonself discrimination in computer (1994)
12. Wang, C., Zhao, Y.: A new fault detection method based on artificial immune systems. Asia-Pacific Journal of Chemical Engineering 3(6), 706–711 (2008)
13. Sajesh, T.A., Srinivasan, M.R.: Outlier detection for high dimensional data using the Comedian approach. Journal of Statistical Computation and Simulation 82(5), 745–757 (2011)
14. Hopkins, M., Reeber, E., Forman, G., Jaap, S.: Spam Base Dataset. Hewlett-Packard Labs (1999)

Random Forest of Oblique Decision Trees for ERP Semi-automatic Configuration

Thanh-Nghi Do[1,2,3], Sorin Moga[1,3], and Philippe Lenca[1,3]

[1] Institut Mines-Telecom; Telecom Bretagne
UMR 6285 Lab-STICC
Technopôle Brest-Iroise - CS 83818 - 29238 Brest Cedex 3 - France
FirstName.LastName@telecom-bretagne.eu
[2] College of Information Technology, Can Tho University, Viet Nam
[3] Université Européenne de Bretagne

Abstract. Enterprise Resource Planning (ERP) is one of the most important parts of company's information system. However few ERP implementation projects are delivered on time. Configuration of ERP based on questionnaires and/or interviews is time consuming and expensive, especially because many answers should be checked and corrected by ERP consultants. Supervised learning algorithms can thus be useful to automatically detect wrong and correct answers. Comparison done on real free open-source ERP data shows that random forest of oblique decision trees is very efficient.

Keywords: ERP configuration, free text classification, random forest of oblique decision trees, ERP5.

1 Introduction

Enterprise resource planning (ERP) systems or enterprise systems are software systems for business management, encompassing modules supporting most of (if not all) functional areas in companies [1]: planning, manufacturing, sales, marketing, distribution, accounting, financial, human resource management... An ERP can thus be the most important part of companies information system. Its software architecture should facilitate integration of modules, providing flow of information between all functions within the enterprise in a consistently visible manner [2].

In the actual context of global competition and the raise of customers expectancy, the companies must have the capability to adapt quickly to market changes and the use of an ERP is one of the solutions. Depending on their implementation, the ERP systems will positively or negatively impact the companies business in a challenging world. Unfortunately, ERP implementation is not always successful. A recent study reports that 56% of ERP project implementations have gone over budget and only 34% of the projects were delivered on the time [3].

J. Sobecki, V. Boonjing, and S. Chittayasothorn (eds.), *Advanced Approaches to Intelligent Information and Database Systems*, Studies in Computational Intelligence 551,
DOI: 10.1007/978-3-319-05503-9_3, © Springer International Publishing Switzerland 2014

The factors that determine the success of an ERP system implementation is a constant research question [4,5,6]. Many factors may influence an ERP implementation success (e.g. cultural factor [7], users training [8], alinement's between the ERP and the company in terms of data, process, and user interface [9], fine-tuning [10], fit between company type and ERP type [11], knowledge sharing and cooperation [12], standardization of business process and organizational structure [2], configuration team organization [13]). Among these factors, fine-tuning, configuration/customization of the ERP implementation, i.e. the process of setting up the ERP in order to correspond to the organisational structure and business process [13], is a key factor [14,10,2,3]. Some effort is thus done to enhance ERP configuration. In order to improve the flexibility of configuration process, [15] proposed to extract an atom decomposition of a business process. The authors use workflows logs when running the ERP on which frequent patterns can be found. The frequent patterns are used later to improve the ERP platform configuration. Another method, based on Material Requirements Plannning (MRP) approach [16] is proposed in [17] for automatic SAP inventory management module configuration. The authors have successfully applied their proposal for stocks management. However MRP requires many complex decisions to be properly implemented and this approach seems suitable for large companies only.

Yet, Small and Medium size Enterprises (SMEs) adopt more and more ERP systems. Indeed, the price of ERP implementation in SMEs becomes acceptable and the benefit is more clearly established. In addition, there are open-source ERP solutions (*OpenERP, Compiere, ERP5, ...*) available for the SMEs and the major vendors of ERP solutions (e.g. SAP, Microsoft, ...) develop their offers to SME sector [12,6]. The implementation cost for an ERP (proprietary or Open Source) solution could be especially high comparing with the price of ERP licence [18,19,20]. The free, open-source, ERP solutions are viable alternative for SMEs as they tackle their specific problems. Thus, for SMEs using off-the-shelf ERP packages, where the implementation is done mainly by configuration, reducing the configuration cost and automation are opportunities for their businesses [21,22,20]. In this context, supervised machine learning approach is a track that can be explored in order to build a semi-automatic configuration tool for free open-source *ERP* as initiated in [22] for the ERP5 system (see www.erp5.org). The ERP5 system will be our case study.

The paper is organized as the following. Section 2 briefly presents the first stage of ERP5 configuration and the data preparation. It is based on the analysis of a questionnaire on the company information (business, organization, etc.). One main issue is to identify questions with wrong answers. When done manually, this task is expensive (in money and in time). Supervised learning approaches are presented in Section 3 to automatically classify the responses. Results are presented and discussed in Section 4. A conclusion is drawn in Section 5.

2   Data Analysis of Questionnaire's Answers

In this section we present one of the main step used during the ERP5 configuration. It consists of fulfilling a questionnaire about the company's information and its business process.

Questionnaire on the Company

The first stage of ERP5 configuration consists in fulfilling a questionnaire. The questionnaire contains questions about the company's business, organisation, customers, etc. The chief executive officer (CEO) is typically in charge of that task.

Some questions allow free text answers (for example *What does the implementation field sell, offer or produce?* with a free text answer zone) and other questions are presented with predefined answers (for example *What types of clients do you have?* with a list such as *consumer, business, administration, not-for-profit*).

In many case, the answers are ill formulated (either wrong, vague, incomplete, out of topic, etc.) or empty. These imperfections are detected later in a post-processing step by the ERP5 consultant when each question is analysed (a questionnaire contains $q = 42$ questions at the date of our study). As a consequence, a costly and time consuming iterative process between the ERP5 consultant and the CEO is started to manually correct the questionnaires.

One way to reduce the time of this iterative process, it to help the ERP5 consultant to only focus on answers that certainly need to be corrected. An automatic tool that detects if an answer has to be corrected or not can thus be very useful. With the help of an ERP5 consultant, we then build a learning set of questionnaires and thus propose to study supervised classifiers efficiency for the binary task of deciding if an answer is correct or has to be corrected. The data set and data preparation are presented below.

Data Sets

The initial real data set contains 3696 lines in a triplet format <*Question*; *Answer*; *Correction*>. Examples of such triplets are presented in Table 1. The *Correction* field is empty if the *Answer* was considered as correct. Otherwise, *Correction* field contains the ERP5 consultant's correction. The *Correction* field is thus the target variable and an *Answer* is considered as good (w.r.t *Question*) if the *Correction* field is empty. The binary target variable can be *correct* (the answer is correct) or *to_correct* (the answer needs a correction).

Obviously, such an unstructured data set with free text needs an important pre-processing step (data cleaning and data representation). In addition such a data set contains many identical *Questions* with different *Answers*.

Table 1. Example of <*Question*; *Answer*; *Correction*> triplet

Question	Answer	Correction
What does the implementation field	MAGIX is a leading international	no marketing
sell, offer or produce?	provider of high-quality software,	
	online services and digital con-	
	tent for multimedia communica-	
	tions. Since 1993, MAGIX ...	
What kind of clients does the com-		check for up-
pany have?		date
Explain what the organization ex-	The 2nd priority is to computerize	specify
pects from an ERP as 2nd priority	their budget management.	budgeting
to improve its own management?		improve-
		ments
...	...	...

Data Cleaning

- non ASCII symbols are replaced by ASCII ones (e.g. <*Ctrl-Enter*> character is replaced by the <space> character);
- duplicate triplets are removed keeping only one instance of each triplet; we then obtain 972 unique triplets <*Question*; *Answer*; *Correction*>.

Data Representation

The bag-of-words (BoW) model [23] is used in order to transform the *clean* unstructured text data set to a structured data set. After applying BoW, the data set contains 4007 different words. Six triplets contain very few words and were deleted at this stage. Last, a feature selection was applied to select the n most discriminant words using mutual information with the LibBow library [24].

The data set is then represented by a $966 \times (n + 1)$ matrix *BoW* where:

- the first n columns represent a discriminant word (n words);
- the $n + 1^{th}$ column represents the class label (*correct* or *to_correct*);
- $BoW[i][j]$ is the frequency of the word j of the <*Question*; *Answer*; *Correction*> triplet with the $ID = i$ (there are 966 rows and the ID is the row number).

There are 112 (12%) <*Question*; *Answer*; *Correction*> triplets in the category *correct* and 854 (88%) in the category *to_correct*. The initial data set has also the following characteristics: it is a high-dimensional data set (large number of features w.r.t number of examples), it is unbalanced (large difference in target variable distribution) and it contains few examples. These issues are challenging for supervised learning.

3   Supervised Classification

Several supervised algorithms can be used to learn the binary target variable (*correct* or *to_correct*). Let's recall that we face a high-dimension and unbalanced data set with few examples in $\Re^n$. In order to save as much time as possible when the ERP consultant has to examine the questionnaires, one of the main objective is to have the best accuracy as possible. However simplicity should also be investigated.

Thus the k nearest neighbour algorithm (kNN) [25] has been tested due to its simplicity. The well-know C4.5 [26] decision tree algorithm and its extended version C4.5-OCE with the Off-Centred-Entropy OCE [27] which has shown to be very efficient for unbalanced data [28] have been also tested. We also investigate the performance of support vector machines [29] and random forests [30] for which many comparative studies (see for example [31]) showed that they are very accurate in case of high dimensional data, especially random forests of oblique decision trees [32,33].

Performance of all algorithms to detect answers that have to be corrected are very similar but there are large difference on *correct* answer accuracy (see Fig. 1). The 3NN ($k = 3$ was the best number of neighbours) and univariate decision trees (C4.5, C4.5-OCE) make much more wrong decision when the answer was *correct* (i.e. the ERP consultant will waste time while checking a *correct* response). Thus these algorithms are no more discussed. We here report the results of the two best approaches, SVM and ensemble of decision trees.

3.1   Support Vector Machines (SVM)

SVM algorithms [29] try to find the best separating plane to separate the class labels. For that purpose it maximizes the distance (or margin) between the supporting planes for each class. Thus a point falling on the wrong side of its supporting plane is considered to be an error. Therefore, a SVM algorithm tries to simultaneously maximize the margin and minimize the error. The SVM algorithms thus pursue these two goals by solving quadratic programs (using several classification functions e.g. linear, polynomial, radial basis or a sigmoid function, and a cost constant to tune the errors and margin size).

3.2   Random Forest

Random forests, introduced in [30], are an ensemble learning method that aim to create a collection of decision trees (the forest). It combines bagging (bootstrap aggregating) idea [34] and the random selection of features [35]. Each tree is grown to maximum depth, i.e without pruning, considering a bootstrap replica of the original data set, i.e. a random sampling with replacement, as training set. The splitting function only consider a randomly chosen subset of the dimensions. These strategies ensure to keep low bias (i.e. high performance classification) and low dependence (i.e. high diversity) between trees in the forest. Each individual decision are then aggregated to classify a new individual. A popular aggregation function is a (unweighted) majority vote of the trees.

Classical random forests (RF) uses a single variable for node splitting. We tested RF-C4.5 with the Shannon entropy like in C4.5 [26] and RF-C4.5-OCE with the Off-Centred-Entropy OCE [28]. However in case of high-dimensionality, i.e. when dependencies among variable are more likely to occur, multivariate splitting can be preferred. We thus also used random forests of oblique decision trees (RF-ODT) that perform multivariate node splitting with linear SVMs [33].

4   Results

All results presented in this section are obtained using our own software[1] based on generally available software library: decision tree toolkit [28], RF-C4.5 [33], RF-ODT [33] and LibSVM [36]. In order to evaluate the performance of classification algorithms, a hold-out cross-validation protocol is used: the data set is randomly partitioned into a *training set* (646 rows, 67 %) to learn the models and a *testing set* (320 rows, 33 %) to test the models. This process (random partition/learn/evaluate) is repeated 10 times and the results are averaged.

The influence of the output of the BoW representation, i.e. the size n of input vector, has to be optimized. We tested n from 50 (very few number of word) up to 1000 (about 25% of the 4007 words in the initial dictionary). RFs need to tune the number of the trees in the forest and the size of the random subset of attributes used for splitting. The number of trees, nb_trees, has been tuned from 50 to 2000 and the size of random subset of attributes, dim_size, from 50 to 1000. The SVM algorithm has been tested with a linear and a non-linear kernel both with a cost constant $C = 10^i, \quad i = 1, \ldots, 8$.

The common classification performance measures (based on the confusion matrix) are used to evaluate the algorithms: $F1$ measure (Eq. 1), $Gmean$ (Eq. 2) and $Accuracy$ (Eq. 3) (global accuracy, accuracy for *correct* and accuracy for *to_correct*) [37]. The $Gmean$ measure represents a trade-off between positive accuracy (class *correct*) and negative accuracy (class *to_correct*).

$$F1 = \frac{2 \cdot TP}{2 \cdot TP + FP + FN} \tag{1}$$

$$Gmean = \sqrt{\frac{TP}{TP + FN} \cdot \frac{TN}{TN + FP}} \tag{2}$$

$$Accuracy = \frac{TP + TN}{TP + FN + TN + FP} \tag{3}$$

Where

- TP (true positive): the number of examples correctly labelled as belonging to the positive class;

---

[1] Our software was developed under General Public License V3.

- FN (false negative): the number of positive examples labelled as negative class;
- TN (true negative): the number of examples correctly labelled as belonging to the negative class;
- FP (false positive): the number of negative examples labelled as positive class.

The data set is highly imbalanced and $F1$ measure is often used, in this case. But, in our application, the user attempts is also to have a high rate of "true positive". So we also focus here on *Accuracy* and *Gmean* measures.

Based on these measures of performance and extensive experimentation, the optimal parameters have been fixed as the following: $nb\_tree = 200$ and $random\_dim = 150$ for RFs and $C = 10^5$ with a linear kernel for SVM.

Performances are presented in Table 2 and in Fig. 1. The global *Accuracy* and the *Accuracy to_correct* are very similar for all algorithms. However they are large differences for the *correct* class. Only RF-ODT and SVM provide very good results. RF-C4.5 and RF-C4.5-OCE will make many mistake while considering *correct* answers and then ask the ERP to check the corresponding questions.

RF-ODT and SVM have very similar performance (differences are not significant). But at this stage we can recommend to use the RF-ODT algorithm. RF-ODT does no dominates SVM on all measures but it is slightly better for *F1* and *GMean* and better for *Accuracy correct*. RF-ODT outperforms SVM in predicting correct answers and thus ask less the ERP consultant to check the corresponding questions, saving more consultant's time. Taking into account these performances and after a discussion with ERP consultants it has been decided to select the RF-ODT algorithm.

Table 2. Performance measure (in %)

Algorithm	Acc. correct	Acc. to_correct	F1	Gmean	Accuracy
RF-C4.5	91.77	99.75	94.74	95.67	98.88
RF-C4.5-OCE	92.89	99.68	95.10	96.21	98.81
RF-ODT	98.16	99.50	97.20	98.82	99.34
SVM	96.83	99.72	97.08	98.26	99.38

The RF-ODT algorithm is thus implemented in the Classification and Suggestion System (C.A.S.), a semi-automatic ERP configuration tool. C.A.S has been evaluated in real situation with real data provided by the Nexidi company (http://www.nexedi.com/) and first results are excellent leading to a gain of productivity [38]. Two groups of experts were in charge of correcting 750 questionnaires. The first group (7 experts) was assisted by C.A.S while the second group (8 experts) proceeded as usual i.e. without help. With C.A.S the first group corrected 93% of wrong answers in less than one minute while the second group achieved only 83%.

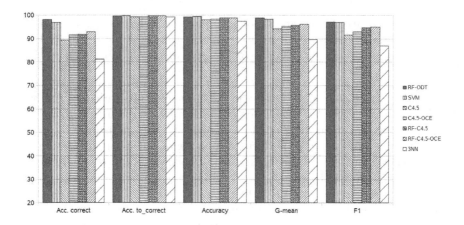

Fig. 1. Performance measure. Graphic representation of Table 2. We integrated the results obtained with kNN ($k = 3$) and C4.5 algorithms for comparison purpose.

5 Conclusion

Entreprise Ressource Planning systems are an important part of companies information system. A well-configuration is a key factor for the success of an ERP implementation. Typically, the configuration based on interviews and/or questionnaires is costly and time consuming. Indeed questionnaires contain many inappropriate responses. They thus need to be corrected by an ERP expert. In this paper, we explore supervised learning to improve the process of correcting questionnaires by comparing several supervised learning algorithms. The Random Forest of Oblique Decision Trees model gives the best and very high performance. It has been implemented in a configuration assisting tool and successively tested with real data. It has provided a gain of productivity.

Acknowledgements. This work was done as a part of Cloud Consulting project of Eureka Eurostars innovation program (project A1107003E). We thank our project partners Nexedi, Technische Universität Dresden and BEIA Consulting for providing us access to the data of the ERP5 Questionnaire System. Special thanks to Klaus Wölfel for valuable comments.

References

1. Hossain, L., Rashid, M.A., Patrick, J.D.: The evolution of ERP systems: A historical perspective. In: Hossain, L. (ed.) Enterprise Resource Planning Global Opportunities and Challenges, pp. 1–16. Idea Group Pub., Hershey (2002)

2. Velcu, O.: Strategic alignment of ERP implementation stages: An empirical investigation. Information & Management 47(3), 158–166 (2010)
3. Solutions, P.C.: Panorama consulting solutions research report - 2013 ERP Report. Technical report, Panorama Consulting Solutions (2013)
4. Nah, F.F.H., Lau, J.L.S., Kuang, J.: Critical factors for successful implementation of enterprise systems. Business Process Management Journal 7(3), 285–296 (2001)
5. Motwani, J., Subramanian, R., Gopalakrishna, P.: Critical factors for successful ERP implementation: Exploratory findings from four case studies. Computers in Industry 56(6), 529–544 (2005)
6. Ahmad, M.M., Pinedo Cuenca, R.: Critical success factors for ERP implementation in SMEs. Robotics and Computer-Integrated Manufacturing 29(3), 104–111 (2013)
7. Soh, C., Kien, S.S., Tay-Yap, J.: Enterprise resource planning: cultural fits and misfits: is ERP a universal solution? Communications of the ACM 43(4), 47–51 (2000)
8. Markus, M., Tanis, C.: The enterprise systems experience-from adoption to success. In: Zmud, R.W. (ed.) Framing the Domains of IT Management: Projecting the Future Through the Past, pp. 173–207 (2000)
9. Hong, K.K., Kim, Y.G.: The critical success factors for ERP implementation: an organizational fit perspective. Information & Management 40(1), 25–40 (2002)
10. Kim, Y., Lee, Z., Gosain, S.: Impediments to successful ERP implementation process. Business Process Management Journal 11(2), 158–170 (2005)
11. Morton, N.A., Hu, Q.: Implications of the fit between organizational structure and ERP: a structural contingency theory perspective. International Journal of Information Management 28(5), 391–402 (2008)
12. Dittrich, Y., Vaucouleur, S., Giff, S.: ERP customization as software engineering: Knowledge sharing and cooperation. IEEE Software 26(6), 41–47 (2009)
13. Robert, L., McLeod, A., Davis, A.R.: ERP configuration: Does situation awareness impact team performance? In: 44th Hawaii International Conference on System Sciences, Kauai, HI, USA, pp. 1–8. IEEE (2011)
14. Olhager, J., Selldin, E.: Enterprise resource planning survey of swedish manufacturing firms. European Journal of Operational Research 146(2), 365–373 (2003)
15. Yi, J., Lai, C.: Research on reengineering of ERP system based on data mining and MAS. In: International Symposium on Knowledge Acquisition and Modeling, Wuhan, China, pp. 180–184. IEEE (2008)
16. Waldner, J.B.: CIM, principles of computer-integrated manufacturing. Wiley, Chichester (1992)
17. Bucher, D., Meissner, J.: Automatic parameter configuration for inventory management in SAP ERP/APO. Working Paper, Kuehne Logistics University, Hamburg, Germany (2010)
18. Timbrell, G., Gable, G.: The SAP ecosystem: A knowledge perspective. In: Nah, F.F.-H. (ed.) Enterprise Resource Planning Solutions and Management, pp. 209–220. IRM Press, Hershey (2002)
19. Snider, B., da Silveira, G.J., Balakrishnan, J.: ERP implementation at SMEs: analysis of five canadian cases. International Journal of Operations & Production Management 29(1), 4–29 (2009)
20. Wölfel, K., Smets, J.P.: Tailoring FOS-ERP packages: Automation as an opportunity for small businesses. In: de Carvalho, R.A., Johansson, B. (eds.) Free and Open Source Enterprise Resource Planning: Systems and Strategies, pp. 116–133. IGI Global (2012)

21. Brehm, L., Heinzl, A., Markus, M.: Tailoring ERP systems: a spectrum of choices and their implications. In: Proceedings of the 34th Annual Hawaii International Conference on System Sciences, pp. 1–9. IEEE Comput. Soc. (2001)
22. Wölfel, K.: Automating ERP Package Configuration for Small Businesses. Diploma thesis, Techniche Universität Dresden, Dresden, Germany (2010)
23. Sebastiani, F., Ricerche, C.N.D.: Machine learning in automated text categorization. ACM Computing Surveys 34, 1–47 (2002)
24. McCallum, A.K.: Bow: A toolkit for statistical language modeling, text retrieval, classification and clustering (1996), http://www.cs.cmu.edu/mccallum/bow
25. Fix, E., Hodges, J.: Discriminatoiry Analysis: Small Sample Performance. Technical Report 21-49-004, USAF School of Aviation Medicine, Randolph Field, USA (1952)
26. Quinlan, J.R.: C4.5: Programs for Machine Learning. Morgan Kaufmann (1993)
27. Lenca, P., Lallich, S., Vaillant, B.: Construction of an off-centered entropy for the supervised learning of imbalanced classes: Some first results. Communications in Statistics - Theory and Methods 39(3), 493–507 (2010)
28. Lenca, P., Lallich, S., Do, T.-N., Pham, N.-K.: A comparison of different off-centered entropies to deal with class imbalance for decision trees. In: Washio, T., Suzuki, E., Ting, K.M., Inokuchi, A. (eds.) PAKDD 2008. LNCS (LNAI), vol. 5012, pp. 634–643. Springer, Heidelberg (2008)
29. Vapnik, V.: The Nature of Statistical Learning Theory. Springer (2010), Softcover reprint of hardcover, 2nd edn. (2000)
30. Breiman, L.: Random forests. Mach. Learn. 45(1), 5–32 (2001)
31. Caruana, R., Karampatziakis, N., Yessenalina, A.: An empirical evaluation of supervised learning in high dimensions. In: Proceedings of the 25th International Conference on Machine Learning, pp. 96–103. ACM, New York (2008)
32. Simon, C., Meessen, J., De Vleeschouwer: Embedding proximal support vectors into randomized trees. In: European Symposium on Artificial Neural Networks, Advances in Computational Intelligence and Learning, Bruges, Belgium, pp. 373–378 (2009)
33. Do, T.-N., Lenca, P., Lallich, S., Pham, N.-K.: Classifying very-high-dimensional data with random forests of oblique decision trees. In: Guillet, F., Ritschard, G., Zighed, D.A., Briand, H. (eds.) Advances in Knowledge Discovery and Management. SCI, vol. 292, pp. 39–55. Springer, Heidelberg (2010)
34. Breiman, L.: Bagging predictors. Technical Report 421, Department of Statistics University of California, Berkeley, California 94720 USA (1994)
35. Ho, T.K.: Random decision forests. In: Proceedings of the Third International Conference on Document Analysis and Recognition, Montreal, Canada, vol. 1, pp. 278–282. IEEE Comput. Soc. Press (1995)
36. Chang, C.C., Lin, C.J.: LIBSVM: a library for support vector machines. ACM Trans. Intell. Syst. Technol. 2(3), 27:1–27:27 (2011)
37. Rijsbergen, C.J.V.: Information Retrieval, 2nd edn. Butterworth-Heinemann, Newton (1979)
38. Wölfel, K.: Suggestion-based correction support for moocs. Summited to Multikonferenz Wirtschaftsinformatik, Paderborn, February 26-28 (2014)

A Novel Conflict Resolution Strategy in Multi-agent Systems: Concept and Model

Ghusoon Salim Basheer[1], Mohd Sharifuddin Ahmad[1], Alicia Y.C. Tang[1],
Azhana Ahmad[1], Mohd Zaliman Mohd. Yusoff[1], and Salama A. Mostafa[1]

[1] Universiti Tenaga Nasional, Jalan IKRAM-UNITEN,
43000 Kajang, Selangor, Malaysia
rawagy2013@gmail.com, semnah@yahoo.com,
{sharif,aliciat,azhana,zaliman}@uniten.edu.my

Abstract. In this paper, we present the concept and model of a conflict resolution strategy for a multi-agent system that covers all aspects of conflict processing, from collecting agents' opinions, recognition of possible conflict status, and through a joint final decision. Our approach is to specify a novel structure for classifying conflict states in decision-making in which related factors, such as number of conflicting agents, agent's confidence level and strength of conflict play essential roles in guiding and selecting the conflict resolution strategies. We provide an example scenario as a proof of concept to show the model's applicability in multiple conflict situations.

Keywords: Software Agent, Multi-agent System, Conflict Resolution.

1 Introduction

In multi-agent systems, agents must communicate with each other and resolve conflicts between them [1]. To do so, the agents must be able to select one of multiple strategies to eliminate conflicts between them. Conflict resolution in multi-agent systems entails a comprehensive investigation of factors that relate to the cause, identification and resolution of conflicts [2].

In this paper, we examine conflicts between agents that have different confidence levels about a particular problem domain. We assume that the confidence levels of agents are given, and that each agent has a different opinion about the domain that generates the conflicts. We seek an optimal algorithm to classify the conflict states by considering our model that resolves conflicts between two agents discussed in [3]. In this paper, we extend the model to include conflicts between three or more agents.

While we assume that the confidence value for each agent in the domain is known, the confidence value is different from other agents. The objective is to develop an optimal solution to the conflict classification problem, which provides a procedural approach to the classification and resolution of conflicts leading to a final decision. To this end, we construct a model that considers three factors: number of conflicting agents, strength of conflict, and confidence level of agents. We firstly consider the states of two-party conflicts, and followed by more than two-party conflicts.

J. Sobecki, V. Boonjing, and S. Chittayasothorn (eds.), *Advanced Approaches to Intelligent Information and Database Systems*, Studies in Computational Intelligence 551,
DOI: 10.1007/978-3-319-05503-9_4, © Springer International Publishing Switzerland 2014

The first part of this paper contains a review of the previous classifications of conflicts in humans' and agents' societies. The review discusses the proposed model of conflict classification. We then propose our technique, in which we suggest a new classification of agents' conflicts based on their confidence and numbers. We organize the rest of the paper as follows: Section 2 reviews the type of conflicts in humans and multi-agent systems. In Section 3, we propose a framework for classification and resolution of conflicts in multi-agent systems. Section 4 elaborates an example scenario of multiple conflict resolutions for our proposed framework and Section 5 concludes the paper.

2   Types of Conflict Resolution Strategies

2.1   Conflicts in Humans

Conflicts in humans occur for many reasons. However, resolving conflicts depends on the nature of the conflicts, the number of stakeholders involved in the conflicts and the interest of each stakeholder has on the outcome of the conflicts. Crawford et al. [4] proposed that stakeholders have two choices, either continue with the conflict or resolve it. Tessier et al. [2] classified the types of conflict in human as shown in Figure 1. He mentioned that only parts of these types have been modeled in multi-agent systems environment due to two reasons. Firstly, the basis of these conflict types is emotions, but the realization of emotional agents is so limited. Secondly, these conflicts types need reflection content in the agents, which current theoretical models of multi-agent system lack.

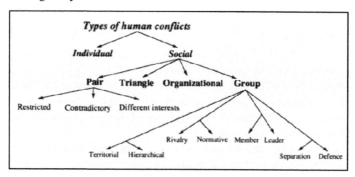

Fig. 1. Human Conflict Types [2]

Rummel [5] detected power as a source of social conflict, and he classified conflicts into three types. The first type is when two individuals are interested in the same thing and to solve this conflict one of them must be excluded (conflict of appropriate interests). The second type of conflict includes i wants x that j does not wants (inverse interest). The third type involves two individuals i and j, and i wants a, but j wants b, where a and b are opposed (incompatible interests).

2.2   Conflicts in Multi-agent Systems

Conflicts between agents occur in multi-agent environment in many instances and are resolved depending on their types and dimensions. Many factors can lead to conflicts in multi-agent environments, like differences in goals, disputes in preferences, changes in expectations about behaviors of others, and conflicts in mental attitudes [5].

Many researchers have proposed different strategies to resolve conflicts in multi-agent systems. Some of these typical strategies include negotiation by Sycara [6], which provided a model for conflict resolution of goals. She proposed a program that resolves labor disputes. Her model performs the negotiation through proposal and modification of goal relaxations, which she proposed by using Case-Based Reasoning with the use of multi-attribute utilities to portray tradeoffs. Barber et al. [7] produced multiple strategies for conflict resolution such as negotiation, self-modification and voting. Selecting each one of these strategies depends on several characteristics like cost and required time.

Ioannidis et al. [8] studied the problem of resolving conflicts of rules that assign values to virtual attributes. They assumed that the set of rules defined by a user is consistent, which means that there is no fact that can be obtained by the rules. They proposed a new model that subsumes all previously suggested solutions, and suggests additional solutions. Jung [9] attempted to solve agents' conflict problem by implementing a new system called CONSA (Collaborative Negotiation System based on Argumentation) based on agent negotiation strategy. Through negotiation, agents propose arguments as justifications or elaborations to explain their decisions.

Tessier et al. [2] classified conflicts into several types: conflicts of ideas, facts, practices, and goods. Müller et al. [10] classified conflicts into three types:

- Conflicts within an individual when he/she is torn between incompatible goals.
- Conflicts between individuals when they want different things, and they must reach an agreement about the same thing.
- Conflicts between individuals when they want the same thing and must reach an agreement of selecting a different thing.

2.3   Classification of Conflicts in Multi-agent Systems

Classifying conflicts is an essential part of realizing and understanding the nature of conflicts. Understanding the nature of conflicts reduces the search space of potential resolution strategies and enables agents to focus on behaviors that are most important for the type of conflict they are attempting to resolve [10].

Tessier et al. [2] classified conflicts into two main classes: Physical Conflicts - conflicts of external resources; and Knowledge Conflicts - agents conflict in beliefs, knowledge and opinions. Liu et al. [11] opined that agents should select an appropriate strategy for conflict resolution depending on three factors: type of conflict, agent's rule, and preference solution. They classified conflicts into three classes: Goal conflicts, Plan conflicts, and Belief conflicts.

3   The Proposed Conflict Classification Framework

In our earlier work [3], we analyzed the social theory of conflict and proposed a conflict resolution model depending on conflict classification. In our approach, agents assign confidence values to their opinions from domain specific pre-defined factors. Our model is developed based on six types of conflict resolution model identified by Tessier et al. [2] as follows:

1. **Flight**: Represent fleeing by one of two opponents.
2. **Destruction**: Takeover one of opponents.
3. **Subservience**: Gives up by one of opponents.
4. **Delegation**: Add a third party to judge between opponents.
5. **Compromising:** Obtain the result of negotiation.
6. **Consensus**: Obtain the agreement of opponents.

Inspired from human's conflict resolution strategies, we proposed a framework for conflict resolution as shown in Figure 2 [3].

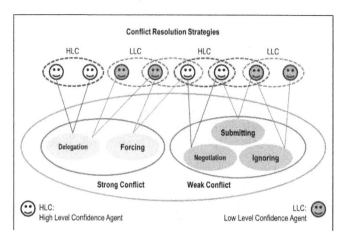

Fig. 2. Selecting Appropriate Conflict Resolution Strategy [3]

- **Forcing**: Corresponds to **Destruction/Flight** in some conflict state. We recognize that there is no chance to resolve the conflict.
- **Submitting/Ignoring**: Corresponds to **Subservience**. In this case, there is no force, but inducement between both sides.
- **Delegation**: Corresponds to Delegation when the conflict cannot be resolved, both opponents request a third party that has deep knowledge to judge.
- **Negotiation**: Corresponds to **Compromising** through negotiation when one of the opponents is willing to yield. This state includes an agreement in a different style.
- **Agreement**: Corresponds to **Consensus**. Each opponent must give all details about its decision to a third party. For this reason, this process comes as a result of a delegation process.

In this paper, we expand the framework of conflict classification in multi-agent systems based on three dimensions:

- Number of conflicting agents, which includes:
  - Conflicts between two agents.
  - Conflicts between three or more agents.
- Confidence level of conflicting agents, which includes:
  - Conflicts between two agents that have the same level of confidence, e.g., both agents have high level of confidence (HLC/HLC), or both have low level of confidence (LLC/LLC).
  - Conflicts between two agents that have different level of confidence, e.g., a high level of confidence against a low level of confidence (HLC/LLC).
- Conflict strength between agents, which includes:
  - Strong Conflict (SC), when more than 50% of agents conflicts in opinions with another agent.
  - Weak Conflict (WC), when less than 50% of agents conflicts in opinions with another agent.

Figure 3 depicts the analytical process of classifying a three-dimensional conflict. The framework consists of three stages:

- Collect agents' opinions and detect conflicting agents in the system.
- Use confidence table to detect confidence value for each agent.
- Classify conflicts states based on three factors:
  - Number of conflicting agents.
  - Conflict strength.
  - Confidence value of conflicting agents.

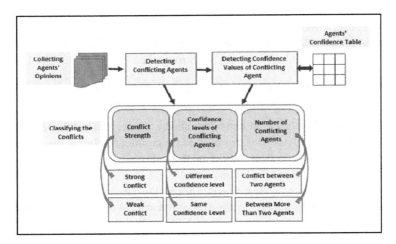

Fig. 3. A Framework for Classifying Conflicts in MAS

3.1   Classification of Conflicts

The conflict classification model proposed in this paper is based on the following definitions:

Definition 1: Given a set of agents, $A=\{a_1, a_2, \ldots, a_n\}$, each agent $a_i \in A$ has a set of specification, S_i, that includes Opinion, O_i, and Confidence, C, i.e., $S_i=(O_i, C)$. An agent's opinion, O_i may conflict with another agent's opinion, O_j, or a set of other agents' opinions $\{O_k, \ldots, O_x\}$.

Definition 2: Let a_i be an agent, such that $a_i \in A$. Each agent in A has an *agent Confidence Value*, C, as a positive integer in an *Agent Confidence Table*, ACT that represents the confidence levels which are determined by the agents themselves.

Definition 3: A Conflicting Agent Set, CAS, is a set of pairs of conflicting agents, i.e., if a_i conflicts with a_j, then CAS=$\{(a_i, a_j)\}$.

Definition 4: For each pair of conflicting agents $(a_i, a_j) \in$ CAS, their *Conflict Strength* is represented by CS_{ij} with two levels of agent's confidence, e.g., High Level Confidence (HLC) and Low Level Confidence (LLC). Three situations are apparent: $C_i = C_j$, or $C_i > C_j$, or $C_i < C_j$.

Definition 5: Referring to Figure 2, for each pair of conflicting agents $(a_i, a_j) \in$ CAS, we define six possible Conflict Resolution Strategies, CRS_{ij}, by detecting the conflict strength, CS_{ij}, and the agent's confidence level (C_i, C_j).
If $C_i = C_j$ and both are HLC agents and CS_{ij}=Strong, then call Evidence Function, EF, and third party Mediator to judge (CRS_{ij}=Delegation).
If $C_i > C_j$ and CS_{ij} = Strong, then (CRS_{ij}=Forcing).
If $C_i = C_j$ and both are HLC agents and CS_{ij}=Weak, then CRS_{ij}=Negotiation).
If $C_i < C_j$ and CS_{ij} = Weak, then (CRS_{ij}=Submitting).
If $C_i = C_j$ and both are LLC agents and CS_{ij}=Weak, then (CRS_{ij}=Ignoring).
If $C_i = C_j$ and both are LLC agents and CS_{ij}=Strong, then (CRS_{ij}=Delegation).

3.2   The Proposed Algorithm

The conflict resolution algorithm considers all the agents and conflict states in the system. Classifying conflicts states can be used for conflict resolution enhancement.
For each agent a_i, in the system, we define the following:
```
Define a set of agents' opinions, O,
Define a set of Conflicting Agent in the system, CAS,
Define a set of Conflict Resolution Strategies as the set of all
possible strategies that include {Delegation, Ignoring, Forcing,
Submitting, Negotiation},
Define the Conflict Strength as the set of two levels {Strong
Conflict (SC), Weak Conflict (WC)},
Define Confidence Level as an array of two levels {High Level
Confidence (HLC), Low Level Confidence (LLC)} for each agent in CAS,
```

```
Evaluate the Confidence values for each agent in CAS,
Classifies conflicting agents array into groups depending on conflict
points,
Calculate a confidence value for each group in the CAS,
While CAS is not empty, Do
 Find two groups conflicts weakly, then
 If the conflict state is between HLC and HLC Then Return Delegation
 If the conflict state is between HLC and LLC Then Return Forcing
 If the conflict state is between LLC and LLC Then Return Delegation
  Delete these two groups from CAS and add a dominant agent
  (depending on the    result from each selecting strategy) to CAS
  For each two groups in CAS
  If conflict strength is strong, Then
  If the conflict state is between LLC and LLC Then Return Ignoring
  If  the  conflict  state  is  between  HLC  and  HLC  Then  Return
Negotiation
  If the conflict state is between HLC and LLC Then Return Submitting
   Delete these two groups from CAS and add a dominant agent
   (depending on the result from each selected strategy) to CAS
```

4   An Example Scenario of Conflict Classification and Resolution

We clarify our approach through a scenario in which our proposed model helps to coordinate and manage conflicts states. In this scenario, we consider a commercial company, in which a Manager meets with his/her Head of Departments to decide on a strategic plan for the company's development. He/She requests their opinions to select an appropriate plan for each stage (of five stages). What strategies could be adopted if there are conflicting views?

In this case, we assume that the Manager has high confidence level due to his experience and knowledge, while the other members of the meeting (i.e., Head of Departments) have varying levels of confidence depending on their confidence factors.

Initialized in the first stage with five agents and five choices, at each stage, the algorithm determines: a set of conflicting agents, the total of its confidence value, and the conflict strengths. Let a_M be the Manager agent and a_1, a_2, a_3, and a_4 be the Head of Departments agents and that the agents select the five plans for the five stages. Figure 4 shows the five agents (a_M, a_1, a_2, a_3, a_4) and their selected plans (P1–P5) in five stages. Notice that the paths from Stage 1 to Stage 5 may have several strong conflicts (SC), e.g., a conflict state at Stage 4 when one agent (a_M) conflicts with four other agents (a_1, a_2, a_3, a_4), or a weak conflict (WC), e.g., the conflict state at Stage 1 when two agents (a_1, a_3) conflicts with three other agents (a_M, a_2, a_4).

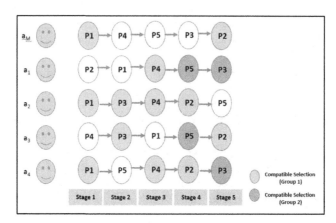

Fig. 4. Five Agents with its Selected Plans

In Stage 1, as shown in Table 1 and Figure 4, in iteration 1, there are two conflict cases, i.e. CAS={(a_1, a_3); $(a_1, [a_M, a_2, a_4])$; $(a_3, [a_M, a_2, a_4])$}. In the first case, agents a_1 conflicts with a_3 and the confidence value for a_1 is 4 and for a_3 is 3. Since the conflict between them is weak, the selected strategy based on the conflict resolution model of Figure 2 is **Submitting** leaving the selected plan as P2. Agent a_3 is then removed from the set, eliminating the third conflict case.

In the second iteration, the second conflict case is resolved, in which agent a_1 conflicts with the agent group (a_M, a_2, a_4). As shown in Table 1, the confidence value for the group is equal to $(6+3+2)=11$. Referring to the conflict resolution model that is proposed in Figure 2, we discovered that the conflict is strong, dictating the selected strategy as Forcing, leaving the selected plan in Stage 1 as **P1**.

Table 1. Selecting a Conflict Resolution Strategy

Stage No.	Iteration No.	Conflicting Agents	Confidence Value of Conflicting Agents	Conflict Type	Conflict Resolution Strategy	Contents of CAS at each Iteration	Selected Plan
1	1	a_1 a_3	4 3	Weak	Submitting	a_1, a_3,(a_M, a_2, a_4)	P2
	2	a_1 a_M, a_2, a_4	4 6+3+2 =11	Strong	Forcing	a_1, (a_M, a_2, a_4)	P1
2	1	a_M a_1	6 4	Weak	Submitting	a_M , a_1,a_4,(a_2,a_3)	P4
	2	a_M a_2, a_3	6 3+3=6	Weak	Negotiation	a_M, a_4, (a_2, a_3)	P3, P4
	3	a_M a_4	6 2	Weak	Submitting	a_M, a_4	P4
3	1	a_M a_1, a_2, a_4	6 4+ 3+2 = 9	Strong	Forcing	a_M,(a_1, a_4 a_2), a_3	P4
	2	a_1, a_2, a_4 a_3	9 3	Strong	Submitting	(a_1, a_4 a_2), a_3	P5
4	1	a_M a_1, a_3	6 4+3=7	Weak	Submitting	a_M,(a_1, a_3), (a_4,a_2)	P5
	2	a_1, a_3 a_2, a_4	4+3=7 3+2=5	Weak	Forcing	(a_1,a_3), (a_4,a_2)	P5
5	1	a_M, a_3 a_1, a_4	6+3= 9 4+2=6	Weak	Submitting	(a_M,a_3),(a_4, a_1),a_2	P2
	2	a_2 a_M, a_3	3 6+3= 9	Weak	Submitting	(a_M, a_3), a_2	P2

In the second stage, there are six conflicting cases, i.e. CAS={(a_M, a_1), $(a_M, [a_2, a_3])$, (a_M, a_4), $(a_1, [a_2, a_3])$, (a_1, a_4), $([a_2, a_3], a_4)$}. Each of a_M, a_1, a_4 agents selects a different plan and the other two agents, (a_2, a_3) select the same plan.

In the first iteration of the algorithm, the conflict between agents a_M and a_1 is resolved by detecting the conflict strength and the confidence value. Since a_M has a higher confidence value and the conflict is weak, the resolution strategy is **Submitting,** leaving **P4** as the selected plan. Agent a_1 is removed from the set, thus eliminating the fourth and fifth conflicts.

In the second iteration, the conflict is between a_M and the agent group (a_2, a_3). The confidence values for each of them are equal (i.e. 6 for a_M and (3+3)=6 for a_2 and a_3). Since the conflict is weak, the strategy to resolve the conflict is **Negotiation,** leaving the selected plan as either **P3** or **P4**.

In the third iteration of Stage 2, the conflict between a_M and a_4 is similarly resolved. Since the confidence value of a_M (i.e. 6) is higher than that of a_4 (i.e. 2), and the conflict between them is weak, then the selected strategy is **Submitting**. Agent a_4 is removed from the set, thus eliminating the sixth conflict.

In the third stage, there are three conflicting cases, i.e. CAS={$(a_M, [a_1, a_2, a_4])$, (a_M, a_3), $([a_1, a_2, a_4], a_3)$}. In the first iteration of stage three, conflict is between a_M and the agent group (a_1, a_2, a_4). In this case, the total confidence value of the triplet (4+3+2= 9) is greater than confidence value of a_M (6), and the conflict is strong. The resolution model suggests that the strategy to resolve this conflict is **Forcing,** so that the selected plan is **P4** favoring the agent group. Agent a_M is removed from the set, eliminating the second conflict case.

In the second iteration, the conflict is between the agent group (a_1, a_2, a_4) and a_3. Since the confidence of the agent group is higher and the conflict is strong, the strategy used is **Submitting,** leaving the selected plan as **P4**.

In the fourth stage, there are three cases of conflicts, i.e. CAS={$(a_M, [a_1, a_3])$, $(a_M, [a_2, a_4])$, $([a_1, a_3], [a_2, a_4])$}. Clearly, the confidence value of a_M (6) is less than the confidence value of (a_1, a_3) (i.e. 4+3=7) and the conflict between them is weak, so that the strategy to use is **Forcing** and the selected plan is **P5**. Agent a_M is removed from the set, eliminating the second conflict case.

In the second iteration, the conflict is between two agent groups, (a_1, a_3) and (a_2, a_4). Since the conflict between these two groups is weak and the first group has a confidence value (4+3=7) that is higher than confidence value of the second group (3+2=5), the **Forcing** strategy resolves the conflict and the selected plan is **P5**.

In the last stage, there are three conflict cases, i.e. CAS={$([a_M, a_3], [a_1, a_4])$, $([a_M, a_3,] a_2)$, $([a_1, a_4,], a_2)$}. In the first iteration, the conflict is between two agent groups (a_M, a_3) and (a_1, a_4). The conflict between them is weak and the first group, (a_M, a_3), has a higher confidence value (6+3=9) than that of the second group, (a_1, a_4), (4+2=6). The selected conflict resolution strategy is **Submitting** and a selected plan for this stage is **P2**. The agent group (a_1, a_4) is removed from the set. In the second iteration, the conflict between the agent group (a_M, a_3) and a_2 is weak, and the confidence value of the group (6+3=9) higher than the confidence value of agent a_2, so the strategy to resolve the conflict is **Submitting** leaving the selected plan as P2.

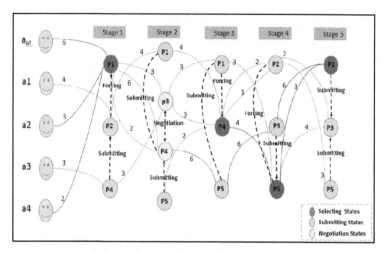

Fig. 5. The Conflict Resolution Model

From the above analysis, it is clear that the most effective plan for the company after resolving all the conflicts is: (Stage 1: P1, Stage 2: P3 or P4, Stage 3: P4, Stage 4: P5, Stage 5: P2).

5   Conclusion

In this paper, we propose a conflict classification algorithm in a multi-agent environment to resolve conflicts between two or more agents or groups of agents. We analyze the social theory of conflict and propose a conflict resolution strategy based on the new conflict classification scheme. The proposed conflict resolution algorithm considers all the conflict states of the agents in a multi-agent system. Classifying conflict states is used for conflict resolution enhancement. We demonstrate that by considering the confidence table as an input to the proposed algorithm, our classification technique is able to resolve multiple conflict situations and arrive at a decision.

References

1. Wagner, T., Shapiro, J., Xuan, P., Lesser, V.: Multi-Level Conflict in Multi-Agent Systems. LNAI, vol. 4386. Springer, Heidelberg (2007)
2. Tessier, C., Chaudron, L., Muller, H.J.: Conflict agents, Conflict management in Multi Agent System, vol. 1. Springer, Heidelberg (2000)
3. Basheer, G.S., Ahmad, M.S., Tang, A.Y.C.: A Framework for Conflict Resolution in Multi-agent Systems. In: Bădică, C., Nguyen, N.T., Brezovan, M. (eds.) ICCCI 2013. LNCS, vol. 8083, pp. 195–204. Springer, Heidelberg (2013)
4. Crawford, D., Bodine, R.: Conflict Resolution Education A Guide to Implementing Programs in Schools. Youth-Serving Organizations, and Community and Juvenile Justice Settings Program Report (1996)

5. Rummel, R.J.: Understanding Conflict and War. In: The Conflict Helix. Conflict in the Sociocultural Field, vol. 2, ch. 27. Sage Publications, California (1976)
6. Sycara, K.P.: Resolving Goal Conflicts via Negotiation. In: AAAI 1988 (1988)
7. Barber, K.S., Liu, T.H., Han, D.C.: Strategic Decision-Making for Conflict Resolution in Dynamic Organized Multi-Agent Systems. In: GDN 2000 PROGRAM. WAP Cellular Walkietalkie (2000)
8. Ioannidis, Y.E., Sellis, T.K.: Conflict resolution resolution of rules assigning values to virtual attributes. ACM 18(2) (1989)
9. Jung, H., Tame, M.: Conflict in Agent Team. Multiagent System, Artificial Intelligent, and Simulated Organizations, vol. 1. Springer, Heidelberg (2002)
10. Müller, H.J., Dieng, R. (eds.): Computational Conflicts: Conflict Modeling for Distributed Intelligent Systems With Contributions by Numerous Experts. Springer (2000)
11. Liu, T.H.A., Goel, M.C.E., Barber, K.S.: Classification and Representation of Conflict in Multi-Agent Systems. The Laboratory for Intelligent Processes and Systems, The University of Texas at Austin (1989)

Cost-Driven Active Learning with Semi-Supervised Cluster Tree for Text Classification

Zhaocai Sun[2], Yunming Ye[2], Yan Li[1], Shengchun Deng[3], and Xiaolin Du[2]

[1] Shenzhen Graduate School, Harbin Institute of Technology
[2] School of Computer Engineering, Shenzhen Polytechnic
[3] Department of Computer Science, Harbin Institute of Technology
zhcsun@hotmail.com, yeyunming@hit.edu.cn, liyan@szpt.edu.cn,
dengshengchun@hit.edu.cn, duxiaolin@gmail.com

Abstract. The key idea of active learning is that it can perform better with less data or costs if a machine learner is allowed to choose the data actively. However, the relation between labeling cost and model performance is seldom studied in the literature. In this paper, we thoroughly study this problem and give a criterion called as cost-performance to balance this relation. Based on the criterion, a cost-driven active SSC algorithm is proposed, which can stop the active process automatically. Empirical results show that our method outperforms active SVM and co-EMT.

Keywords: Active Learning, Semi-supervised Learning, Cluster Tree.

1 Introduction

In intelligent information and data mining, less training data but with a higher performance is always desired[11,20,17]. However, it is empirically shown in most cases that the performance of a classifier is positively correlated to the number of training samples. Active learning is then proposed to solve this contradicting but meaningful problem[9,11]. Because trained on the most informative data, the active classifier can also achieve a high performance, even though less training data is used than ordinary classifiers. This famous phenomenon is called "less is more"[9].

In general, the procedure of an active learner is summarized as follows[11,17]. 1, Only the most uncertain data (e.g. near to the class margin) are sampled from the unlabeled data set; 2, The sampled data are labeled by the "oracle" and added into the labeled data set; 3, On the new and enlarged labeled data set, the active classifier is retrained. Through repeating the above three steps T times, the active classifier becomes stronger and stronger. It is noticed that, in most active algorithms, T must be pre-set (e.g. T=50). That is, the drawback of current active algorithms is that there is no mechanism to stop the active process automatically. In other words, there is no criterion to compare two classifiers in the process of active learning. Although performing better than the prior classifier generally, the posterior classifier used more labeled data for training, thus has the higher cost. So, it needs a criterion to balance the performance and the cost of the active classifier.

J. Sobecki, V. Boonjing, and S. Chittayasothorn (eds.), *Advanced Approaches to Intelligent Information and Database Systems*, Studies in Computational Intelligence 551,
DOI: 10.1007/978-3-319-05503-9_5, © Springer International Publishing Switzerland 2014

In this paper, we thoroughly study the relation of performance and cost, and propose a criterion for assessing the active classifier, which is termed as cost-performance. Moreover, based on the classification model of active SSC (Semi-Supervised Cluster tree), we give a method to estimate the cost-performance and design a mechanism for automatically stopping the active process.

The rest of this paper is organized as follows. Section 2 reviews related works about active learning, in which the active SSC algorithm is emphasizing. Section 3 gives the theoretical analysis of the cost-performance. Section 4 proposes our cost-driven active SSC algorithm, which can automatically find out the classifier with the highest cost-performance. Section 5 and Section 6 are experiments and conclusions respectively.

2   Related Works

2.1   Active Learning Methods

The key idea of active learning is that if a machine learner is allowed to choose the data actively then it can perform better with less costs. So as far, it has widely applied to the field in which labeled samples are hard or expensive to extract, such as speech recognition[4], image retrieval[5], web page categorization[19,16], and so on. In general, based on a classification model (e.g. SVM or logistic regression), active learning focuses on the strategy of sampling the most informative data. It includes Uncertainty Sampling[3], Query-By-Committee[13], Expected Model Change[12], Expected Error Reduction[8], Variance Reduction[2], Density-Weighted Methods[10], etc. In recent years, some researchers also combined semi-supervised learning and active learning to obtain a better learner[20].

2.2   Active Semi-Supervised Cluster Tree Review

SSC (Semi-Supervised Cluster) algorithm is a semi-supervised method, which build a tree classifier (illustrated in Fig.1) with both labeled data and unlabeled data. Not like C4.5 [7] or CART [14], a clustering algorithm, specially k-means, is adopted to

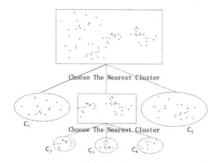

Fig. 1. Semi-supervised cluster tree. Black dots denote unlabeled data, others are labeled data.

group close data (including both labeled samples and unlabeled samples) into different sub sets. With the help of unlabeled data, the semi-supervised tree has more stable nodes and a higher performance. Empirical results have shown that SSC have a better generalization ability and suits to deal with the data with very few labeled samples. For more details, please refer to our past work [15].

In [16], we also proposed an active SSC algorithm based on batch mode active strategy. For most active algorithms, only one sample is sampled in each active step. This one-one strategy may induce inefficiency, especially on complex classification models[6]. Hence, in each active step, we sampled unlabeled data from all leaf nodes.

$$\mathcal{Q}(\mathbb{T}) = \bigcup \mathcal{Q}(C_k) \tag{1}$$

where C_k is a leaf node of tree $\mathbb{T}$.

Because our active strategy balances clusters of tree, active SSC is an effective and stable active approach.

3 The Analysis of Cost-Performance

Given a classifier and N labeled training data, $\Theta(N)$ denote the classification accuracy of learned classifier, and we have the following properties.

Property 1. $\Theta(x)$ gradually increases when the labeled data is few and is concave over the whole data space.

This property inherits the general decision tree's property that with the increase of training examples, the model accuracy increases. However, if overwhelming training data are provided, then the classifier tree is overfitting and thus is concave.

Property 2. $0 \leq \Theta(x) \leq 1$ and $\Theta(x) = 0$ when $x = 0$.

If there is no labeled samples, then there is no class information. In this case, the classifier is invalid and the corresponding model accuracy equals to zero, i.e. $\Theta(0) = 0$.

The term "cost" we used in this paper especially refers to human's effort to label on a given data sample.

Assumption 1. The cost of training a classifier only relies on the number of labeled training samples.

Denoting the cost as $\pi(N)$, our study is to balance $\Theta(N)$ and $\pi(N)$ to get a better classification system.

Property 3. $\pi(x)$ is an increasing and convex function.

The cost $\pi(x)$ is increasing obviously. For convex property, it is caused by two reasons. 1, With the scale enlarging, the human will becomes bored to label the sample more and more. Since working efficiency decreases, the cost of labeling one sample increases. 2, In active learning, the earlier queried samples are simple to label relatively. If an active classifier performs very well, the queried sample has little differentiable characteristics actually. With the number of labeled samples increasing, it is harder and harder to classify the queried samples for the human.

Based on Property 3, we give a formulation of cost function.

Assumption 2. $\pi(N) = e^{\rho N}$

Note that, ρ is a parameter. It relies on the users and/or the difficulty of labeling data. For example, Tagging a long blog costs different to tagging a short blog.

In active learning, the higher performance and the lower cost are both desired. Thus, we define a new criterion to balance them in this paper.

Definition 1. The cost-performance of a classifier is the ratio of performance to cost.

$$\Psi(N) = \frac{\Theta(N)}{\pi(N)} \tag{2}$$

The functions of performance, cost and cost-performance are shown in Fig.2. When

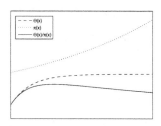

Fig. 2. The functions of performance, cost and cost-performance

$x \geq 0$,

$$0 \leq \Theta(x) \leq 1, \pi(x) > 1, \tag{3}$$

For cost-performance function, there is,

$$0 \leq \Psi(x) < 1 \tag{4}$$

Theorem. In $(0, \infty)$, there is one and only one x_0 satisfies $\Psi'(x_0) = 0$.
Proof. Suppose there are $x_0, x_1 > 0$ satisfies

$$\Psi'(x_0) = 0, \Psi'(x_1) = 0 \tag{5}$$

That is,

$$\frac{\Theta'(x_0)\pi(x_0) - \Theta(x_0)\pi'(x_0)}{\pi^2(x_0)} = 0, \tag{6}$$

and,

$$\frac{\Theta'(x_1)\pi(x_1) - \Theta(x_1)\pi'(x_1)}{\pi^2(x_1)} = 0, \tag{7}$$

Let $F(x) = \Theta'(x)\pi(x) - \Theta(x)\pi'(x)$, there exists x^*, $x^* \in (x_0, x_1)$ or $x^* \in (x_1, x_0)$, that satisfies

$$F'(x^*) = 0 \tag{8}$$

That is,

$$F(x^*) = \Theta''(x^*)\pi(x^*) - \Theta(x^*)\pi''(x^*) = 0 \tag{9}$$

Otherwise, $\Theta(x)$ is strictly concave and $\pi(x)$ is strictly convex.

$$\Theta''(x) < 0, \pi''(x) > 0 \tag{10}$$

There is,

$$F(x) = \Theta''(x)\pi(x) - \Theta(x)\pi''(x) < 0 \tag{11}$$

Eq.9 and Eq.11 are conflicting. So, the supposing does not stands. There is only one point x_0 that satisfies $\Psi''(x_0) = 0$. □

This Theorem tells the uniqueness of extremum. If the performance dose not increase, the active process stops.

4   Cost-Driven Active Semi-Supervised Cluster Tree

4.1   The Estimation of Performance

As discussed, the performance of a classifier usually is verified by the classification accuracy. But in training of an active classifier, the classifiers are blind to the test data set. The classification accuracy is unable to get. Hence, it is important to estimate the accuracy only by the training data.

An intuitive idea is the generalization error. For example, SVM reduces error by controlling the upper bound of VC-dimension[18]. But for tree classifier (e.g. C4.5,CART), it is hard to estimate a general error. However, our cluster uses clustering to generate a tree. Samples in the same node are similar very much. The quality of a cluster can be estimated by labeled samples in it.

In this paper, we use the information entropy to estimate the performance of a classifier. Suppose the leaf set of cluster tree $\mathbb{T}$ is $\mathcal{L}$. For each cluster C_k in $\mathbb{T}$, it includes labeled set C_k^L and unlabeled set C_k^U. Suppose there are M classes, $\omega_1, \omega_2, \cdots, \omega_M$, on C_k^L. Let the frequency for ω_m is q_m. The entropy of C_k is defined as,

$$E(C_k) = \sum_{m=1}^{M} q_m \log(q_m) \tag{12}$$

The entropy $E(C_k)$ embodies the quality of C_k. The smaller the entropy, the higher the quality. Summing all entropies of leaves with weights, the entropy of tree $\mathbb{T}$ is get.

$$E(\mathbb{T}) = \sum_{C_k \in \mathcal{L}} \frac{|C_k|}{|X|} E(C_k) \tag{13}$$

Using the entropy $E(\mathbb{T})$, the performance of our semi-supervised cluster tree is estimated. According to the relation of performance and entropy, we can estimate the cost-performance by cost and entropy.

Definition 2. The cost-entropy is defined as

$$\tilde{\Psi}(\mathbb{T}, N) = \frac{1 - E(\mathbb{T}, N)}{\pi(N)} \tag{14}$$

We expect cost-performance is equivalent to cost-entropy, $\Psi \approx \tilde{\Psi}$.

4.2   The Cost-Driven Active SSC Algorithm

Summarizing the above content, this section proposes a cost-driven active SSC algorithm. Comparing to current active algorithms, there is a mechanism to stop the active precess automatically. We exam the cost-entropy $\tilde{\Psi}(\mathbb{T}, N)$ in each step of the active process. If the cost-entropy of the posterior classifier is lower than that of the prior classifier, $\tilde{\Psi}_{t+1} < \tilde{\Psi}_t$, it is considered that the active classifier reaches the maximum of cost-performance, thus the active process is stopped. In addition, to our experience, in the beginning of the active process, the performance may fluctuate sharply. Hence, we set a threshold T_0 to avoid the fluctuation of cost-entropy. That is, Only when the active process becomes stable, the cost-entropy is compared to finding the best active classifier. The detail of our algorithm is listed in Algorithm 1.

Algorithm 1. The cost-driven active SSC algorithm

Input: Unlabeled training examples D^U
Output: Semi-supervised cluster tree $\mathbb{T}$
  1: initialize factors, $\tilde{\Psi}_0 = -1, \tilde{\Psi}_1 = 0, t = 1$.
  2: initialize the labeled pool, $D^L = \emptyset$.
  3: randomly query examples from D^U and add them into D^L.
  4: **while** $\tilde{\Psi}_{t-1} < \tilde{\Psi}_t$ and $t < T_0$ **do**
  5:     generate a semi-supervised tree $\mathbb{T}$ with D^L and D^U;
  6:     query examples from D^U with N2other strategy and add them into D^L;
  7:     $\tilde{\Psi}_{t-1} = \tilde{\Psi}_t$;
  8:     update $\tilde{\Psi}_t$ (as Eq.14);
  9:     $t = t + 1$;
 10: **end while**

---

5   Experiments

In our experiment, the public Reuters-21578 data set is chosen as the data set. The original Reuters-21578 data set consists of 21578 web pages from 135 different categories. We selected the first 10 categories for the classification task. The detailed characteristics is reported Table.1

Table 1. Characteristics of Reuters 10 Classes Data Set

Category	Earn	Acq	Money-fx	Grain	Crude	Trade	Interest	Ship	Wheat	Corn
Documents	3713	2055	321	298	245	197	142	114	110	90

Our experiments include two parts. In Section 5.1, we verified the feasibility of the proposed cost-entropy strategy. The trend of entropy and classification accuracy is given to illustrate their relation. Based on cost-entropy, an auto-stoping strategy in active learning is shown. In Section 5.2, we compared our algorithm to some famous active classifiers. The evaluation index includes cost-performance, classification accuracy, F-measure, etc.

5.1 Empirical Studies on Cost-Entropy Strategy

The purpose of this experiment is to evaluate the effectiveness of the active SSC of which the active learning is controlled by the proposed proposed cost-performance criterion. We expected to find from the experimental results that: (1)the trend of model performance increases, whereas the trend of entropy decreases when the labeled samples are enough; (2) the cost-entropy can be used to help to find the best tradeoff between cost and performance.

Table 2. Details of 4 subsets used in the experiment

sub set	classes	instances	categories contained
Set R2	2	5768	Earn & Acq
Set R3	3	314	Ship & Wheat & Corn
Set R5	5	648	Trade & Interest & Ship & Wheat & Corn
Set R10	10	7285	Earn & ... & Corn

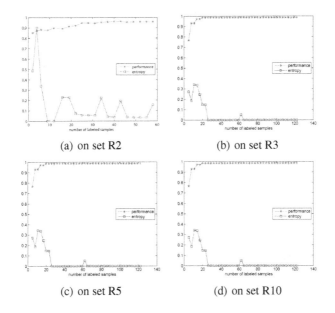

(a) on set R2

(b) on set R3

(c) on set R5

(d) on set R10

Fig. 3. Experiments of Performance V.S. Entropy

From the data shown in Table 1, we generate four group of data sets for this experiment. The data sets are called as R2;R3;R5;R10, which include two, three, five and all ten classes respectively. More information about these data sets can be seen in Table 2. For the model training process, only one web page per class is randomly selected out to form the initial labeled data set, and the class label of the rest web pages are removed to form the unlabeled data set. Then active SSC is run with this experimental settings. At each active learning step, the corresponding model accuracy (stands for performance) as well as the entropy calculated are reported in Fig.3. Each active SSC per data set is plotted in a sub figure, in which the model performance and entropy are plotted in red line and blue line, respectively.

From the results on all 4 sub data sets, the performance of active SSC increases gradually with the increase of labeled samples. The entropy cure of each active SSC fluctuates when the labeled samples are few, which is reasonable. In the largest data set R10, the number of web pages is over 7000, and more labeled samples are used in this setting. It can be seen that both the performance and entropy of three active SSCs are steadily converged. Similar observations could be found in the data set R3 and R5, which is reported in Fig.3 (b) and (c). Although R2 contains more than 5000 web pages, a comparably smaller number of labeled samples are used in this data set, and the corresponding entropy curve fluctuates greater than that of the rest settings. However, the entropy trend gradually decreases. These empirical results meet our expectation very well. This is reasonable that the sum of accuracy and entropy is approximately equal to 1, which is expected.

As discussed in previous section, the cost-entropy ratio is the best candidate to approximate the cost-performance ratio, which is used to seek the best tradeoff between cost spent and model accuracy. The study on the relationship between cost-performance and costentropy is performed on two datasets and the results are reported in Fig.4. In this experiment, the cost coefficient ρ was set to 0.001. It is well noticed that the cost-entropy curve fluctuates a lot at the beginning stage of the learning process. After this fluctuating stage (the left area of the dotted vertical line), the cost-entropy curve becomes more stable. As discussed in section 4, the selection of this stage is an empirical value. The circled part in all sub figures of Fig.4 is reported as the maximum value of the cost-performance acquired after the fluctuating stage. In this experiment, we simply set this value to be in proportion to the number of labeled samples used to learn the active SSC.

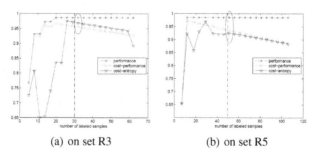

(a) on set R3   (b) on set R5

Fig. 4. Experiments of Cost-Performance V.S. Cost-Entropy

5.2 Comparison Experiment

In this experiment, we compared the proposed cost-driven active SSC to active SVM[1] and co-EMT[20]. We first randomly grouped the sub sets of Reuters to form 5 different experimental data sets consist of web pages from 3, 4 and 5 classes, respectively, to form a multi-class environment. Description about the data set refers to Table 3. The active SSC learning process will be terminated if the cost-entropy reaches its maximum value after the fluctuating stage. Because active SVM and co-EMT do not have an auto-termination mechanism, they reach the maximum learning steps. Only results with the best cost-performance is reported. The evaluation criteria includes precision, recall, and F-Score. The calculated results are reported in Table 4.

Table 3. Details of multi-class datasets

subset	classes	instances	categories
S1	4	456	Interest, Ship, Wheat, Corn
S2	4	563	Trade, Interest, Ship, Wheat
S3	3	864	Money-fx, Grain, Crude
S4	3	740	Grain, Crude, Category
S5	5	653	Category, Trade, Interest, Ship, Wheat, Corn

Table 4. Results of comparing with Co-EMT

Subset	Method	labeled samples	cost-performance	TP Rate	FP Rate	Precision	Recall	F-Measure
S1	Co-EMT	49	.8998	.9450	.0196	.9465	.9450	.9447
	active SVM	57	.8524	.9645	.0057	.9611	.9697	.9703
	active SSC	48	**.9426**	.9890	.0021	.9896	.9890	.9890
S2	Co-EMT	57	**.9270**	.9814	.0192	.9821	.9814	.9814
	active SVM	65	.9115	.9887	.0156	.9895	.9885	.9887
	active SSC	44	.9037	.9444	.0339	.9515	.9444	.9450
S3	Co-EMT	36	.9382	.9726	.0110	.9749	.9726	.9726
	active SVM	73	.8445	1	0	1	1	1
	active SSC	52	**.9493**	1	0	1	1	1
S4	Co-EMT	61	.9046	.9615	.0123	.9627	.9615	.9603
	active SVM	58	.9263	1	0	1	1	1
	active SSC	32	**.9685**	1	0	1	1	1
S5	Co-EMT	82	.7834	.8503	.0810	.8588	.8503	.8425
	active SVM	89	.7812	.8621	.0669	.8635	.8570	.8514
	active SSC	56	**.8362**	.8979	.0488	.8987	.8979	.8968

It is well noticed that active SSC achieves the highest cost-performance value on four data sets: S1; S3; S4andS5. And on these data sets, precision, recall and F-measure of active SSC outperforms the Co- EMT every well. Moreover, the number of labeled samples of active SSC is much lower than that of Co-EMT on S4andS5, and exceeds that of Co-EMT in S3. When the Co-EMT outperforms the active SSC in S2, the number of labeled samples of active SSC is much fewer than that of Co-EMT. This might be caused by the inappropriate empirical set up for the fluctuating stage. For active SVM, The F-measure is close to active SSC. But it uses more labeled samples than Co-EMT and active SSC. The reason is that both active SSC and Co-EMT are active+semi-supervised learning algorithms ,while active SVM is an active learning algorithm. Semi-supervised learning uses unlabeled samples to help train the classifier, which will reduce the number of labeled samples and save the cost. To summarize, based on these observed results, we can conclude that the proposed cost-sensitive strategy works well in terms that, in most cases, it can acquire a better model performance but with a much lower labeled samples.

6   Conclusions

To further reveal the relation between labeling cost and model performance, the cost-performance is proposed in this paper which is then approximated by cost-entropy to be as the cost-sensitive strategy. This cost-sensitive strategy could be used to terminate the active learning process as well as to find the best tradeoff between labeling cost and model performance. Experimental results show the effectiveness of our method.

Acknowledgment. This research is supported in part by NSFC under Grant No. 613031 03 and No.61073051, and Shenzhen Science and Technology Program under Grant No. JCY20130331150354073, No. CXY201107010163A and No.CXY201107010206A.

References

1. Baram, Y., El-Yaniv, R., Luz, K.: Online choice of active learning algorithms. The Journal of Machine Learning Research 5, 255–291 (2004)
2. Geman, S., Bienenstock, E., Doursat, R.: Neural networks and the bias/variance dilemma. Neural Computation 4(1), 1–58 (1992)
3. Lewis, D., Gale, W.A.: A sequential algorithm for training text classifiers. In: ACM SIGIR Conference on Research and Development in Information Retrieval, pp. 3–12 (1994)
4. Nallasamy, U., Metze, F., Schultz, T.: Active learning for accent adaptation in automatic speech recognition. In: IEEE SLT 2012, pp. 360–365 (2012)
5. Niu, B., Cheng, J., Bai, X., Lu, H.: Asymmetric propagation based batch mode active learning for image retrieval. Signal Processing (2012)
6. Patra, S., Bruzzone, L.: A cluster-assumption based batch mode active learning technique. Pattern Recognition Letters 33(9), 1042–1048 (2012)
7. Quinlan, J.R.: C4. 5: programs for machine learning. Morgan Kaufmann (1993)
8. Roy, N., McCallum, A.: Toward optimal active learning through monte carlo estimation of error reduction. In: ICML 2001 (2001)
9. Schohn, G., Cohn, D.: Less is more: Active learning with support vector machines. In: ICML 2000, pp. 839–846 (2000)
10. Settles, B.: Active learning literature survey. University of Wisconsin, Madison (2010)
11. Settles, B.: Active learning. Synthesis Lectures on Artificial Intelligence and Machine Learning 6(1), 1–114 (2012)
12. Settles, B., Craven, M., Ray, S.: Multiple-instance active learning. In: Advances in Neural Information Processing Systems, pp. 1289–1296 (2007)
13. Seung, H.S., Opper, M., Sompolinsky, H.: Query by committee. In: Proceedings of the Fifth Annual Workshop on Computational Learning Theory, pp. 287–294 (1992)
14. Steinberg, D., Colla, P.: Cart: classification and regression trees. Salford Systems, San Diego (1997)
15. Su, H., Chen, L., Ye, Y., Sun, Z., Wu, Q.: A refinement approach to handling model misfit in semi-supervised learning. In: Cao, L., Zhong, J., Feng, Y. (eds.) ADMA 2010, Part II. LNCS, vol. 6441, pp. 75–86. Springer, Heidelberg (2010)

16. Sun, Z., Ye, Y., Zhang, X., Huang, Z., Chen, S., Liu, Z.: Batch-mode active learning with semi-supervised cluster tree for text classification. In: WI 2012, pp. 388–395 (2012)
17. Tong, S., Koller, D.: Support vector machine active learning with applications to text classification. The Journal of Machine Learning Research 2, 45–66 (2002)
18. Vapnik, V.N.: The nature of statistical learning theory. Springer (2000)
19. Zhu, J., Ma, M.: Uncertainty-based active learning with instability estimation for text classification. ACM Transactions on Speech and Language Processing 8(4), 5 (2012)
20. Zhu, X., Lafferty, J., Ghahramani, Z.: Combining active learning and semi-supervised learning using gaussian fields and harmonic functions. In: ICML Workshop on the Continuum from Labeled to Unlabeled Data in Machine Learning and Data Mining, pp. 58–65 (2003)

Parameter Estimation of Non-linear Models Using Adjoint Sensitivity Analysis

Krzysztof Łakomiec and Krzysztof Fujarewicz

Silesian University of Technology
Akademicka 16, 44-100 Gliwice, Poland
{krzysztof.lakomiec,krzysztof.fujarewicz}@polsl.pl

Abstract. A problem of parameter estimation for non-linear models may be solved using different approaches, but in general cases it can be always transformed to an optimization problem. In such a case the minimized objective function is a measure of the discrepancy between the model solution and available measurements. This paper presents the ADFIT program — a tool for numerical parameter estimation for models that contain systems of non-linear ordinary differential equations. The user of the program provides a model in a symbolic form and the experimental data. The program utilizes adjoint sensitivity analysis to speed up gradient calculation of the quadratic objective function. The adjoint system generating the gradient is created automatically based on the symbolic form of the model. A numerical example of parameter estimation for a mathematical model arising in biology is also presented.

Keywords: Identification, parameter estimation, nonlinear systems, ordinary differential equations, sensitivity analysis, automatic differentiation.

1 Introduction

The problems of numerical parameter estimation for nonlinear mathematical models are not trivial. Although today available numerical software, like for example Matlab, supports this task by using specific built-in toolboxes, in many scientific areas this task is still a challenge and cannot be solved by general-purpose tools. For example, identification of models of biological processes, such as cell signaling pathways or metabolic reactions, is still a difficult task. These difficulties arise mainly from complicated models (non-linear and high-dimensional) and limitations in available measurements (noisy and rare in time).

There are many possible approaches to parameter estimation of linear and non-linear models [5], [6]. In the general case a problem of parameter estimation for non-linear models may be solved by transforming it to an equivalent optimization problem. In such a case the minimized objective function is a measure (usually quadratic) of a discrepancy between the model solution and the available measurements. The optimization process is non-trivial and slow because computation of the value of the objective function requires numerical simulation (integration of ODEs) of the model in each iteration of the optimization

J. Sobecki, V. Boonjing, and S. Chittayasothorn (eds.), *Advanced Approaches to Intelligent Information and Database Systems*, Studies in Computational Intelligence 551,
DOI: 10.1007/978-3-319-05503-9_6, © Springer International Publishing Switzerland 2014

procedure. The time of computation rises greatly with the number of estimated parameters of the model. In such a case it is very useful to obtain the information about the gradient of the objective function. This gradient may be calculated approximately using a finite difference approach or precisely (apart from the numerical errors) by using so-called forward sensitivity analysis. In both situations the time for the gradient computation rises with the number of estimated parameters. Fortunately, there is another possible way to calculate the gradient — by using so-called adjoint sensitivity analysis which requires only one simulation of the adjoint system in order to obtain the whole gradient and the time of computation is independent of the number of estimated parameters. Such an approach was proposed in our former papers [1], [2], [3].

This paper presents a computer software — the ADFIT program — written in Matlab that uses adjoint sensitivity analysis for non-linear model identification. The key element of the program is a fully automated process of adjoint system construction based on the model given by the user in symbolic form.

The paper also presents an example of application of the ADFIT program to parameter estimation for the mathematical model of DNA damage repair described in [4].

2   The ADFIT Tool

The ADFIT tool is a set of scripts in Matlab environment, which create the C code from the symbolic form of the model. This C code can be compiled directly to MEX a file and used in internal Matlab functions. This tool can be used for parameter estimation (based on discrete-time experimental data) of mathematical models containing a system of ordinary differential equations (the current version can load up to 30 model equations). The graphical user interface (GUI) of this tool is shown in Fig. 1.

2.1   ADFIT Features

The main features of the ADFIT tool are:

- Parameter estimation for models containing system of up to 30 ordinary differential equations,
- Two built-in numerical methods for solving differential equations: the Euler method and the 4-order Runge-Kutta with a constant step of integration,
- Ability to estimate the initial conditions for each equation in the system
- Two methods of gradient calculations: adjoint sensitivity analysis or the finite difference method.

The program fits parameters of the model to experimental data by minimization of the scalar objective function:

$$J = \sum_{i=1}^{n} \sum_{j=1}^{N_i} (y_i(t_j) - d_i(t_j))^2 \tag{1}$$

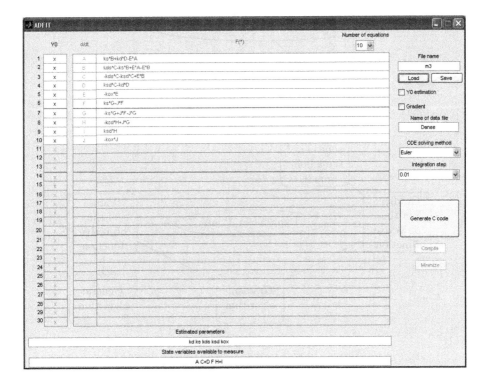

Fig. 1. Graphical user interface of the ADFIT tool

where: n is the number of measured variables, N_i is the number of measurements of the ith variable, $y_i(t_j)$ and $d_i(t_j)$ are the ith outputs of the model and the corresponding experimental data at time t_j.

In order to minimize the objective function the ADFIT program utilizes a trust-region method. However the results of this method are strongly dependent on the initial point of estimation. Hence, the ADFIT program uses a genetic algorithm to find a suboptimal starting point for the trust-region method that finally uses the gradient obtained by the adjoint sensitivity analysis. See Fig. 2 for a complete dataflow diagram of the ADFIT tool.

2.2   Graphical User Interface of the Program

The graphical user interface (GUI) of the ADFIT tool, presented in Figure 1, contains a set of options:

Number of equations – means the number of equations in the system,
Y0 – column of values that means the initial conditions for specific equations (if the checkbox Estimation Y0 is enabled then this column is disabled),
d/dt – column of state variables,

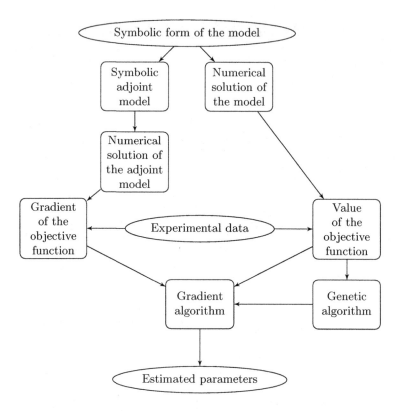

Fig. 2. Dataflow of the ADFIT tool, in ellipses are data provide/retrieve by user

F(*) – right side of ordinary differential equations (symbolic form),

Estimated parameters – names of estimated parameters (separated by spaces),

Measured state variables – names of state variables (separated by spaces) that
    are given by experimental data,

File name – name of the file containing the output C code,

Load – load the the file that contains the previously saved model,

Save – save to the the file with the current model,

Y0 estimation – if enabled, the program also estimates the initial conditions for
    every variable in the system, otherwise the user must provide a column Y0
    with all initial conditions,

Gradient – if enabled, the program utilizes adjoint sensitivity analysis in order
    to obtain a gradient, otherwise a finite difference method is used,

Name of data file – name of the file that contains a matrix of the experimental
    data,

ODE solving method – allows the user to choose the method for differential equa-
    tion solving (Euler or 4-order Runge-Kutta method),

Integration step – selects the integration step,

Generate C code – generates the C code of the function that calculates the objective function (and the gradient if the option Gradient is enabled),

Compile – compiles the generated C file to a MEX file,

Minimize – executes the internal script to minimize the objective function using the generated MEX file.

The experimental data in the ADFIT tool is provided by a matrix named Data which contains in the first column the times of measurements and in the next columns the values of state variables (or combinations of these variables) in the order that they are entered in the field Measured state variables. The experimental data should include values sampled at the same times — the program is not suitable for data where different variables are sampled at different times.

2.3   Availability

The ADFIT program may be downloaded from the web page cellab.polsl.pl.

3   Example of Application

The ADFIT tool was used to estimate the parameters of a model of repair of irradiated DNA repair [4]. Ionizing radiation creates two types of breaks in double stranded DNA: single strand breaks when breaks are created in only one strand, or double strand breaks when breaks are created close to each other in both opposite strands. The model presented in this section was made to describe the behavior of a supercoiled circular viral DNA called a minichromosome in cells. After irradiation DNA forms in analyzed cells a specific supercoiled structure called minichromosome. After irradiation of cells, two forms of damaged minichromosome DNA are obtained:

- the circular form which contains only single strand breaks,
- the linear form which contains a double strand break and single strand breaks.

The circular form is repaired directly to the supercoiled form by repair of single strand breaks, but the linear form may be repaired by two possible paths:

- First, the single strand breaks are repaired to obtain the linear form and next the double strand break is repaired to produce the circular supercoiled form, or
- First, the double strand break is repaired to produce the circular form and next the single strand breaks are repaired to produce the supercoiled form.

Considering the above information the model may be described using a system of ordinary differential equations:

$$\begin{cases} \dfrac{\mathrm{d}S}{\mathrm{d}t} = k_s \cdot CssB + INH \cdot k_d \cdot L \\[2mm] \dfrac{\mathrm{d}CssB}{\mathrm{d}t} = INH \cdot k_d \cdot LssB - k_s \cdot CssB \\[2mm] \dfrac{\mathrm{d}LssB}{\mathrm{d}t} = -INH \cdot k_{ds} \cdot LssB - k_{sd} \cdot LssB \\[2mm] \dfrac{\mathrm{d}L}{\mathrm{d}t} = k_{sd} \cdot LssB - INH \cdot k_d \cdot L \end{cases} \tag{2}$$

To estimate the parameters of this system, data from experiments done in Laval University on the repair of irradiated Epstein-Barr virus DNA present in Raji cells were used. In this experiment the repair of irradiated Epstein-Barr virus DNA present in Raji cells was examined. Additional information about these experiments is presented in [4].

The experimental data contains information about the quantities of the supercoiled DNA form (variable S in the model) and of the total linear form ($LssB + L$ in the model). Each sample is measured under two different experimental conditions: normal and with inhibition of repair of double strand breaks (logic variable in the model INH takes 1 in normal condition, and 0 otherwise). We don't have any information about the quantity of circular forms of minichromosome DNA (variable $CssB$ in the model).

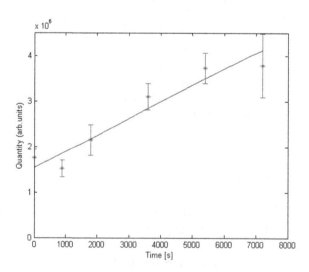

Fig. 3. Parameter estimation results: the supercoiled form (S variable in the model). Solid line – output of the model, points – the mean of experimental data, vertical lines – standard deviation of three separate measurements.

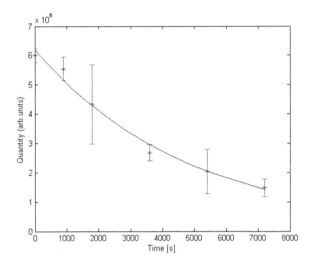

Fig. 4. Parameter estimation results: the total linear form (variable $L + LssB$ in model)

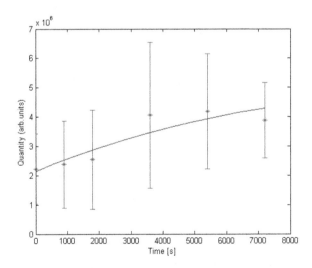

Fig. 5. Parameter estimation results: the supercoiled form (S variable in the model) with inhibition of double strand breaks repair

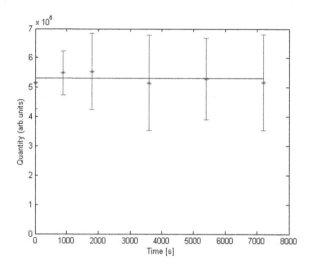

Fig. 6. Parameter estimation results: the total linear forms (variable $L + LssB$ in model) with inhibition of double strand breaks repair

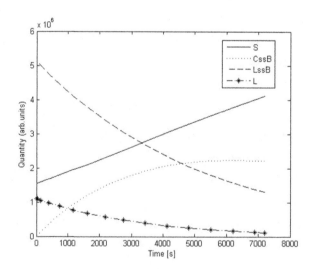

Fig. 7. Simulation of the model after parameter estimation in normal condition – variable INH in model is set to 1. Variables: S means the supercoiled form, $CssB$ means the circular form with single strand breaks, $LssB$ means the linear form with single strand breaks and L means the linear form without single strand breaks.

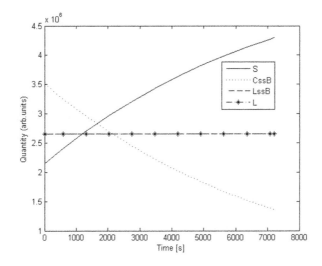

Fig. 8. Simulation of the model after parameter estimation with arrested double strand breaks repair – variable INH in model is set to 0. Variables: S means the supercoiled form, $CssB$ means the circular form with single strand breaks, $LssB$ means the linear form with single strand breaks and L means the linear form without single strand breaks.

Four parameters K_s, K_d, K_{sd}, K_{ds} (which are kinetic constants for particular repair pathways) and four initial conditions for all variables are estimated. In order to do this, we put the data into the ADFIT tool (as in Fig. 1) and started minimization of the objective function encoded in the output MEX file. After the minimization process we obtained the estimated parameters of this model and initial conditions. The results of fitting the model after this estimation process to the experimental data are shown in Figures 3, 4, 5, 6. Vertical lines represent the standard deviation of three independent measurements. We see that the optimized model describes the data well. In Figures 7, 8 we can see the simulation of the optimized model showing the change of particular fractions of DNA during repair. Additionally, in these Figures we see the changes of the circular DNA forms that we cannot obtain by experimental methods.

4   Conclusion

The ADFIT tool presented in this paper is a universal tool for automatic parameter estimation of models containing systems of ordinary differential equations. Through the use of C language a significant acceleration of the process determining the model parameters has been obtained. The built-in method for obtaining the gradient, which is based on sensitivity analysis, makes the process faster and efficient in comparison to the finite difference method (the procedure converges to the solution after fewer algorithm iterations).

The ADFIT program is being constantly developed, and we plan to improve it by elimination existing limits and introduction of the ability to estimate parameters of models containing delayed differential equations.

Acknowledgement. This work was supported by the Polish National Science Center under grant DEC-2012/05/B/ST6/03472. The authors would also like to thank Ronald Hancock from Laval University for reading the manuscript and valuable comments.

References

1. Fujarewicz, K., Galuszka, A.: Generalized Backpropagation through Time for Continuous Time Neural Networks and Discrete Time Measurements. In: Rutkowski, L., Siekmann, J.H., Tadeusiewicz, R., Zadeh, L.A. (eds.) ICAISC 2004. LNCS (LNAI), vol. 3070, pp. 190–196. Springer, Heidelberg (2004)
2. Fujarewicz, K., Kimmel, M., Swierniak, A.: On Fitting of Mathematical Models of Cell Signaling Pathways Using Adjoint Systems. Mathematical Biosciences and Engineering 2(3), 527–534 (2005)
3. Fujarewicz, K., Kimmel, M., Lipniacki, T., Swierniak, A.: Adjoint Systems for Models of Cell Signaling Pathways and Their Application to Parameter Fitting. IEEE/ACM Transactions on Computational Biology and Bioinformatics 4(3), 322–335 (2007)
4. Kumala, S., Fujarewicz, K., Jayaraju, D., Rzeszowska-Wolny, J., Hancock, R.: Repair of DNA Strand Breaks in a Minichromosome In Vivo: Kinetics, Modeling, and Effects of Inhibitors. PLoS One 8(1) (2013)
5. Ljung, L.: System Identification: Theory for the User. Pearson Education (1998)
6. Nelles, O.: Nonlinear System Identification: From Classical Approaches to Neural Networks and Fuzzy Models. Engineering Online Library. Springer (2001)

The Effect of Electronic Word-of-Mouth, Customer Expectations, and Emotions on Intention to Take Low Cost Airlines

Chia-Ko Lee[1], Yi-Chih Lee[2,*], Yuh-Shy Chuang[2], and Wei-Li Wu[2,*]

[1] Feng Chia University Ph.D. Program in Business, Feng Chia University
and Department of Administration, Cheng Ching General Hospital, Taiwan
[2] Department of International Business,
Chien Hsin University of Science and Technology, Taiwan
lyc6115@ms61.hinet.net, wuweili0709@yahoo.com.tw

Abstract. With the rapid development of low cost airlines in the Taiwan's airlines market, there are both positive and negative critiques of low cost airlines. Inexpensive prices are the way for low cost airlines to attract the public; however, unstable service quality is criticized by travelers. This study seeks to incorporate the effect of word-of-mouth, customer expectations, and emotional contagion into the dimensions to be explored, focusing on whether such factors affect the behavioral intentions of travelers to take low cost airlines. Results show that online word-of-mouth for low cost airlines would not affect consumers' intention to take low cost airlines. However, the expectations and emotional contagion of consumers would affect the intention to take low cost airlines.

Keywords: Electronic word-of-mouth, Customer Expectation, Emotional contagion, Behavior intention.

1 Introduction

The economic growth in Taiwan has led the public to demand better quality of life. Furthermore, the increasing conveniences of international travel help the rapid development of leisure activities overseas. According to the statistics of the Tourism Bureau of Taiwan in 2011, 19.1% of Taiwanese people have traveled abroad, which increased to 20.6% in 2012; the total person-times of outbound travel increased from 9.5 million in 2011 to more than 10 million in 2012 [1]. The Taiwanese tourism market has evolved from the early tour groups to semi- or even full-backpacking trips as people are familiar with outbound travel, and now seek new and different types of travel.

In order to enjoy more travel destinations or save on total travel costs, tourists strive to effectively lower transportation costs. In recent years, a number of low cost airlines have been established worldwide, which is conducive to the promotion of outbound travel. Early on, low cost airlines first started in the U.S., later expanding to Europe, Australia, and Asia. The first low cost airline that provided service in Taiwan was Jetstar Asia Airways of Singapore, which was established in 2004 [2]. Over the next few years, Air Asia of Malaysia, Cebu Pacific Air of the Philippines, Qatar

---

* Corresponding authors.

J. Sobecki, V. Boonjing, and S. Chittayasothorn (eds.), *Advanced Approaches to Intelligent Information and Database Systems*, Studies in Computational Intelligence 551,
DOI: 10.1007/978-3-319-05503-9_7, © Springer International Publishing Switzerland 2014

Airways, Tigerair, Air Busan of Asiana, Eastar Jet, Jin Air, t'way, Flyscoot, and Peach also entered the Taiwanese market.

With the rapid development of low cost airlines in the Taiwanese airline market, there are positive and negative critiques of low cost airlines. Inexpensive prices are the way for low cost airlines to attract the public; however, unstable service quality is criticized by travelers. Thus, this study seeks to incorporate the effect of word-of-mouth, customer expectations, and emotional contagion, into the dimensions to be explored, and focuses on whether such factors affect the behavioral intention of travelers to take low cost airlines.

2   Literature Review

2.1   Electronic Word-of-Mouth

The transmission of information between people originally relied on face-to-face communication; however, due to advancement of Internet technology, the exchange of online information has become an important channel for people of the new generation. Electronic word-of-mouth can deliver large amounts of diverse data, can be operated at low cost [3] and is very convenient to use; therefore, it is utilized by many people online. Blogs and social networking sites (SNSs) are important forms for the transmission of electronic word-of-mouth, where people can express their thoughts, share their feelings, and comment on various matters [4]. Meanwhile, through social networking, people can quickly transmit information to various corners of the world, and even become a powerful format for social issues. Research has shown that, in the online world, electronic word-of-mouth can affect the purchase intentions of consumers [5]; therefore, this study incorporates the factor of electronic word-of-mouth into exploration, deducing Hypothesis 1:

H1: For consumers, more positive electronic word-of-mouth has a greater effect on intentions to take low cost airlines

2.2   Customer Expectations

Scholars have pointed out that customer expectations are consumer beliefs regarding products or services, the expected benefit of the product prior to purchase, and that customer expectations can be predicted and observed in terms of overall feelings, individualized services, and reliability [6]. Consumers frequently use word-of-mouth, experience, and mass media to acquire their expectations for products or services. After they have personally experienced the product or service, they would compare their actual experiences with expected experiences. When satisfaction with their actual experiences is greater than the expected experience, they would feel satisfied; if they are equal, then there is no positive or negative feeling; however, when the expected experience is greater than the satisfaction with the actual experience, there would be complaints or dissatisfaction. Therefore, the establishment of customer expectation relies on continuous communication between corporations and consumers, in order to benefit positive information transfer. This study deduces Hypothesis 2:

H2: When expectations are low, the intention to take low cost airlines is greater

2.3   Emotional Contagion

Scholars have defined emotional contagion as the automatic imitation of the other person's expressions, gestures, voices, postures, and actions in the process of interacting with other people, and they are also inclined to capture the feelings of others; this process is called emotional contagion [7]. The topic of emotional contagion has be explored from several dimensions, with research pointing out that people would be under the effect of contagion, resulting in negative emotional changes; the stronger the negative event, the stronger the feelings, and the stronger the individuals negative emotions [8]. In the service industry, when service employees show a cold countenance, dispassionate language, and flippant service actions, these would all be perceived by customers. When customers receive negative service attitudes, their brains would quickly respond with negative emotions, resulting in poor overall service impression [8]. Therefore, positive or negative information and commentary regarding low cost airlines on the Internet all have considerable effect on potential consumers who intend to use low cost airlines. When customers shop, they hope to gain good emotional experiences; therefore, the emotional effect of reading website commentaries would deeply affect their consumption intentions. This study deduces Hypothesis 3:

H3: For consumers, emotional contagion positively affects the intention to take low cost airlines

3   Methods

This study explores to how customers are affected by eWOM, Customer Expectation, and Emotional Contagion when customers use blogs or websites to search low cost carrier information. Thus, the construct measurement, and reliability and validity analysis of the variables of the measurement models in this study are described, as follows:

The independent variables in this study are eWOM, Customer Expectation, and Emotional Contagion. In eWOM construct, revised from network News. In the Customer Expectation dimension, this study modified Turel and Serenkom (2006), with 7 points measured using a 7-point Likert scale (7=highly disagree, 1=highly agree) for testing[9]. Emotional Contagion primarily referred to questionnaires by Doherty (1997); there were a total of fifteen questions, using a Likert 7-point scale (1=highly disagree, 7=highly agree)[10]. Regarding the dimension of Behavioral Intention, the scale is modified from Zeithaml et al.(1996), with a total of 5 questions[11]. This study uses Partial Least Square (PLS) to conduct reliability and validity testing, as well as overall model analysis. The software used is smartpls, with overall reliability and validity as shown in Table 1. The dimensions of the measurement model have good reliability, with composite reliability greater than 0.7, and average variance extracted (AVE) over 0.5, both of which reach the standard for convergent validity.

Table 1. The fit index of the model

Construct	Composite Reliability	AVE	Cronbach's Alpha	eWOM	Customer Expectation	Emotional Contagion	Behavioral Intention
eWOM	1.00	1.00	1.00	1.00			
Customer Expectation	0.92	0.78	0.86	0.11	0.88		
Emotional Contagion	0.94	0.73	0.92	-0.06	0.41	0.85	
Behavioral Intention	0.92	0.75	0.83	0.13	0.62	0.43	0.87

4   Results

This study uses interviews to collect information in Taiwan. There are 104 valid questionnaire samples in this study.

This study established 3 hypotheses. The hypotheses are confirmed using PLS for structural equation analysis, with the structural model results as shown in Table 2. Standardized coefficient and significance testing show that customers who accept positive electronic word of mouth have higher behavioral intentions (Estimate value=0.089, p=0.116), but as this is not statistically significant, H1 is not supported; however, when customers have less positive customer expectations, it would actually elevate their behavioral intentions (Estimate value=0.524, p=0.005), thus, H2 is supported. Research results show that when customers are susceptible to emotion, it would affect intention to take the low cost carrier (Estimate value=0.221, p<0.001), H3 is supported. The model in this study has Average R^2=0.43, which is significantly greater than 0.35[12], GOF=0.59, thus, the model has good fit. The framework of this study is shown in figure 1.

Table 2. The results of the structural equation model

	Structural Path	Estimate Value	t value	p value	Verify
H1	eWOM→Behavioral Intention	0.089	1.584	p=0.116	Not Support
H2	Customer Expectation→Behavioral Intention	0.524	7.521	p=0.005	Support
H3	Emotional Contagion→Behavioral Intention	0.221	2.895	p<0.001	Support

p<0.05 significant

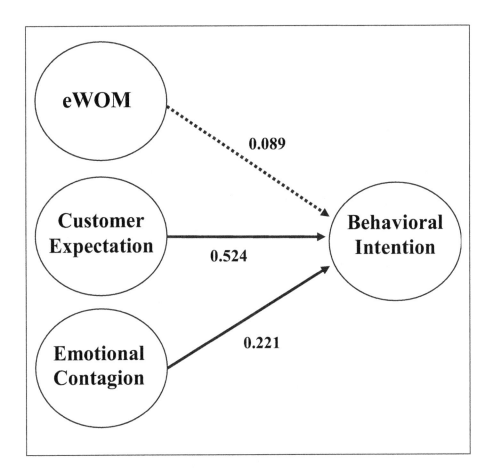

Fig. 1. The framework of this study

5   Discussions and Conclusion

Travelling abroad has become an important way for people to maintain quality of life, engage in leisure and entertainment activities, and conduct business. The main purpose of this study is to use electronic word-of-mouth, customer expectations, and the personal emotions of consumers to explore the intentions of customers to take low cost airlines. The results show the following:

Electronic word-of-mouth regarding low cost airlines does not affect consumers' intention to take low cost airlines. A possible reason is that, even though the media frequently reports on delays of low cost airlines, or cancellations due to short-notice adjustments of flights, as low cost airlines have clear stipulations posted online, and people post reminders on SNSs, traveler unhappiness is quickly replaced by the other conveniences provided by low cost airlines; therefore, the commentary quality of low cost airlines on the Internet would not directly affect the intention of consumers to take low cost airlines.

Next, this study shows that the expectations of consumers would affect their intention to take low cost airlines. A possible reason is that travelers do not expect low cost airlines to provide good service quality, may have delays or cancellations, and do not provide convenience goods on the plane; however, compared to regular airlines, airfare is very inexpensive, and can be conveniently purchased online without going through travel agencies. For consumers who expect reduced travel costs, low cost airlines are indeed very attractive modes of transportation. Therefore, this study shows that when customers have lower expectations regarding the quality of low cost airlines, they would have greater intentions to take low cost airlines.

In terms of individual consumers, when their emotions are affected by factors and their surroundings, intentions to take low cost airlines are more likely to change. Research shows that, when consumers feel stronger emotional contagion, they are more effective in their intention to take low cost airlines. When a traveler has friends and relatives who have positive attitudes regarding low cost airlines, then a traveler with strong emotional contagion would more likely take the low cost airline; and vice versa. Therefore, while the service quality of airlines would affect the usage intentions of consumers, the personalities of customers would also have considerable effect on usage intentions.

5.1   Limitation

The explanatory model proposed by this study has room for improvement. In addition, measurement items for the constructs can incorporate more areas for measurement. There were constraints of research issues and design. It is suggested that future studies explore different low cost airlines, as they have different service qualities, thus, future researchers can incorporate different brands for analysis. Furthermore, this study did not compare customers of different age groups. As different age groups have different demands and levels of tolerance for travel quality, customers should be divided into groups for further analysis.

References

1. Tourism Bureau, Republic of China: 2012 Survey of Travel by R.O.C. Citizens (2012), http://admin.taiwan.net.tw/public/public.aspx?no=137 (retrieved)
2. Jou, R.C., Liu, Y.H., Wang, W.C.: A Study On International Airline Choice Behavior Between LCC And FSC - From Taipei To Singapore. Transportation Planning Journal 36(3), 307–332 (2007)
3. Dellarocas, C.: The digitization of word of mouth: Promise and challenges of online feedback mechanisms. Management Science 49(10), 1407–1424 (2003)
4. Chen, C.Y., Lu, T.Y.: The Implications of Travel Blogs for Destination Marketing: An Example of Tanshui. Journal of Tourism and Leisure Studies 14(2), 135–159 (2008)
5. Chen, Y.F.: The Influence of e-WOM on Consumer Purchase Intention with Prospect Theory. Journal of e-Business 12(3), 527–546 (2010)
6. Anderson, E.W., Fornell, C.: Foundations of the American customer satisfaction index. Total Quality Management 11(7), 869–882 (2000)

7. Hatfield, E., Cacioppo, J.T., Rapson, R.L.: Emotional contagion. Cambridge University Press, New York (1994)
8. Du, J.G., Fan, X.C.: Multiple Emotional Contagions and Its Dynamic Impact on Consumer's Negative Emotion under Service Encounters. Acta Psychologica Sinica 41(04), 346–356 (2009)
9. Turel, O., Serenkom, A.: Satisfaction with Mobile Services in Canada: An Empirical Investigation. Telecommunications Policy 30(5/6), 314–331 (2006)
10. Doherty, R.W.: The emotional contagion scale: a measure of individual differences. Journal of Nonverbal Behavior 21(2), 131–154 (1997)
11. Zeithaml, V.A., Berry, L.L., Parasuraman, A.: The Behavioral Consequences of Service Quality. Journal of Marketing 60, 31–46 (1996)
12. Cohen, J.: Statistical Power Analysis for the Behavioral Sciences. Lawrence Erlbaum, Hillsdale (1988)

Part II
Intelligent Systems Advanced Applications

ETOMA: A Multi-Agent Tool
for Shop Scheduling Problems

S. Bouzidi-Hassini, S. Bourahla, Y. Saboun, F. Benbouzid-Sitayeb, and S. Khelifati

Laboratoire LMCS, Ecole nationale Supérieure d'informatique, Alger, Algérie
{s_hassini,s_bourahla,y_saboun,f_sitayeb,s_khelifati}@esi.dz

Abstract. In this paper, we present ETOMA a multi-agent framework dedicated to developing and testing floor shop production schedules. It is applied for both production and joint production and maintenance scheduling as well. ETOMA architecture is composed of three modules: Develop, Test and Blackboard. The first one defines all agents that model the floor shop and their behavior. The second one takes care of testing the scheduling solution by using Taillard benchmarks depending on floor shop types. The considered types are Flow-shop, Job-shop and Open-shop. Finally, the Blackboard insures communication between the two predefined modules. ETOMA allows developing and testing any scheduling solution without imposing a specific architecture for agents. Moreover, ETOMA provides at the end of simulations a report composed of a recapitulative table or a curve according to user choice.

Keywords: Multi-Agent Framework, Simulation, Scheduling, Production, Maintenance, Taillard Benchmarks.

1 Introduction

Scheduling is an important activity for manufacturing systems. It allocates tasks to machines over time by satisfying constraints related to tasks (precedence constraints, due dates ...) or to machines (availability, cummutativity, renewability....) and optimizing objective functions (minimizing Cmax[1], minimizing jobs' tardiness....) [1]. Floor shop organizations vary according to products to be manufactured. Flow-shop, job-shop and open-shop are principally the theoretical types of floor shops.

In production floor shops, and after a certain period of use, machines can break down and then provoke production disturbance. Preventive maintenance has been introduced to carry out maintenance operations in machines and equipments before the failure takes place. The objective is then to prevent failures before they happen and thus to minimize the probability of failure. As a consequence, preventive maintenance activities have to be scheduled with respect to production schedules because they use the same machines. Joint production and maintenance schedules are composed of production and maintenance tasks scheduled simultaneously in order to prevent machine failure and thus propose realistic deadlines.

---

[1] Cmax : completion date of the last operation of the last job.

J. Sobecki, V. Boonjing, and S. Chittayasothorn (eds.), *Advanced Approaches to Intelligent Information and Database Systems*, Studies in Computational Intelligence 551,
DOI: 10.1007/978-3-319-05503-9_8, © Springer International Publishing Switzerland 2014

Multi-agent technology has been widely applied to resolve scheduling problems [2] [3] [4]. Indeed reactivity, distributed decision making and collaboration are proprieties that make agents a good support for modeling floor shops. Agents that may represent any actors in the floor shop (jobs, machines, tasks... etc.) negotiate to get an agreement on how to allocate tasks and deadlines to machines. It is obvious that for implementing multi-agent solutions, we have to use a multi agent framework. This latter is a tool that offers services (agents' creation, destruction and communication) for developers in order to implement their solutions. Nowadays, many frameworks exist for this aim like JADE [6], MaDKit [7], Magique [8], etc. They are general frameworks each one has its specificity over the others but not dedicated to the scheduling problem. All existing tools that treat the scheduling problem [14] are only simulators of floor shops i.e. they illustrate how jobs pass through machines.

Implementing and testing scheduling solutions is essential for proving their efficiency by comparing them with existing ones. Several experimentations have to be done. It is about varying the problem data, calculating objective functions and comparing it with existing benchmarks. This test method is called "the benchmarking". According to [5] "a benchmarking is a learning process structured so as to enable those engaging in the process to compare their services-activities-processes-products-results in order to identify their comparative strengths and weaknesses".

In order to facilitate developing and testing multi-agent scheduling solutions, we propose in this paper a tool called ETOMA (Environnement de Test des Solutions d'Ordonnancement Multi-Agents). It is composed of three modules: the first one called "Develop" allows defining proposed solutions by specifying agents and their behavior. The second one called "Test" generates automatically tests by comparing results with existing benchmarks. Finally, "the Blackboard" is the third module which insures communication and data exchange between the two previous components.

¶The rest of the paper is organized as follows. In section 2 we present the ETOMA architecture and detail all its components. Section 3 illustrates through an execution example the ETOMA services. The last section enumerates the ETOMA advantages and presents some work perspectives.

2   ETOMA Architecture

As mentioned previously, ETOMA is composed of three main modules *Develop*, *Test* and *Blackboard* (figure1). In what follows we detail them.

2.1   Develop Description

This module offers necessary services for defining user's solution. Services are about agents' creation, destruction and communication. It respects FIPA[2] [9] norms regarding agents' identification, life cycle and provided services.

---

[2] FIPA : The Foundation for Intelligent Physical Agents.

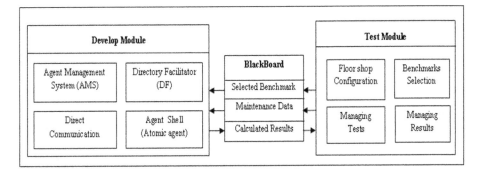

Fig. 1. Block diagram of ETOMA

2.1.1 Agent Characterization: Develop agent can offer one or many calculation services. It has its own state, behavior, control thread and a capacity for interacting with other agents. We propose an agent model based on the concept of *competence* as it is identified in Magique platform [8]. An agent is an empty shell in which we insert competences. These latter can vary from simple behaviors as responding to messages to realization of complex calculations. The idea is to separate the agent from what it can do. Thus, an agent is defined as « an entity talented with competences » [10]. We fix minimal set of competences for an agent. We admit that two initial competences are necessary for defining an atomic agent: "the ability of communicating" and "the ability of getting new competences".

2.1.1.1 Agent's Identification: Each agent is identified by a unique identifier called Agent IDentifier (AID) so it can be distinguished on the platform. The AID is attributed to the agent when it is created; it is composed of two fields:

- *Globally Unique IDentifer (GUID) :* local name of the agent at the platform
- *The platform name:* by default, it is machine host's name.

Exemple: machine_agent@user-pc

2.1.1.2 Agent Competences: A competence is defined as a sequence of operations needed to execute a specific task. It can be perceived as a java class that can be added to an agent. In programming, this approach favors reusability and modularity. Once a competence is developed, it operates in different contexts. It can be considered as a reusable software component. An agent behavior is composed of several competences that interpret agent's behavior in the environment.

2.1.1.3 Agent's Education: At the beginning, all agents have similar behavior. They have only two competences: communication and new competences acquisition. Consequently, the difference between agents is generated by their different education.

Their competences can be fixed by the programmer at their creation or learned dynamically due to interactions with other agents.

2.1.1.4 Machine Agent Description: From literature review [11] [12], we find that no generic methods exists for production scheduling based on multi-agent systems. However, all of these define a *Machine Agent* that represents the production resource. For this reason, we define a *"Machine Agent"* with elementary competences needed for users. The defined competences are:

- *Proposing Duration:* a Machine Agent has to propose processing times for tasks it can execute. It brings these processing times from current test's scenario.

- *Getting Maintenance Benchmark:* for each production scenario, our system generates a maintenance plan (section 2.2.4). In order to propose production task's processing time, Machine Agent has to take into consideration maintenance tasks programmed on the same machine. For this, it retrieves the maintenance benchmark from the blackboard.

2.1.2   Develop Services: We identify some FIPA agents for the framework management. We retain the Agent Management System (AMS) and the Directory Facilitator (DF).

2.1.2.1 Agent Management System (AMS): In general, this agent insures the following functionalities:

- *Inscribing a new agent:* every agent has to be inscribed at the AMS in order to save its identifier (AID) in the AMS directory.

- *Modifying existing AID:* the AMS verifies first that no agent is identified by the new AID.

- *Deleting existing AID:* when an agent exists no more at the platform, its AID has to be removed.

- *Searching an agent.*

- *Creating, throwing, suspending, waking and killing other agents:* every agent's father can effect theses actions by a request to the AMS.

2.1.2.2 Directory Facilitator (DF): The DF is an agent that offers yellow pages service. It allows the correspondence between agents and their competences or services. It can do two actions: *inscribing agents at the directory with their services* and *finding agents that can perform searched services.*

2.1.2.3 Inter-agents Communication: Inter-agent communication is direct by sending and receiving asynchronous messages. Agents can communicate directly without need to an intermediate agent like the FIPA ACC (Agent Communication Channel). In order to keep received messages, each agent is endowed with a mail box called a BAL. The BAL operates like a post mail and it follows a FIFO strategy. We propose three communication models:

- *Mono-diffusion (one-to-one):* there is only one receiver of the message.

- *Multi-diffusion (one-to-many):* It exist many receivers for the posted message.

- *Broadcasting (one-to-all):* all platform agents must receive the message.

We have implemented many structures of messages. We propose a library of basic messages. These structures allow exchanging strings, objects, competences or messages in accordance with FIPA specifications. All messages must contain parameters that vary according to exchanged data. The obligatory ones are: *performative, sender identifier, receiver identifier* and the *content*.

2.2   Test Description

This module offers the following functionalities:

2.2.1   Saving and Configuring a Scheduling Solution: Before starting a simulation, user has to specify the floor shop to be represented and the objective function that it wants to optimize. After that, the system chooses adequate benchmark for tests.

- *Floor shop types:* the developed system allows testing solutions for *job-shop*, *flow-shop* and *open-shop* configurations.

- *Objective function:* This information serves only for displaying results. It represents the function that the user wants to optimize.

- *Used benchmarks:* we use Taillard benchmarks [13] for several reasons. First, these benchmarks are frequently used in production scheduling. Second, they include the three principal floor shop types (*flow-shop, job-shop* and *open-shop*). Finally, their good documentation makes them easy to understand.

After defining a scheduling solution i.e. agents and their behaviors, we can save it under an XML format in order to reuse it later. Also, our system can configure a production shop floor from an XML description.

2.2.2   Interface Agent: *Test* module contains an agent called *"Interface Agent"*. It is a subclass of the atomic agent class. It is a mediator between user, Develop and Test Module (fig. 2). It identifies one or two test scenarios according to floor shop configuration selected by user. Then it puts down it on the blackboard. Finally, it retrieves results after the solution execution in order to display it. Interface Agent competences are:

- *Putting a benchmark on the blackboard.*
- *Putting maintenance plan*: this point will be explained at section 2.2.4.

- *Results retrieval:* allows results recuperation from the blackboard in order to display it.
- *Running Simulation.* Interface Agent is responsible of starting simulations.

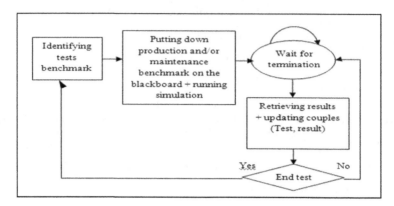

Fig. 2. The Interface Agent behavior

2.2.3   Managing Results: We include the following functionalities:

- *Tests report generation:* by generating a report, tests and all information about studied shop floor can be grouped. It summarizes floor shop type, ran tests number and final results i.e. the objective function value.
- *Graphical visualization:* graphical visualization is preferred for comparing results with those of the benchmark and so evaluating the objective function. In this order, our system offers two graphical visualizations: curves and tables.

2.2.4   Generating Maintenance Benchmark: Our system supports production and maintenance joint scheduling. By lack of benchmarks dedicated to this kind of scheduling, we have proceeded as follows: several maintenance tasks can be planned for one machine. Their processing time takes its value randomly from an interval fixed by user. Example: [10, 30]. Actually, we cannot specify randomly maintenance tasks because it depends on machines' operating duration. Consequently, we admit that a period between two maintenance tasks can be calculated by equation 1.

$$Ti = MP_i * NTM . \tag{1}$$

Where MP_i is the operating duration average of the selected production benchmark and NTM is the number of production tasks between two maintenance tasks. It's a parameter that can be fixed by user.

2.3   The Blackboard

It's an important component within the platform. It insures communication between Develop and Test modules. Indeed, the user solution needs data from the test module and this later need at its turn data from the former. In plus, as no generic scheduling method exists, we don't know which agent is responsible for solution calculation. So this structure allows results recuperation in order to display it by the test module. It is composed of three components. The first one is used for benchmarks deposition and retrieval. The second one, for results deposition and retrieval and the last one, for maintenance plan deposition and retrieval.

3   Experimentation

In order to illustrate all above functionalities, we consider a solution example (figure 3) and we run it on our framework. Let consider a multi agent scheduling for a flow-shop where the objective function is the Cmax minimization. In plus of platform's Machine Agent and Interface Agent, we identify two other agents:

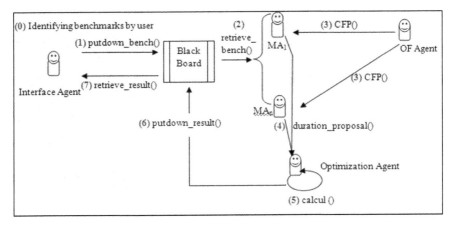

Fig. 3. Interaction among agents for scheduling

- *OF Agent:* is responsible of inializing the negotiation process for scheduling by sending a call for proposals (CFP) to Machine Agents.
- *Optimization Agent:* its role is to calculate resulting schedules.

As illustrated in figure 3, after user selection of benchmarks, the Interface Agent puts down every benchmark on the blackboard while creating Machine Agents. The OF Agent sends a CFP to existent Machine Agents. These latter, send their proposals according to the selected benchmark to the Optimization Agent. Optimization Agent calculates the resulted Cmax by executing the "calcul()" competence and puts down it on the blackboard. Finally, the interface Agent retrieves the result in order to save it.
We run the precedent example on our framework and we illustrate agents' behavior and framework functionalities via some screen captures. Figure 4 shows the ETOMA main GUI[3]. It is composed of four parts:

---

[3] GUI : Graphique User Interface.

Part one: describes defined agents for the scheduling solution. At the beginning, there is only the AMS, the Interface Agent and the DF. After that, the user can create agents corresponding to its solution (figure 5). In the previous example, we create OF Agent and Optimization Agent. Finally, we add competences to the created agents. For OF Agent, we add the "Ask_Proposal()" competence and for Optimization Agent, we add the "Calcul()" competence. Notate that Machine Agents are created automatically when the benchmark is selected (part three).

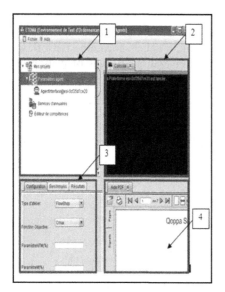

Fig. 4. The main GUI of ETOMA

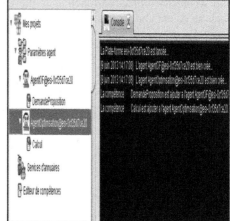

Fig. 5. Creating agents defined by user's solution

Part two: a console panel that keeps track of solutions execution is showed. Indeed, it is very important for users to follow their software execution. So, this part displays each performed task by agents principally exchanged messages.

Part three: Allows users to define floor shops type (flow-shop, job-shop or open-shop), the objective function (Cmax, Lmax and Rmax) and maintenance parameters (NMT and M). In our example, we select (figure 6) a flow-shop configuration, Cmax as an objective function and we do not consider maintenance parameters. In the same part, we can identify tests benchmark at the "benchmark" tag. As mentioned previously, we use Taillard banchmarks. Notate that if we want to use other benchmarks, they must have the same structure of Taillard benchmarks. When we identify a benchmark type, we create corresponding Machine Agents (figure 7) (the number of Machine Agents depends on selected benchmark). We select for example "tail 20_20_4". That means that the flow shop is composed of 20 Jobs which pass through 20 machines. Number 4 corresponds to the identification of the instance. Finally, we select a set of benchmarks for execution and we run the simulation.

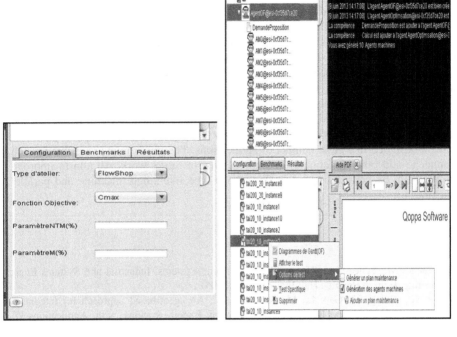

Fig. 6. Selecting floor shop characterizes

Fig. 7. Creating agent's machine according to benchmark's selection

Part four: After simulation's termination, we can view results as a curve (figure 8) or a table (figure 9). Furthermore, we can generate a recapitulative report that contains all information about the executed simulation.

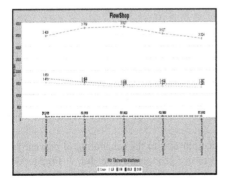

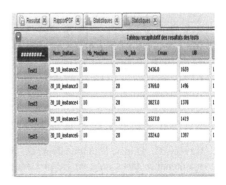

Fig. 8. The resulted curve

Fig. 9. The statistics table

4   Conclusion and Future Work

In this paper, we presented a new tool called ETOMA for developing and testing multi-agent schedules. It is composed of three modules. The first one called *Develop* defines the user multi-agents' solution by defining all system agents and their behavior. The second one called *Test* manages tests and the third one called *Blackboard* insures indirect communication between the two previous modules. ETOMA does not impose any architecture for the system that represents the floor shop nether for agents to be defined by users. The concept of competence adopted by our system insures flexibility of agents' behavior and reusability of agents' functionalities. ETOMA covers the three existent floor shops' configuration: *Flow-shop*, *Job-shop* and *Open-shop* what make it designed to a large number of applications. The platform is extensible in such manner to add other functionalities and options. In future work, we aim to improve the platform by considering other flow shops type like single machine and parallel machines.

References

1. Pinedo, M.: Scheduling, Theory, Algorithms and Systems. Industrial and Systems Engineering. Edition Prentice Hall International (1995)
2. Li, X., Zhang, C., Gao, L., Li, W., Shao, X.: An agent-based approach for integrated process planning and scheduling. Journal of Expert Systems with Applications 37, 1256–1264 (2010)
3. Kouiss, K., Pierreval, H., Mebarki, N.: Using multi-agent architecture in FMS for dynamic scheduling. Journal of Intelligent Manufacturing 8, 41–47 (1997)
4. Owliya, M., Saadat, M., Goharian, M., Anane, R.: Agents-based Interaction Protocols and Topologies in Manufacturing Task Allocation. In: Proceeding of the 5th International Conference on System of Systems Engineering (2010)
5. Jackson, N., Lund, H.: Benchmarking for Higher Education. Society for Research into Higher Education & Open University Press, Buckingham (2000)
6. JADE, Java Agent DEvelopment Framework, http://jade.tilab.com
7. MaDKit, http://www.madkit.org
8. Magique, Multi-agent hiérarchique,
   http://www.lifl.fr/MAGIQUE/presentation/index.html
9. FIPA, The Foundation for Intelligent Physical Agents, http://www.fipa.org
10. Mathieu, P., Routier, J.C., Secq, Y.: Multi-Agent hiérarchique,
    http://www2.lifl.fr/SMAC/projects/magique/presentation/presentationContent.html#intro
11. Coudert, T., Grabot, B., Archimède, B.: Système multi agents et logique floue pour un ordonnancement coopératif production/ maintenance. Journal of Decision Systems 13(1), 27–62 (2004)
12. Cavory, G., Dupas, R., Goncalves, G.: A genetic approach to the scheduling of preventive maintenance tasks on a single product manufacturing production line. Int. J. Production Economics 74, 135–146 (2001)
13. Taillard, E.: Benchmarks for basic scheduling problems. European Journal of Operational Research 64(2), 278–285 (1993)
14. Masc platform, http://www.irit.fr/MASC

Forecasting Stock Market Indices Using RVC-SVR[*]

Jing-Xuan Huang and Jui-Chung Hung[**]

Department of Computer Science, University of Taipei, Taipei 100, Taiwan
juichung@gmail.com

Abstract. This paper addresses stock market forecasting indices. Generally, the stock market index exhibits clustering properties and irregular fluctuation. This paper presents the results of using real volatility clustering (RVC) to analyze the clustering in support vector regression (SVR), called "real volatility clustering of support vector regression" (RVC-SVR). Combining RVC and SVR causes the parameters of estimation to become more difficult to solve, thus constituting a highly nonlinear optimization problem accompanied by many local optima. Thus, the genetic algorithm (GA) is used to estimate parameters.

Data from the Taiwan stock weighted index (Taiwan), Hang Seng index (Hong Kong), and NASDAQ (USA) were used as the simulation presented in this paper. Based on the simulation results, the stock indices forecasting accuracy performance is significantly improved when the SVR model considers the RVC.

Keywords: Support vector regression, Forecasting index of stock market, Genetic algorithm, Real volatility clustering.

1 Introduction

Forecasting the stock market is a prevalent research topic [1], as are forecasting methods such as the adaptive network-based fuzzy inference system (ANFIS) in Taiwan's stock market forecasting [2], generalized autoregressive conditional heteroskedasticity (GARCH) in econometrics [3], the support vector machine (SVM) in Taiwan's stock market forecasting [4], and support vector regression (SVR) in the forecasting research on stock indices [5]. Using these forecasting methods is appropriate when the fluctuation of stock market indices is regular; however, when it is irregular, the performance degrades.

To overcome this degradation, many researchers in recent years have proposed using robust methods, such as SVR, in the stock market [6]. Using this model for forecasting is effective, but it loses its efficiency in strong-fluctuation stock markets. Therefore, this paper proposes a method in which the stock data is clustered according

---

[*] This work was supported in part by the National Science Council under grant number NSC 102-2221-E-845-001.

[**] Correspond author.

to its degree of real volatility, and then combines the robust characteristics of SVR to forecast a stock index. The problem in this model was complex and nonlinear; thus, the genetic algorithm (GA) was used for parameter estimation.

The GA, which was proposed in 1975 [7], emulates the process of natural evolution to achieve optimization. Undergoing selection, copying, crossover, and mutation, similar to genes, yields the next optimal generation; therefore, the GA is suitable for solving nonlinear problems [8].

2   Proposed Method

SVR implements the structural risk minimization principle to achieve optimization [9]. In linear regression function $f(\mathbf{x}_i) = \mathbf{w}^T \mathbf{x}_i + b$, where $\mathbf{x}_i \in R^D$ for $i = 1, 2, ...N$. $\mathbf{x}_i$ is the input attribute values of the ^{i}th training data in D dimensional space, N is the size of training data, $\mathbf{w}$ is normal to the hyperplane, and $-b/\|\mathbf{w}\|$ is the perpendicular distance from the hyperplane to the origin.

Regarding nonlinear functions, the mapping function was adopted to transfer the functions to a hyper-surface to be considered as linear functions. In SVR, the mapping function is also called the "kernel function." The radial basis function (RBF) used in this study is expressed as

$$K(\mathbf{x}_i) = exp(-\frac{\|\mathbf{x}_i - \mathbf{x}\|^2}{\sigma^2})$$

(1)

where σ is the kernel bandwidth parameter.

Thus, the nonlinear function through the kernel function is described as $f(\mathbf{x}_i) = \mathbf{w}^T K(\mathbf{x}_i, \mathbf{x}) + b$, where $\mathbf{x}_i \in R^D$ for $i = 1, 2, ...N$, and the SVR problem is expressed as [6]

$$\min \frac{1}{2} \mathbf{w}^T \mathbf{w} + C \sum_{i=1}^{N} (\xi_i + \xi_i^*)$$

$$\text{subjected to } \mathbf{y}_i - \left(\mathbf{w}^T K(\mathbf{x}_i, \mathbf{x}) + b \right) \leq \varepsilon + \xi_i$$

(2)

$$\left(\mathbf{w}^T K(\mathbf{x}_i, \mathbf{x}) + b \right) - y_i \leq \varepsilon + \xi_i^*$$

$$\xi_i, \xi_i^* \geq 0 \quad \forall i = 1...N$$

where ξ_i , $\xi_i^{(*)}$ is the slack variable that specify the upper and the lower training errors subject to an error tolerance ε , ε indicates the loss function, and including the cost function C as trade-off between the flatness of $f(\mathbf{x}_i)$ and the training error. The role of the slack variables, loss function and support vectors are shown in Fig. 1.

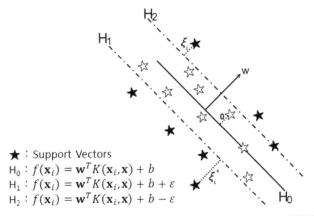

Fig. 1. The slack variables, loss function, and support vectors of SVR

As Eq. (2) is a quadratic optimization problem with inequality constraints, it is solved by the following Lagrange form [6]

$$\max_{\alpha} -\frac{1}{2}\sum_{i=1}^{N}\sum_{j=1}^{N}\left(\alpha_i - \alpha_i^*\right)\left(\alpha_j - \alpha_j^*\right)K(\mathbf{x}_i,\mathbf{x})\cdot K(\mathbf{x}_j,\mathbf{x}) - \varepsilon\sum_{i=1}^{N}\left(\alpha_i + \alpha_i^*\right) + \sum_{i=1}^{N}\mathbf{y}_i\left(\alpha_i - \alpha_i^*\right) \quad (3)$$

$$\text{subject to}\sum_{i=1}^{N}\left(\alpha_i^* - \alpha_i\right) = 0 \text{ and } 0 \le \alpha_i^{(*)} \le C$$

where α, α_i^* are Lagrange multipliers. After the α, α_i^* have been determined, the parameter b can be estimated under Karush-Kuhn-Tucker as follow

$$f(\mathbf{x}_i) = \sum_{i=1}^{N}\left(\alpha_i - \alpha_i^*\right)K(\mathbf{x}_i,\mathbf{x}) + b$$

$$\max\{-\varepsilon + \mathbf{y}_i - \sum_{i=1}^{N}\left(\alpha_i - \alpha_i^*\right)K(\mathbf{x}_i,\mathbf{x}) \mid \alpha_i < C \text{ or } \alpha_i^* > 0\} \le b \quad (4)$$

$$\le \min\{-\varepsilon + \mathbf{y}_i - \sum_{i=1}^{N}\left(\alpha_i - \alpha_i^*\right)K(\mathbf{x}_i,\mathbf{x}) \mid \alpha_i > 0 \text{ or } \alpha_i^* < C\}$$

Three parameters are in SVR: cost function C, ε-insensitive loss function, and the bandwidth σ of the kernel function. The ε-insensitive loss function makes the model robust. If the difference between estimated values and objective values is less than ε, the difference is considered as 0. When the value of ε is high, the stock situation cannot be determined. By contrast, when ε is too low, it is not robust and does not provide an accurate model for strong-fluctuation stock markets. Mandelbrot observed the volatility clustering characteristic, stating that "large changes tend to be followed by large changes, of either sign, and small changes tend to be followed by small changes" [10-11]. Therefore, the stock closing index's real volatility was adopted in this study for clustering.

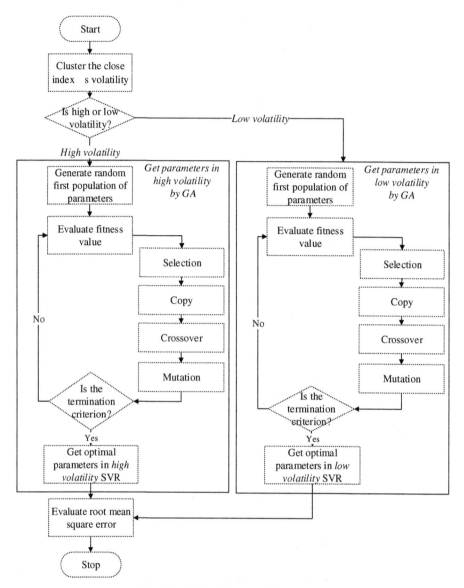

Fig. 2. RVC-SVR's manual flow chart

The rate of a weighted stock price return and its real volatility is obtained using

$$r_i = \ln(p_i) - \ln(p_{i-1})$$
$$v_i = r_i^2 \qquad (5)$$

where p_i is the stock index at time i, r_i is the rate of stock return at time i, and v_i is the volatility at time i [10].

The proposed method for forecasting stock indices by using the SVR model is expressed as

$$\min \frac{1}{2}\mathbf{w}_{\gamma}^{T}\mathbf{w}_{\gamma}+C_{\gamma}\sum_{i=1}^{N}(\xi_{\gamma i}+\xi_{\gamma i}^{*})$$

$$\text{subjected to } \mathbf{y}_{\gamma i}-\left(\mathbf{w}_{\gamma}^{T}K(\mathbf{x}_{\gamma i},\mathbf{x}_{\gamma})+b_{\gamma}\right)\leq\varepsilon_{\gamma}+\xi_{\gamma i}$$

$$\left(\mathbf{w}_{\gamma}^{T}(\mathbf{x}_{\gamma i},\mathbf{x}_{\gamma})+b_{\gamma}\right)-\mathbf{y}_{\gamma i}\leq\varepsilon_{\gamma}+\xi_{\gamma i}^{*} \quad (6)$$

$$\xi_{\gamma i},\xi_{\gamma i}^{*}\geq 0 \quad \forall i=1...N$$

$$\gamma\in\{0,1\}$$

where γ indicates either high or low volatility, which is clustered using k-means [12]. When γ equals 0, it refers to high volatility, and when γ equals 1, it refers to low volatility.

In general, the optimization problem is a very highly nonlinear function. Many local extreme values may exist. Therefore, we will use GA to specify these parameters, C_{γ}, ε_{γ}, and σ_{γ} to solve the RVC-SVR problem. In this paper, a simple genetic algorithm is composed of four operators, which are select, copy, crossover, and mutation [13] to search the parameter set $\{C_{\gamma},\varepsilon_{\gamma},\sigma_{\gamma}\mid\gamma\in\{0,1\}\}$ in Eq. (6). **Fig. 2** describes the manual flow chart of the proposed-method, in which the unknown parameters in Eq. (6) are estimated using the GA.

3   Simulation

Taiwan stock weighted indices, Hang Seng indices, and NASDAQ indices from October 13, 2009 to December 22, 2011 were adopted in the experiment in this study, including October 13, 2009 to May 27, 2011 for training and May 30, 2011 to December 22, 2011 for testing. Fig. 3 represents the sample autocorrelation and sample partial autocorrelation for the stock index volatility in Taiwan, Hong Kong, and NASDAQ respectively, all with a 95% confidence interval. Based on Fig. 3, most lags were outside the boundary, meaning that the stock index volatility had significant autocorrelation. The idea of using the autocorrelation and partial autocorrelation of volatility for determining time series behavior was proven using the GARCH model [14].

The estimation of error in the experiment is a root mean square error (RMSE), which is obtained using

$$\text{RMSE} = \sqrt{\frac{\sum_{i=1}^{M}\left[\mathbf{y}_{i}-f(\mathbf{x}_{i})\right]^{2}}{M}} \quad (7)$$

where $f(\mathbf{x}_{i})$ is the forecasting value of the ith data, y_{i} is the true value of the ith data, and M is the size of the forecasting data.

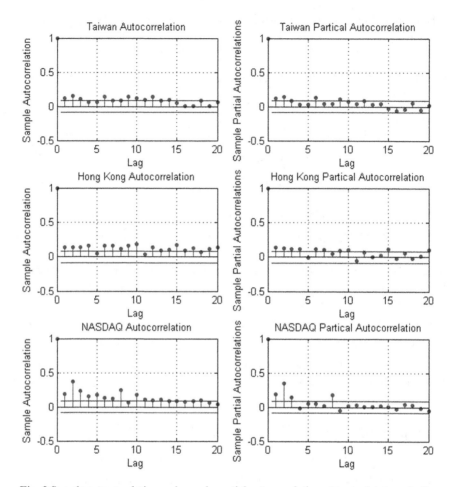

Fig. 3. Sample autocorrelation and sample partial autocorrelation of the stock index volatility

Table 1. Test results of the forecasting for Taiwan, Hong Kong, and NASDAQ data based on SVR [5], RVC-SVR, and ANFIS

Stock	Method	RMSE
Taiwan	SVR	84.336
	RVC-SVR	79.808
	ANFIS	331.976
Hong Kong	SVR	233.294
	RVC-SVR	215.676
	ANFIS	5945.617
NASDAQ	SVR	36.227
	RVC-SVR	34.691
	ANFIS	61.194

Table 2. The parameters for Taiwan, Hong Kong, and NASDAQ data based on SVR and RVC-SVR

Stock	Method	Class	Parameters		
			C	ε	$g=1/\sigma^2$
Taiwan	SVR		2^{15}	$2^{(-3)}$	$2^{(-4)}$
	RVC-SVR	$\gamma=0$	$2^{14.957}$	$2^{(-0.680)}$	$2^{(-5.713)}$
		$\gamma=1$	$2^{14.909}$	$2^{6.264}$	$2^{(-0.851)}$
Hong Kong	SVR		2^{15}	2^{6}	$2^{(-4)}$
	RVC-SVR	$\gamma=0$	$2^{14.994}$	$2^{2.027}$	$2^{(-2.695)}$
		$\gamma=1$	$2^{14.663}$	$2^{2.383}$	$2^{(-5.649)}$
NASDAQ	SVR		2^{15}	$2^{(-15)}$	$2^{(-5)}$
	RVC-SVR	$\gamma=0$	$2^{14.998}$	$2^{4.737}$	$2^{(-4.045)}$
		$\gamma=1$	$2^{14.996}$	$2^{4.636}$	$2^{(-0.273)}$

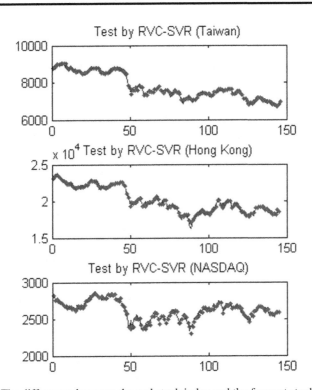

Fig. 4. The differences between the real stock index and the forecast stock index

Table 1 lists the test results of SVR [5], ANFIS [2], and RVC-SVR for the forecasting stock market indices. Fig. 4 illustrates the differences between the real stock index and forecast stock index in the RVC-SVR experiments, and it demonstrates that the RVC-SVR model determines the trend of the stock index. Table 2 lists the detailed parameters of the RVC-SVR model. We further observed that the optimal parameters are significantly different in each cluster of data set clustered by real volatility, meaning that RCV-SVR reduces its heterogeneity and enlarges each data set to obtain a more robust model for forecasting.

4   Conclusion

Because of clustering properties and irregular fluctuation in stock market indices, the designed method, RCV-SVR, for estimating stock market indices, combines real volatility characteristics for clustering to reduce the behavioral heterogeneity in stock indices with SVR. The estimated performance improved significantly. Therefore, the proposed method is effective for obtaining accurate estimations when encountering strong fluctuation.

References

1. http://apps.webofknowledge.com/summary.do?product=WOS&doc=1& qid=5&SID=S2J3Qbhu8ziLCdlprey&search_mode=AdvancedSearch
2. Wei, L.T., Chen, T.L., Ho, T.H.: A hybrid model based on adaptive-network-based fuzzy inference system to forecast Taiwan stock market. Expert Systems with Applications 38, 13625–13631 (2011)
3. Engle, R.: GARCH 101: The Use of ARCH/GARCH Models in Applied Econometrics. Journal of Economic Perspectives 15(4), 157–168 (2001)
4. Chen, S.M., Kao, P.Y.: TAIEX forecasting based on fuzzy time series, particle swarm optimization techniques and support vector machines. Inf. Sci. 247, 62–71 (2013)
5. Hsieh, H.I., Lee, T.P., Lee, T.S.: A Hybrid Particle Swarm Optimization and Support Vector Regression Model for Financial Time Series Forecasting. International Journal of Business Administration 2(2), 48–56 (2011)
6. Vapnik, V.N., Golowich, S.E., Smola, A.: Support Vector Method for Function Approximation, Regression Estimation and Signal Processing. Advances in Neural Information Processing System 9, 281–287 (1997)
7. Holland, J.H.: Adaptation in Natural and Artificial Systems: An Introductory Analysis with Applications to Biology, Control, and Artificial Intelligence. University of Michigan Press (1975)
8. Whitley, D.: A Genetic Algorithm Tutorial: Statistics and Computing (1993)
9. Vladimir, N.V.: An Overview of Statistical Learning Theory. IEEE Transactions on Neural Networks 10, 988–999 (1999)
10. Hung, J.C.: Applying a combined fuzzy systems and GARCH model to adaptively forecast stock market volatility. Applied Soft Computing 11, 3938–3945 (2011)
11. Mandelbrot, B.: The variation of certain speculative prices. The Journal of Business 36(4), 394–419 (1963)
12. Kanungo, T., David, M.M., Nathan, S.N., Christine, D.P., Silverman, R., Angela, Y.W.: An Efficient k-Means Clustering Algorithm: Analysis and Implementation. IEEE Transactions on Pattern Analysis and Machine Intelligence 24(7), 881–892 (2002)
13. Pratibha, B., Manoj, K.: Genetic Algorithm – an Approach to Solve Global Optimization Problems 1(3), 199–206 (2010)
14. Bollerslev, T.: Generalized Autoregressive Condition Heteroskedasticity. Journal of Econometrics 31(3), 307–327 (1986)

Particle Swarm Optimization Feature Selection for Violent Crime Classification

Mohd Syahid Anuar, Ali Selamat, and Roselina Sallehuddin

Faculty of Computing, Universiti Teknologi Malaysia, Johor Bahru, Malaysia
syah2105@yahoo.com, {aselamat,roselina}@utm.my

Abstract. Crime prevention is one of the important roles of the police system in any country. One of the components of crime prevention is crime rate classification. Thus, this study proposed a crime classification model by combining Artificial Neural Network (ANN) model and Particle swarm optimization (PSO) model. PSO is used as feature selection to select the significant features that affects the capability of ANN as classifier. This combination is expected to generate more accurate classification result with minimum error. To evaluate the performance of the proposed model, comparison with ANN model without PSO is carried out on the Communities and Crime dataset. The proposed model is found to produce better classification accuracy as compared to ANN model alone in classifying crime rates. Besides improving the classification accuracy, the proposed model has reduced the learning convergence time in training phase.

Keywords: crime, classification, prediction, particle swarm optimization.

1 Introduction

Recently, increasing volumes of crime had brought serious problems in the community. Therefore, crime prevention is one of the important roles of the police system in any country. One of the components of crime prevention is crime rate classification. Crime classification is one of a prime focus in the field of Criminology. Research on crime forecasting has increased because of the potential and effectiveness of classification in crime prevention programs. Crime classification can assist the police to make a proper operational and tactical strategies in the future, such as to allocate police patrols in the right area, install CCTV in the right place and plan other operations. A common practice is to identify hot spots in the preceding period based on their geographical location and assume these hot spots will persist into the next period [1].

Some researches in crime classification have been done by several researchers [2, 7, 9, 11, 12]. Several methods have been applied in crime classification research such Naïve lag, exponential smoothing methods and classical decomposition, Decision Tree, Regression and ARIMA. Even though the application of Artificial intelligence (AI) techniques especially artificial neural networks (ANNs) have shown a great classification performance, their applications in crime dataset are still rare. ANN is a model inspired by brain. ANN with back propagation learning algorithm is typically

J. Sobecki, V. Boonjing, and S. Chittayasothorn (eds.), *Advanced Approaches to Intelligent Information and Database Systems*, Studies in Computational Intelligence 551, DOI: 10.1007/978-3-319-05503-9_10, © Springer International Publishing Switzerland 2014

used as a benchmark model for any classifier. However, ANN is a black-box learning approach where it cannot determine automatically the significant input features. To overcome the limitation of ANN in choosing relevant features as input, feature selection is needed. Particle Swarm Optimization (PSO) is one of the popular optimization techniques that have capability to perform the feature selection task [4]. Thus, the objective of this paper is to propose a new classifier that combines PSO and ANN to classify the crime rates. Therefore, in this paper we intend to combine ANN and PSO as a classifier to classify crime categories.

2   The Proposed Classification Model, PSO-ANN

Figure 1 shows the overview of classification process using ANN and PSO as feature selection in classifying violent crime categories. There are four phases involve namely data preparation, feature selection, classification and evaluation. The detail explanation of each phase will be given in the following section.

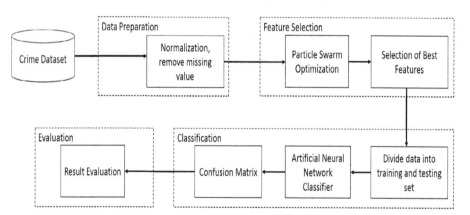

Fig. 1. Classification Process for of Crime Rate Categories

2.1   Crime Dataset

The Communities and Crime data set is obtained from the UCI Machine Learning Repository. This dataset focuses on communities in United States of America (USA). The data comprises of socio-economic data from the '90 Census, law enforcement data from the 1990 Law Enforcement Management and Admin Stats survey and crime data from the 1995 FBI UCR. The dataset consists of 147 number of attributes and 2215 number of instances with missing values [16].

2.2   Data Preparation

Data preparation is essential for successful data classification. Poor quality data typically result in incorrect and unreliable data classification results. Thus, the following data preparation mechanisms were carried out to obtain the final set of attribute in an appropriate form for further analysis and processing.

- All the attributes with large number of missing values were removed
- The newly added nominal attribute named 'Categories' is created from the attribute named '*violentPerPop*' – if the value is less than 25% than the Categories is 'Low'. If the value is equal to or greater than 25% than the Categories is 'Medium'. If the value is equal to or greater than 40% than the Categories is 'High' [6]. There are 1871 instances classified as 'Low', 242 instances classified as 'Medium' and 102 instances classified as 'High'.
- All attributes are set to numeric except 'Categories' which is set as nominal.
- The final number of attributes after data preparation are implemented is 104. This number of attribute will undergo feature selection process.
- All the data will be normalize into [0, 1] using min-max method by using (1).

$$V_i = \frac{v_i - \min v_i}{\max v_i - \min v_i} \qquad (1)$$

Where V_i is normalized variable at i_{th} row. *Min* v_i is minimum value for variable v. and max v_i is maximum value of v.

3   Experiment

3.1   Feature Selection

PSO is used as feature selection tool to select a subset of relevant features to be used in model construction of ANN classifier. PSO will examine and then recognize the irrelevant and redundant features. Redundant features are those which provide no more information than the currently selected features, and irrelevant features provide no useful information in any context. Hence removing these features can improve the classification accuracy and learning times. Based on the review of the existing literature [4, 5, 14, 15], PSO is one of the frequently used feature selection techniques since it provides an accurate classification result and easy to implement. The overview process of PSO feature selection is shown in figure 2 [4, 5].

PSO is initialized with a population of random solutions, called 'particles'. Each particle is treated as a point of an S-dimensional space. The i_{th} particle is represented as $X_i = (x_{i1}, x_{i2}, ..., x_{iS})$. The pBest is the best previous position (best fitness value) and represented as $P_i = (p_{i1}, p_{i2}, ... p_{iS})$. The best among all the particles is represented as 'gBest'. The velocity is the rate of change of the position for particle i and represented

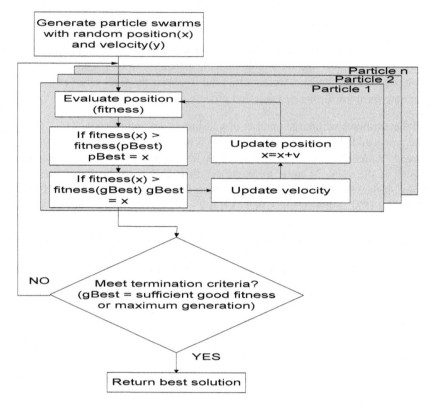

Fig. 2. PSO Process

as $V_i = (v_{i1}, v_{i2}, ..., v_{iS})$. The particles are manipulated according to the fitness function which is a positive linear function of time changing according to the generation iteration and can be represented as equation (2) and (3).

$$V_{id} = \text{w} * v_{id} + c_1 * rand(\ ) * (p_{id} - x_{id}) + c_2 * rand(\ ) * (p_{gd} - x_{id}) \quad (2)$$
$$x_{id} = x_{id} - v_{id} \quad (3)$$

where $d = 1, 2, ..., S$, w is the inertia weight. The learning rate c_1 and c_2 represent the weighting of the acceleration terms that pull each particle toward pBest and gBest positions. Lastly, rand() is a random function in the range [0,1]. Basically, the steps involves in PSO are as follow:

1. Initialize parameters such as the acceleration constants, inertia weight, number of particles, maximum number of iterations, velocity boundaries, initial and constrained velocities and positions, and eventually the error limit for the fitness function.
2. Evaluate the particles' fitness function values, comparing with each other, therefore setting the local best (pBest) and global best (gBest).
3. In accordance with Equations (2) and (3), calculate a particle's new speed and position, and then update each particle.

4. For each particle, compare the current fitness value with the local best. If the current value is better, update the local best fitness value and particle position with the current one.
5. For each particle, compare the current fitness value with the global best. If the current value is better, update the global best fitness value and particle position with the current one.
6. If a stop criterion is achieved, then stop the procedure, and output the results. Otherwise, return to step 2.

Table 1 lists the parameters setting for PSO which are based on suggestions in the previous studies [4-6].

Table 1. PSO Parameters Setting

Experiments	Learning Rates		Number of Particles (n)
	C1	**C2**	
1	1.0	1.0	5
2	1.0	2.0	
3	2.0	1.0	
4	2.0	2.0	
5	1.0	1.0	20
6	1.0	2.0	
7	2.0	1.0	
8	2.0	2.0	
9	1.0	1.0	100
10	1.0	2.0	
11	2.0	1.0	
12	2.0	2.0	

Figure 3 visualize the result of PSO feature selection in term of number of features and fitness value. For, the first graph, the x-axis represents the i_{th} number of experiment and y-axis represents the number of features used in each experiment. In the second graph, x-axis represents the i_{th} number of the experiment and y-axis represents the fitness value used.

Figure 3 shows that the number of features is decreasing and the value of fitness function is increasing when the experiment is iterated. The PSO algorithm will stop searching after it found the optimal solution which is the lowest number of features with the highest fitness value [4]. Result from this study shows that the experiment 10 produces the optimum number of features with highest fitness value. Out of 104 initial features used, only five features are identified by PSO as significant features. The new selected attributes (represented as number) are 17, 51, 72, 100, and 103. Table 2 shows the detail information of selected significant attribute for crime classification after implemented feature selection.

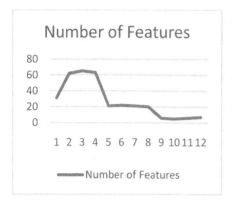 

Fig. 3. Number of features and fitness value for PSO feature selection

Table 2. Attribute description for selected significant feature

Attribute No.	Attribute Name	Description
17	*pctWsocsec*	percentage of households with social security income in 1989
51	*pctKidsBornNevrMarr*	percentage of kids born to never married
72	*houseVacant*	number of vacant households
100	*popDensity*	population density in persons per square mile
103	*violentPerPop*	total number of violent crimes per 100K popuation

3.2   Classification Evaluation

The effectiveness of PSO feature selection is evaluated using ANN classifier. To validate the performance of PSO as feature selection, comparison with ANN classifier without PSO is conducted. 10-fold cross validation methodology was chosen to split data into training and testing. In 10-fold cross validation, the complete dataset is split randomly into 10 mutually exclusive subsets of approximately equal size. Each classification model will be trained and tested 10 times which mean one round of cross validation involves the partitioning of data into two subsets of training and testing data. Validating the analysis for testing subset is based on the training subset. The test outcomes of all folds are compiled into a confusion matrix. Table 3 shows the Confusion Matrix for the classification result. The rows represent the actual categories and the columns represent predicted categories.

The bold numbers represent the number of correct classification produce by classifier. From the classification result obtained, PSO-ANN produced better classification performance compared to ANN. PSO-ANN correctly classifying for both

Table 3. Confusion Matrix

Prediction Method		Confusion Matrix		
		(Predicted Class)		
		Low	Medium	High
PSO-ANN	Low	**1864**	7	0
	Medium	2	**240**	0
	High	0	2	**100**
ANN	Low	**1851**	20	1
	Medium	17	**218**	7
	High	1	14	**87**

categories 'Low' and 'High' classes 13% higher than ANN. For 'Medium' class, PSO-ANN improved the ANN classification ability about 22%. This result indicates that the application of PSO in ANN classifier has successfully removed the irrelevant factors that affect the performance of ANN classifier in crime classification.

To further validate the performance of PSO-ANN model, five statistical performance measurement criteria are also applied. They are average accuracy, error rate, precision and recall [6]. The formula for each measurement is given as follow:

$$Average\ Accuracy = \frac{\sum_{i=1}^{l}\frac{TP_i+TN_i}{TN_i+FN_i+FP_i+TN_i}}{l} \quad (4)$$

$$Error\ Rate = \frac{\sum_{i=1}^{l}\frac{FP_i+FN_i}{TP_i+FN_i+FP_i+TN_i}}{l} \quad (5)$$

$$Precision = \frac{\sum_{i=1}^{l}\frac{TP_i}{TP_i+FP_i}}{l} \quad (6)$$

$$Recall = \frac{\sum_{i=1}^{l}\frac{TP_i}{TP_i+FN_i}}{l} \quad (7)$$

Where TP, TN, FP and FN denote true positive, true negative, false positive and false negative. Average accuracy measure the average per class for the overall number of correct classification of classifier. Error rate measure the average of classification error per class. Precision is an average per class for the positive label given by classifier. Recall is an average effectiveness of a classifier to identify class labels.

4   Result

Table 4 shows the classification results obtained from PSO-ANN model and ANN model. PS0-ANN outperformed ANN in terms of average accuracy (99%, 98%), error rate (0.0033, 0.0181) and processing time (52s, 3862s). In fact PSO-ANN need only five features to beat ANN classification performance. Therefore, it will improve the computational time of PSO-ANN since the network's structure used is smaller. PSO-ANN only need 52 second to correctly classified crime rates category while ANN that used 104 features need 3862 second to do the same task. Table 4 shows the detail statistical results of the classification on crime dataset.

Table 4. Classification Result PSO-ANN vs. ANN

	Number Of Features	Average Accuracy	Error Rate	Time (s)
PSO-ANN	5	0.9917	0.0033	52
ANN	104	0.9819	0.0181	3862

Figure 4 visualize the recall and precision for PSO-ANN and ANN. PSO-ANN improved the capability of ANN's recall and precision about 7% and 6% respectively.

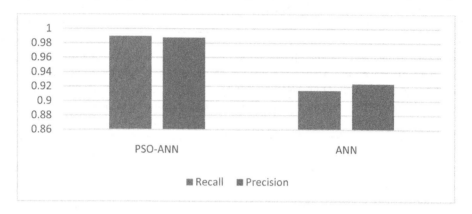

Fig. 4. Recall vs Precision

To wrap up, the application of PSO in ANN classifier has helped ANN to recognize the optimum number of relevant features that can improve the ANN classification performance and reduce the computational cost.

5   Conclusion

In this paper, PSO feature selection was proposed as feature selection to overcome the limitations of ANN classification on the crime dataset. Experimental results shows that the application of PSO on ANN improved the classification accuracy of general ANN, reduce the node amounts of the input and hidden layers (reduce the complex structure of ANN) and decrease the learning time or computational cost of ANN. It offers a promising alternative to improve the application of ANN in knowledge discovery and data dining from a large –scale information pool.

Acknowledgements. The Universiti Teknologi Malaysia (UTM) under research grant 03H02 and Ministry of Science, Technology & Innovations Malaysia, under research grant 4S062 are hereby acknowledged for some of the facilities utilised during the course of this research work.

References

1. Noor, N.M.M., Retnowardhani, A., Abd, M.L., Saman, M.M.Y.: Crime Forecasting using ARIMA Model and Fuzzy Alpha-cut. Journal of Applied Sciences 13(2013), 167–172 (2013)
2. Nasridinov, A., Ihm, S.-Y., Park, Y.-H.: A Decision Tree-Based Classification Model for Crime Prediction. In: Park, J.J., Barolli, L., Xhafa, F., Jeong, H.-Y. (eds.) Information Technology Convergence, vol. 253, pp. 531–538. Springer Netherlands (2013)
3. Sokolova, M., Lapalme, G.: A systematic analysis of performance measures for classification tasks. Information Processing & Management 45, 427–437 (2009)
4. Omar, N., bin Othman, M.S., binti Ibrahim, R., binti Jusoh, F.: Particle Swarm Optimization Feature Selection for Classification of Survival Analysis in Cancer. International Journal of Innovative Computing (2012)
5. Wang, X., Yang, J., Teng, X., Xia, W., Jensen, R.: Feature selection based on rough sets and particle swarm optimization. Pattern Recognition Letters 28, 459–471 (2007)
6. Rahman, S.A., Bakar, A.A., Hussein, Z.A.M.: Filter-wrapper approach to feature selection using RST-DPSO for mining protein function. In: 2nd Conference on Data Mining and Optimization, DMO 2009, pp. 71–78 (2009)
7. Iqbal, R., Murad, M.A.A., Mustapha, A., Panahy, P.H.S., Khanahmadliravi, N.: An Experimental Study of Classification Algorithms for Crime Prediction. Indian Journal of Science and Technology 6 (2013)
8. Iqbal, R., Murad, M.A.A., Sharef, N.M.: Emerging and Prospering Trends in Crime Analysis and Investigation Systems: A Literature Review. Journal of Theoretical & Applied Information Technology 47, 53 (2013)
9. Janeela Theresa, M.M., Joseph Raj, V.: Fuzzy based genetic neural networks for the classification of murder cases using Trapezoidal and Lagrange Interpolation Membership Functions. Applied Soft Computing 13, 743–754 (2013)
10. Phillips, P., Lee, I.: Mining co-distribution patterns for large crime datasets. Expert Systems with Applications 39, 11556–11563 (2012)
11. Tao, W.: Crime Data Mining Based on Extension Classification. In: Wu, Y. (ed.) Software Engineering and Knowledge Engineering: Vol. 2. AISC, vol. 115, pp. 383–390. Springer, Heidelberg (2012)
12. Zou, K.-Q., Zhou, X.-M., Liu, F.-X.: Application of Fuzzy Clustering Method in the Crime Data Analysis. In: Cao, B.-Y., Xie, X.-J. (eds.) Fuzzy Engineering and Operations Research. AISC, vol. 147, pp. 41–50. Springer, Heidelberg (2012)
13. Chung-Hsien, Y., Ward, M.W., Morabito, M., Wei, D.: Crime Forecasting Using Data Mining Techniques. In: 2011 IEEE 11th International Conference on Data Mining Workshops (ICDMW), pp. 779–786 (2011)
14. Zhi-Hui, Z., Jun, Z., Yun, L., Chung, H.S.H.: Adaptive Particle Swarm Optimization. IEEE Transactions on Systems, Man, and Cybernetics, Part B: Cybernetics 39, 1362–1381 (2009)
15. Leu, M.-S., Yeh, M.-F., Wang, S.-C.: Particle swarm optimization with grey evolutionary analysis. Applied Soft Computing 13, 4047–4062 (2013)
16. Asuncion, A., Newman, D.J.: UCI Machine Learning Repository, School of Information and Computer Science, University of California, Irvine, CA (2007),
   `http://archive.ics.uci.edu/ml/datasets/`
   `Communities+and+Crime+Unnormalized`

A Goal-Based Hybrid Filtering for Low-rated Users Recommendation Issue Using Neighborhood Personalized Profile Similarities in E-Learning Recommendation Systems

M.W. Chughtai[*], Ali Selamat, and Imran Ghani

Software Engineering Department, Faculty of Computing,
Universiti Teknologi Malaysia (UTM),
Skudai, 81310, Johor Darul Ta'zim, Malaysia
mr.chughtai88@gmail.com, {aselamat,imran}@utm.my

Abstract. The e-learning recommender systems are based on the users past history, ratings, likes or dislikes. The low-rated (less rated history profile) users may cause of zero or non-relevant recommendations issue in these days, which lose the users interest. This research proposed the goal-based hybrid filtering approach that used to perform the personalized similarities between users personalized profile preferences collaboratively. The aim of this research study is to improve the low-rated user's recommendations by tackling the collaborative filtering and k-neighborhood personalized profile preferences similarities in e-Learning recommendation scenarios. The experiments has been tackled with famous 'Movielens' dataset while the experimental results has been performed with the help of (average mean precision *Pr*: 79.90%) and (average mean recall *Re*: 83.50%) respectively. A conducted result demonstrates the effectiveness of proposed goal-based hybrid filtering in the improvement of low-rated users profile recommendations in e-learning recommendation systems.

Keywords: Goal-based, hybrid filtering, recommender systems, e-learning, collaborative filtering, k-nearest neighbors.

1 Introduction

The intensification of web informational content increased the difficulties of find the relevant content quickly and efficiently. For end-users, Idiosyncratic e-Learning scenarios are diverse in learning information / guidance and content-based electronic learning environment (sometimes called e-Learning) is one of them [2]. Where electronic learning provides many benefits to users, there are some drawbacks too; the one and major drawback is Large Content Mismanagement because traditional learning organizations cover their mismanagement by increasing the number of pages. This page-to-page running environment has not been considered an effective and

---

[*] Corresponding author.

J. Sobecki, V. Boonjing, and S. Chittayasothorn (eds.), *Advanced Approaches to Intelligent Information and Database Systems*, Studies in Computational Intelligence 551,
DOI: 10.1007/978-3-319-05503-9_11, © Springer International Publishing Switzerland 2014

good management strategy for content handling [3]. The reason of discouraging this strategy is that the user spends a lot of time for visiting every page and retrieving the required learning content. This process cuts the users interest. However, Recommender Systems in e-Learning knows as e-Learning Recommender Systems [4]; offers more flexibility for users to overcome the large content management issue. It helps to decrease the content searching time, increase the user's interest, and provide the recommendations relevant to user's goals / interests [5].

1.1   Recommender System Approaches

Recommender or Recommendation System is a branch of information retrieval, gradually it filters the result in three ways; namely Content-based Filtering (CBF), Collaborative Filtering (CF), and Hybrid Filtering (HF) [6]. the Content-based Filtering (CBF) only recommends relevant items/learning contents to users that are similar to the ones they preferred them self's in the past [7], while in Collaborative Filtering (CF), the users recommend relevant items/learning contents that other users with similar interest and preferences liked in the past [8]. The Hybrid Filtering (HF) is a third way of e-Learning Recommender Systems to tackle the filtering results [9]. Simultaneously, these approaches play a controversial role to tackle the users required goals although with respect to traditional research aspects, every filtering approach has its own limitations. The HF basically, hibernates the features of CF and CBF, somehow the encouraged researchers combine some adopted machine learning techniques / approaches with CBF or CF (sometimes with the combination of both) to emerge the Artificial Intelligent aspects in it.

1.2   Low-rated User Profile Recommender Issue

The growing user-based challenges affect the performance of recommender systems in e-Learning, one of them is low-rated (less rated history) user's profile recommendation issue. This issue moderated against the user profile preferences and user's past / history ratings (sometimes called voting or like/dislike) on specific learning contents / items. The contemporary literature induced that the low-rated users (who do not have much past / history profile rating against any learning content / item in the system) may facing the forecasting recommendation issues. In other words, the recommender systems are unable to forecast the recommendations to low-rated user profiles [10, 11]. Critically, this issue has two causes; (1) user is not regular in the system, or (2) user has not rated or visited much learning contents / items in the past / history. In both cases, the recommender systems unable to recommend the required learning content / item to the end user, which loss the user's interest too.

1.3   Research Contribution

Salient aspects of our research work contribution towards e-learning recommender systems are as follows:

— Drive a goal-based hybrid filtering approach to improve the performance of e-learning recommender systems on low-rated profiles recommendations.

— Improve traditional collaborative filtering with hybridization of k-nearest neighbors features to overcome the low-rated user's profile recommendations.
— Operate validation / normalization to improve the traditional dataset driven method.
— The goal-based hybrid filtering approach is flexible with user's profiles; the flexible in terms that it do not need users extra information, the "age, gender, and occupation" profile preferences are enough to compute the recommendations for low-rated users.

2   Related Work

The author [4] combined the collaborative and content-based filtering approaches and used a keyword maps technique for extracting the low-rated content automatically. Researchers are proffered the approaches hybridization (partially used machine learning) to tackle the low-rated users profile recommendation issues on more appropriate and systematic way. For example, Bharadwaj in 2013 proposed user-oriented content-based recommender system (UCB-RS) [1] with two stages. In the first stage, the [1] used fuzzy theoretic content-based filtering to generate the initial population of users' preferences by interactive genetic algorithm (IGA) using reclusive methods (RM's). In second stage, the [1] used k-mean algorithm for clustering the item in order to handle time complexity of interactive genetic algorithm (IGA). Usually, the traditional k-mean is unable to handle time computational complexity with genetic algorithm if K is small [12]. Mostly with large set of data, the traditional k-mean does not give sufficient performance. K.I.Bin Ghauth in 2010 also proposed a hybrid system of recommendation to overcome the low-rated users issue for the e-learning environments.

M.Lee In 2006 proposed a collaborative filtering recommender system combined with the SOM Neural Network for low-rated users profile recommendations. The author [13] categorized the users based on their demographic information, and used a clustering technique to cluster the users in each category according to their preference to items using the SOM Neural Network. M.Jahrer in 2010, used several approaches such as SVD (Singular Value Decomposition), Neighborhood Based Approaches [14], restricted Boltzmann Machine, Asymmetric Factor Model and Global Effects to build recommender systems. The author [15] show that linearly combining these algorithms increases the accuracy of prediction.

Many researchers have been used / adopted different techniques (i.e. k-nearest neighbors "a branch of neural networks," data mining, data clustering and patterns recognition, etc.). These techniques have been categorized in machine learning approaches that have been used in recommender Systems in order to domain requirement and the cause of sorted issue. Nevertheless, many researchers have adopted different machine-learning techniques (i.e. k-nearest neighbors "a branch of neural networks," data mining, data clustering and patterns recognition, etc.) in recommender systems in order to domain requirement and the cause of users profile recommendation issue [16-18]. These techniques increase the efficiency that has ineffectual in tackling the learning contents by analyzing users past preferences.

3   Goal-Based Hybrid Filtering

This research work proposed a Goal-based hybrid filtering approach, which hybridized selected features of collaborative filtering and k-nearest neighbors to improve the low-rated users profile recommendations. Here, neighborhood personalized profile similarities measures the resemblance between the collaborative user's profiles preferences (age, gender, occupation) without using any extra user profile information with the help of famous 'Movielens' dataset [19].

3.1   Dataset in Used

This research used the famous 'Movielens' dataset for experiments [19]. The specialty of 'Movielens' is that it is a real-time validated dataset with and famous for experimenting the constructed hybrid recommendation systems [20-22]. Through the previous literature and our understanding, we noticed some key-points in dataset that mentioned in Table 1.

Table 1. 'Movielens' dataset D key-points

Key-Point 1:	The strength of users 'U' is 943.
Key-Point 2:	Low-rated user's profiles are 872.
Key-Point 3:	High-rated user's profiles are 61.
Key-Point 4:	The strength of learning contents / items 'I' is 1682.
Key-Point 5:	Each user 'U' vote / rate 'R' at least 20 learning contents / items 'I' from 1682
Key-Point 6:	The rating scores from 1 to 5
Key-Point 7:	The rating strength (Ru,i) of 943 users 'U $\in$ u' against 1682 learning contents / items 'I $\in$ i' is around 100,000

Table 1 defines the clear aspects of 'Movielens' dataset. In general, classification, the user U profile is fully depending on the ratings 'R' against each learning content / item I. Although, it is expressed that 1,682 different items 'I' contains 100,000 ratings 'R' (sometimes called vote) by 943 users 'U' [10, 21-23]. The density of the user-item matrix with respect to ratings created from the 'Movielens' dataset is;

$$total\ ratings\ Ru,i\ total\ users\ \ u \in U)\ \times\ total\ items\ \ i \in I)$$

$$(1)$$

$$\therefore \quad \frac{\sum R_{u,i}}{(\sum u \in U) \times (\sum i \in I)}$$

$$= \frac{100,000}{(943) \times (1,682)} = 15.86\%$$

In equation (1), 15.86% considered as appropriate fine in terms of sparsity for the evaluation and validation of the dataset. The validation / normalization 'Movielens' dataset helps to rearrange the dataset as (a) training set: low rated users and (b) testing set: high-rating profiles sequences.

3.2   Testing and Training Set

The rearrangement of 'Movielens' dataset 'D' in training and testing sets has been employed using the users 'U 'profiles as per their ratings 'R' information. In the general method of training / testing set portionalization [1], the dataset have been divided as 50% of training set and 50% testing set [1]. Alternatively, in this proposed work allocate the training and testing sets in a slitter different way. The detail view is shown in Figure 1. The dataset portionalization helps the system to classify the training set (presents as TR) and testing set (presents as TS) from 'Movielens' dataset 'D'. The definitions of these training and testing sets are shown in Table 2.

Table 2. The definitions and purposes of derived training set (T_R) and testing set (T_S).

Subset	Definition	Purpose
TR	Training set of users 'U' profiles with ratings 'R' of range (1 = 'R' $\in$ r $\leq$ 1000).	To improve the low-rated users profile recommendation
TS	Testing set of users 'U' profiles with ratings 'R' of range ('R' $\in$ r > 1000).	

This research study classifies training T_R and testing T_S sets in a slighter different way from the previous literature [1]. Methodologically, training and testing sets (T_R and T_S) has been used in collaborative filtering with k-nearest neighbors features (mentioned in section 3.1) and it works to overcome low-rated users profile recommendations in proposed goal-based hybrid filtering approach for e-Learning recommender systems.

3.3   Hybrid Filtering Approach

This section discusses the inheritance of a famous machine learning (k-neighbors) features with collaborative filtering approach. This inheritance helps to overcome the low-rated user's recommendations issue that shows the usefulness of user' personalized profile similarities using k-neighbors features collaboratively.

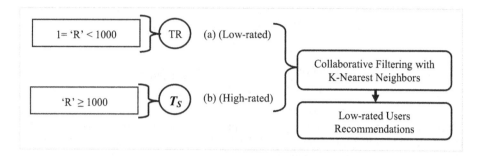

Fig. 1. Operational flow of collaborative filtering with k-neighbor features for low-rated users 'U' recommendations.

Fig 1 shows the operational flow of collaborative filtering with k-nearest neighbor's scheme to the improvement of low-rated users 'U' recommendations. Generally, in the collaborative filtering approaches [11, 21, 24], the system recommend relevant items / learning contents to the target user that other users with similar interest and preferences visited / liked or high rated in the past. This means, the traditional collaborative approaches did not worked well with the low-rated users 'U' if the user is not regular or less visited / liked or rated items in past. To handle this, the research work proposed foremost a slight different way, this work used the total ratings 'R' of per user to classify the (a) low-rated users and (b) high-rated users (discussed in section 3.2) in the system.

4   Implimentation

The implementation of proposed goal-based hybrid approach has been made by following operational steps:

Step 1 - (Initialization): Initialize the (a) user 'U', (b) item / learning content 'I' and (c) ratings 'R' data sets from 'Movielens' dataset 'D'.

$$U = \{u_1, u_2, u_3, \ldots, u_n \mid u \in U\} \qquad (a)$$
$$I = \{i_1, i_2, i_3, \ldots, i_m \mid i \in I\} \qquad (b)$$
$$R = \{r_1, r_2, r_3, r_4, r_5 \mid r \in R\} \qquad (c)$$

Step 2 - (Classification): The system arrange low-rated users which total ratings 'R' (1 ~ 1000) as training set (TR) and high-rated users which total ratings 'R' are (1000 above) as testing set (TS).

Step 3 - (Collection): With respect to the training (TR) and testing set (TS), the system collects the profile preferences (age, gender, occupation) from the users personalized profiles for computing the collaborative similarity measurement.

Step 4 - (Hybridization): Hybridize the collaborative and k-neighborhood features to conduct following parameters:

— *Profile Similarities:* Profile similarity has been calculated between the training set T_R (low-rated users 'u' profile preferences) and testing set T_S (high-rated users 'v' profile preferences) to improve the low-rated user recommendation accuracy. The following similarities have been measured by using Euclidean cosine vector-based similarity method in equation (2) and (3).

$$\sum uv = \sum \frac{dot_{(u,v)}}{(norm_u)_2 \times (norm_v)_2} \tag{2}$$

$$Sim(u,v) = \frac{\sum uv \times (\sum u \sum v)}{\sqrt{(\sum u^2 \times (\sum u)^2)}\sqrt{(\sum v^2 \times (\sum v)^2)}} \tag{3}$$

With the help of equation (2) and (3), the similarity has been tackled between the low-rated user (as vector 'u') and high-rated user (as vector 'v') personalized profile preferences (age, gender, and occupation). The similarities of user 'u' and user 'v' are represented as Sim(u,v). The similarity results are demonstrated in section 4.

— *Similar k-neighborhoods:* Compute the similarities Sim(u,v) of users 'u' and users 'v', choose only those k-neighbor users 'v' profiles, which are highly similar to user 'u' profile preferences.
— *Neighborhoods related items / learning content:* Collects the items i that rated by the k-neighbors similar users of user's 'v'.

Step 5 - (Recommendation): Receives the recommended items / learning-contents 'I' and recommend them to low-rated users. The recommendations are denoted as rm.

5   Results and Discussion

The experimental result has been tackled with the help of equation (2) using cosine vector-based similarity matrix factorization [Un x Vn]. This study randomly elects 10 low-rated users profiles (as mentioned in section 2.3) as (U1, U2, U3, U4, U5, U6, U7, U8, U9, U10) and the N number of users profiles with ratings greater than 1000 identifies as Vn high-rated user's profiles. The low-rated / users 'u' ∈ 'U' have been considered as training set TR and high-rated user's 'v' ∈ 'V' are as testing set TS (discussed in section 2.3). Table 3 shows the resulting matrix of conducted experiment with (U1, U2,...,Un=10 | 'u' ∈ 'U') training and (V1, V2,...,Vn | 'v' ∈ 'V') testing set. The similarity resultant matrix of [U10 x Vn] has been measured (from 0.1 to 1.0) parameters between low-rated users (U1, U2,...,Un=10 | 'u' ∈ 'U') as training set and high-rated users (V1, V2,...,Vn | 'v' ∈ 'V') as testing set from the 'Movielens' Dataset D. This similarity matrix element has computed with the help of following similarity parameters in equation 4:

$$\text{Similarity}_{[U_{10} \times V_n]} = \begin{cases} 0.9 \text{ to } 1.0 & \text{if there is perfect similarity between user } u \text{ and user } v \\ \geq 0.1 \text{ and } < 0.9 & \text{if there is mature similarity between user } u \text{ and user } v \\ 0 & \text{if there is no similarity between user } u \text{ and user } v \end{cases} \qquad (4)$$

With the help of equation (4) similarity parameters of [U10 × Vn] from 0.1 to 1.0 between low-rated users (U1, U2,...,Un=10 | 'u' ∈ 'U') and high-rated users (V1, V2,...,Vn | 'v' ∈ 'V').

Table 3. Similarity matrix of training-set TR_2 and testing-set TS_2 as $U_{n=10} \times V_n$

	V_1	V_2	V_3	V_4	V_5	V_6	V^7	V_8	....	V_n
U_1	0.957	0.938	0.911	0.944	0.938	0.729	0.952	0.944	....	0.949
U_2	0.938	0.957	0.954	0.956	0.892	0.788	0.878	0.752	....	0.948
U_3	0.947	0.982	0.957	0.955	0.934	0.584	0.621	0.913	....	0.499
U_4	0.788	0.878	0.752	0.957	0.956	0.715	0.842	0.949	....	0.701
U_5	0.819	0.899	0.621	0.697	0.957	0.934	0.584	0.621	....	0.913
U_6	0.946	0.715	0.842	0.949	0.701	0.957	0.788	0.878	....	0.752
U_7	0.949	0.715	0.842	0.949	0.701	0.952	0.957	0.915	....	0.955
U_8	0.955	0.955	0.584	0.621	0.913	0.956	0.955	0.957	....	0.938
U_9	0.899	0.621	0.957	0.948	0.947	0.925	0.899	0.935	....	0.499
U_{10}	0.911	0.897	0.893	0.944	0.899	0.621	0.697	0.952	....	0.957

The table 3 stated that low-rated user's profiles similarity results exhibit almost identical accuracy that would be indicated by Precision Pr (equation 5) values. Their recall Re (equation 6) values differ considerably, hinting at similar behavior with respect to the types of users scored profile content similarities.

$$Precision\ (Pr) = \frac{Total\ size\ of\ recommended\ learning\ contents}{Total\ size\ of\ learning\ contents} = \frac{T_{rm}}{T_i} \qquad (5)$$

$$Recall\ (Re) = \frac{Total\ size\ of\ recommended\ learning\ contents}{Total\ size\ of\ relevent\ learning\ contents} = \frac{T_{rm}}{T_{I_{(u.v) \in \hat{r}}}} \qquad (6)$$

The results evaluation has been done with the help of precision equation (5) and recall equation (6) respectively. In equations (5) and (6), T_{rm} embodies the total number of recommended contents / items, T_i represent total contents / items stored in the system and $T_{Iv \in r}$ is considered as total relevant content / items with respect to ratings. The conducted experimental study on the improvement of traditional collaborative filtering features with k-neighborhood to improve the low-rated users profile recommendation issue. The evaluation stage compared the research work of

Table 4. The results evaluation of Proposed Goal-based hybrid filtering, measured by Average Mean of Precision *Pr* and Average Mean of Recall *Re*

	Kant, V. and K.K. Bharadwaj [1]	**Proposed Goal-based Hybrid Filtering**
Average Mean of Precision *Pr* (equation 5)	66.43%	79. 90%
Average Mean of Recall *Re* (equation 6)	78.53%	83.50%

(Kant and Bharadwaj, 2013), the researchers stat that Kant and Bharadwaj's work based on machine learning based recommendations method. The evaluation results of [1] and the proposed goal-based hybrid filtering are given in table 4.

In the table 4, the average mean of precision Pr (equation 5), used to calculate the precision mean average rate of recommendation on behalf of the similarity between training set U and testing set V users profiles similarity measurement with the help of table 3. While the average mean of recall Re (equation 6), used to measures the average rate of desired learning content / items appearing among the low-rated users profile recommendations. The results show that the proposed goal-based hybrid filtering improved results as compared to the work of [1]. Figure 3 shows the results using bar graph diagram.

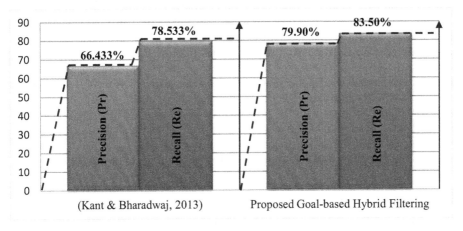

Fig. 2. The results evaluation of proposed Goal-based Hybrid Filtering by Mean of Precision *Pr* and Mean of Recall *Re*

Fig 2 shows the evaluation results of proposed goal-based hybrid filtering and [1] using bar-graph. The results demonstrate that proposed work helps to improve the work if the parameters will be passed accurately. The approach of (Kant and Bharadwaj, 2013) contains (Precision Mean: 66.43%, Recall Mean: 78.53%). The proposed goal-based hybrid filtering combined k-neighborhood features in traditional collaborative filtering features, which improves the low-rated user profile recommendations as revealed in the result (Precision Mean: 79.90%, Recall Mean: 83.50%). The results demonstrate that proposed Goal-based hybrid filtering works well to improve the low-rated users profile recommendations. The improvement has tackled with only one machine learning (k-nearest neighbor) technique for the enhancement and improvement of the content-based filtering technique in the normal way with collaborative features for e-learning recommendation systems. The experiments (Figure 3) demonstrate the importance of proposed goal-based hybrid filtering as through these experiments generates the good recommendation results are achieved from multi user personalized similarity process.

6   Conclusion

Recommender system is a very active research field. A number of works has already mentioned in (section 2) related work of this paper. This study includes a partial element of research with aim to work with collaborative filtering and k-neighbors scheme features to improve the low-rated personalized profile preference similarities scheme. The proposed goal-based hybrid filtering approach works with minimum profile preferences (age, gender, occupation) without using any extra information (highly rated / visited items history) information for low-rated user's recommendation. Methodologically, the proposed approach compute the similarity of low-rated users profile preferences with other high-rated users personalized profile preferences, and recommend the similar high-rated users learning content / item to the low-rated users with respect to their similarity parameters (equation 4). This proposed research has used the famous 'Movielens' dataset for experimental descriptive personalized profile similarities. Table 2 defines the flow of training and testing set and figure 1 shows the operational flow of the proposed work. The equation (3) and (4) helps to compute the profile similarities between low-rated users U10 and high-rated users V_n (shows in table 3). The recommendation average a result of low-rated users has been calculated with the help of average means precision Pr (equation 5) and recall Re (equation 6). The experimental results of proposed approach and results evaluation in table 4, demonstrate that the proposed goal-based hybrid filtering improves the recommendation rates of low-rated user profile collaboratively.

7   Future Work

The user's requirements are increasing with the passage of time, for this, more research is required to be applicable in real world situations in the field of e-Learning recommender systems [25, 26]. Therefore, the work on low-rated users profile recommendations in e-Learning recommender systems should reviewed with an eye toward the next generation and will be experimented on different other datasets too. This future work will help to improve the recommendation approaches / methods / techniques and results to offer more secure and useful appropriate learning contents that are most relevant to the user's goal.

Acknowledgement. The Universiti Teknologi Malaysia (UTM) under research grant 03H02 and Ministry of Science, Technology & Innovations Malaysia, under research grant 4S062 are hereby acknowledged for some of the facilities utilized during the course of this research work.

References

1. Kant, V., Bharadwaj, K.K.: A User-Oriented Content Based Recommender System Based on Reclusive Methods and Interactive Genetic Algorithm. In: Proceedings of Seventh International Conference on Bio-Inspired Computing: Theories and Applications (BIC-TA 2012). AISC, vol. 201, pp. 543–554. Springer, Heidelberg (2013)
2. Salmon, G.: E-moderating: the key to teaching and learning online, 2nd edn. Softcover, ed. L.K. Page. Taylor & Francis, London (2003)
3. Shishehchi, S., et al.: Review of personalized recommendation techniques for learners in e-learning systems. In: 2011 International Conference on Semantic Technology and Information Retrieval (STAIR) (2011)
4. Ghauth, K.I., Abdullah, N.A.: Measuring learner's performance in e-learning recommender systems. Australasian Journal of Educational Technology 26(6), 764–774 (2010)
5. di Bitonto, P., et al.: A recommendation method for e-learning environments: The rule-based technique. Journal of E-Learning and Knowledge Society 6(3), 31–40 (2010)
6. Lara, R.O., Arroyo, S., Lausen, H., Roman, D., Chirita, P.: A semantic web services framework for distributed e-learning environments. Informe técnico, L3S Research Center (2004)
7. Adomavicius, G., Tuzhilin, A.: Toward the next generation of recommender systems: a survey of the state-of-the-art and possible extensions. IEEE Transactions on Knowledge and Data Engineering 17(6), 734–749 (2005)
8. Ghazanfar, M.A., Prugel-Bennett, A.: A Scalable, Accurate Hybrid Recommender System. In: Third International Conference on Knowledge Discovery and Data Mining, WKDD 2010, pp. 94–98. IEEE (2010)
9. Burke, R.: Hybrid recommender systems: Survey and experiments. User Modeling and User-Adapted Interaction 12(4), 331–370 (2002)
10. Mi, Z., Xu, C.: A Recommendation Algorithm Combining Clustering Method and Slope One Scheme. In: Huang, D.-S., Gan, Y., Premaratne, P., Han, K. (eds.) ICIC 2011. LNCS, vol. 6840, pp. 160–167. Springer, Heidelberg (2012)
11. Ge, F.: A User-Based Collaborative Filtering Recommendation Algorithm Based on Folksonomy Smoothing. In: Zhou, M., Tan, H. (eds.) CSE 2011, Part II. CCIS, vol. 202, pp. 514–518. Springer, Heidelberg (2011)
12. Dash, R., Dash, R.: Comparative Analysis of K-means and Genetic Algorith Based Data Clustering. International Journal of Advanced Computer and Mathematical Sciences 3(2), 257–265 (2012)
13. Lee, M., Choi, P., Woo, Y.T.: A hybrid recommender system combining collaborative filtering with neural network. In: De Bra, P., Brusilovsky, P., Conejo, R. (eds.) AH 2002. LNCS, vol. 2347, pp. 531–534. Springer, Heidelberg (2002)
14. Selamat, A., Raza, M.A.: An improvement on genetic-based learning method for fuzzy artificial neural networks. Applied Soft Computing 9(4), 1208–1216 (2009)

15. Jahrer, M., Töscher, A., Legenstein, R.: Combining predictions for accurate recommender systems. In: Proceedings of the 16th ACM SIGKDD International Conference on Knowledge Discovery and Data Mining. ACM (2010)

16. Hsu, M.-H.: A personalized English learning recommender system for ESL students. Expert Systems with Applications 34(1), 683–688 (2008)

17. Bin Ghauth, K.I., Abdullah, N.A.: Building an E-learning Recommender System Using Vector Space Model and Good Learners Average Rating. In: Ninth IEEE International Conference on Advanced Learning Technologies, ICALT 2009. IEEE (2009)

18. Pin-Yu, P., et al.: The development of an Ontology-Based Adaptive Personalized Recommender System. In: 2010 International Conference on Electronics and Information Engineering (ICEIE) (2010)

19. Research, G.: MovieLens Dataset (1997),
    http://movielens.umn.edu/html/tour/index.html

20. Qiu, T., Chen, G., Zhang, K., Zhou, T.: An item-oriented recommendation algorithm on cold-start problem. Euro Physics Letters, 95(5) (2011)

21. He, L., Wu, F.: A Time-context-based Collaborative Filtering Algorithm. In: IEEE International Conference on Granular Computing, GRC 2009. IEEE, Nanchang (2009)

22. Symeonidis, P., Nanopoulos, A., Manolopoulos, Y.: Providing justifications in recommender systems. IEEE Transactions on Systems, Man and Cybernetics, Part A: Systems and Humans 38(6), 1262–1272 (2008)

23. Verbert, K., Drachsler, H., Manouselis, N., Wolpers, M., Vuorikari, R., Duval, E.: Dataset-driven research for improving recommender systems for learning. In: 1st International Conference on Learning Analytics and Knowledge (2011)

24. López-Nores, M.B.-F., Pazos-Arias, Y., Gil-Solla, J.J.: Property-based collaborative filtering for health-aware recommender systems. Expert Systems with Applications 39(8), 7451–7457 (2012)

25. Khribi, M.K., Jemni, M., Nasraoui, O.: Automatic Recommendations for E-Learning Personalization Based on Web Usage Mining Techniques and Information Retrieval. In: Eighth IEEE International Conference on Advanced Learning Technologies, ICALT 2008 (2008)

26. Jiang, L., Elen, J.: Why do learning goals (not) work: a reexamination of the hypothesized effectiveness of learning goals based on students' behaviour and cognitive processes. Etr&D-Educational Technology Research and Development 59(4), 553–573 (2011)

StudyAdvisor: A Context-Aware Recommendation Application in e-Learning Environment

Phung Do[1], Hiep Le[2], Vu Thanh Nguyen[1], and Tran Nam Dung[3]

[1] University of Information Technology
Vietnam National University HoChiMinh City
{phungdtm,nguyenvt}@uit.edu.vn
[2] John von Neumann Institute, Vietnam National University HoChiMinh City
hiep.le.ict@jvn.edu.vn
[3] University of Science, Vietnam National University HoChiMinh City
trannamdung@yahoo.com

Abstract. The explosion of world-wide-web has offered people a large number of online courses, e-classes and e-schools. Such e-learning applications contain a wide variety of learning materials which can confuse the choices of learner to select. In order to address this problem, in this paper we introduce an e-learning application named StudyAdvisor which integrates a context-aware recommender system to suggest suitable learning materials for learners. Particularly, StudyAdvisor provides lessons and exercises or questions related to Fundamental of Database domain. After learners do exercises and take examinations, we then use the study result of learners to determine the learning levels (context) and base on this information to give suggestions by using a context-aware recommendation technique named STI. As a result, lessons to preview and exercises to practice are recommended for learners.

Keywords: e-learning, context, recommender systems.

1 Introduction

E-learning applications play a very important role in supporting online learners in recent decades. These are network-based applications which provide internet users flexible environment to learn actively, every time and everywhere with internet connection. As in the majority of applications in e-commerce, e-learning applications contain the plentiful learning materials namely courses, lessons, references and exercises. As a result, learners face the problem in selecting learning materials which are suitable for their learning levels from the potentially overwhelming number of alternatives.

In order to address this problem, the recommender system (RS) is one of the effective solutions. RSs indicate software tools having techniques to generate suggestions which are suitable items users might prefer [1]. In e-learning, RSs suggest appropriate learning materials (items) for learners (users). Yet RSs do

J. Sobecki, V. Boonjing, and S. Chittayasothorn (eds.), *Advanced Approaches to Intelligent*   119
Information and Database Systems, Studies in Computational Intelligence 551,
DOI: 10.1007/978-3-319-05503-9_12, © Springer International Publishing Switzerland 2014

not take into consideration additional information such as time, place, companion and others which can influence preferences or tastes of user. This additional information is called contextual information or context briefly (more details in section 2). The RSs which deal with contextual information are called context-aware recommender systems (CARSs) [1].

In this paper, we present an e-learning application, StudyAdvisor, which uses one of techniques from CARSs to provide recommendations. This application contains lessons, examinations, practice tests and questions related to the *Fundamentals of Database* domain. It is undeniable that each learner has different competence to study. For instance, an exercise could be easy for a learner, however it may be difficult for others to do. Thus we are motivated to consider the learning level to make recommendations.

We first identify the learning level for each user in the system. This information is determined automatically and considered as context. We then acquire ratings, the rating for a learner (a user) and a question (an item) indicates the probability this learner is able to answer this question correctly. Finally, a method from CARSs is applied to recommend learning materials for learners.

In the rest of this paper, the structure is organized as follow: section 2 presents the related work about CARSs and prior work which applies RSs in e-learning environment. Section 3 describes the main points of StudyAdvisor application, section 4 reveals the technique to infer contextual information implicitly and section 5 describes method to obtain ratings. Next, section 6 and section 7 present the modeling and recommendation processes respectively. Section 8, conclusions and future work, appears last.

2   Related Work

Context is any information that can be used to characterize the situation of an entity. An entity is a person, place, or object that is considered relevant to the interaction between a user and an application [2]. Entity is often a user, an item and the rating from a user over an item in terms of RSs.

There are three ways [3] to obtain contextual information include:
- *Explicitly* from the relevant objects by asking direct questions or eliciting through other means.
- *Implicitly* from the data or the environment such as the change in location automatically detected by devices or the time stamp of a transaction.
- *Inferring* the context using statistical or data mining methods.

CARSs indicates RSs which incorporate contextual information into recommendation process to model and predicting tastes of users. As in RSs, collaborative filtering (CF)[1],[4] is often used in CARSs to make recommendations. Recall that CF uses a 2-dimensions matrix (*the rating matrix $U \times I$*) made up by a list of users (U) and a list of items (I) (as shown in Fig 1(a)). The rating matrix represents preferences' users which are explicit ratings with scale 1-5 or implicit indications such purchasing frequencies or click-throughs [4]. Because of the appearance of context, the rating matrix is extended as a multi-dimensional matrix (denoted $U \times I \times C$) with contexts. Fig 1(b) shows an example of multi-dimensional rating matrix where contextual information is time. There are missing values in the matrix where users did not give their preferences for certain items (in certain contexts) and CF has to predict them. CF is extended to deal with contextual information called context-aware collaborative filtering (CACF).

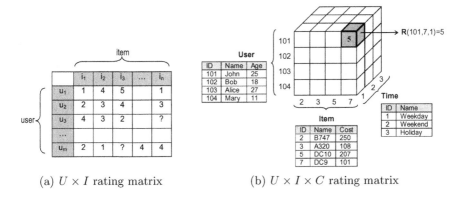

(a) $U \times I$ rating matrix   (b) $U \times I \times C$ rating matrix

Fig. 1. Rating Matrices

The prior work of CARSs are implemented in e-commerce [5], entertainment (music [6], tourism [7], movie [3], [8]) and food [9] domains.

Regarding e-learning domain, recommendation task is implemented with methods in RSs without the using of context. In particular, in [10], neighborhood-based CF is applied to recommend suitable learning materials. Additionally, [11] maps educational data into user×item matrix [4] and applies matrix factorization technique to predict student performances.

Furthermore, recommendation processes use hybrid methods as well. Specially, [12] applies both user-based and rule-based methods to suggest relevant courses for learners. A scientific paper recommendation engine using model-based CF and hybrid techniques is built in [13]. The work in [14] proposes equations to adapt CF into e-learning meanwhile [15] combines RS with ubiquitous computing to enhance learning with memory-based CF and association mining techniques in botanic subject. In [16], a recommender system with ApiroriAll and memory-based CF methods is constructed to suggest java lessons.

E-learning RSs also use item's attributes to generate suggestions. The work in [17],[18] propose hybrid attribute-based methods with CF and content-based techniques to recommend books for learners, attributes consist of subjects, education levels, prices and authors or implicit attributes from history ratings of learners.

In relation to context-aware e-learning systems, [19] analyzes the user's knowledge gap (context) to filter suitable learning contents instead of using a recommendation technique.

Prior works mentioned above do not use any CACF technique. Therefore, the main contribution of this paper is to build an e-learning application with a CACF technique to recommend learning materials for learners.

3   The StudyAdvisor System

The StudyAdvisor system is an e-learning website which provides useful learning resources related to the Fundamentals of Database course for online learners. Such learning resources consist of : (1) a list of lessons related to Fundamentals of Database course, (2) examinations and (3) practice tests. Each examination or practice test includes questions. There are many types of question such as true-false, multiple-choice, matching, completion (as shown in Fig. 3) and others with different hard levels in this application. This application targets Vietnamese students; therefore, we use Vietnamese as the main language for all lessons, questions, examinations and practice tests.

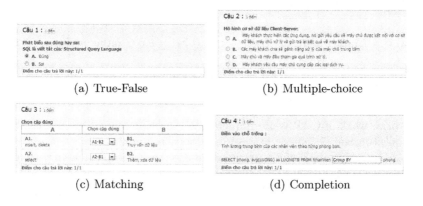

(a) True-False

(b) Multiple-choice

(c) Matching

(d) Completion

Fig. 2. Types of question (Vietnamese GUI)

This application supports four types of user (student, teacher, faculty secretary and administration system) with different functions. Nonetheless, students (or learners) are the main actors of this application. They register, log in and use the application to access lessons, take examinations and do practice tests. In order to highly support for learners, we build a recommender engine based on learning levels, it gives recommendations such as lessons and questions for learners to review and practice.

The recommendation function is the most important in this application and Fig. 3 demonstrates how recommendations are produced. We rely on a CACF algorithm to generate recommendations. The input data for every CACF algorithm must be a rating matrix $(U \times I \times C)$ which is partially filled by ratings for items (I) from users (U) in different contexts (C). To adapt to our application, the terms "user, item and context" are interchangeably used as "learner, question and learning level" respectively. The method to achieve the rating matrix and generate recommendations is now shortly presented.

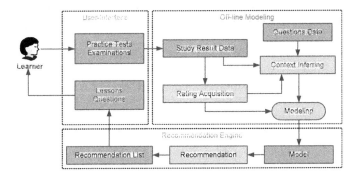

Fig. 3. Method to produce recommendations

First Through user's interfaces, learners take *examinations* or do *practice tests* and study results data is then recorded. The study results data contains information about the timestamp, answers, score, correct answers, incorrect answers, related-lessons, users and others.

Second At server side with offline mode, the *study results data* is regularly processed with three main steps executed consecutively.

- *Context Inferring*: The study results data and *questions data* of system are used to infer context. As a result, the C component of the rating matrix is identified.
- *Rating Acquisition*: The main purpose of this step is to collect ratings extracted from the study results data. This step aims to find out the U, I and ratings components of the rating matrix.
- *Modeling*: A model-based CACF algorithm (STI method [20]) is applied with the rating matrix obtained from two steps above to build the recommender model which is stored/updated in the system database.

Finally *Recommendation* is processed. The model is retrieved to predict ratings and to produce recommendations. The *recommendation list* is normally a list of *lessons* or a list of *questions* displayed in the user's interface. The users continue taking examinations and doing practice tests to enrich the study results data for the next modeling processes.

Details of context inferring, rating acquisition, modeling and recommendation processes are clarified in the next sections.

4   Context Inferring

Context in StudyAdvisor system is the learning level which depicts the academic performance of a learner in the fixed period of time. It means that a learner could have a variety of learning levels which depends on study results of this learner. The learning level is inferred implicitly and periodically in this application. We define the learning level (context) has 7 values: beginner, elementary, pre-intermediate, intermediate, upper-intermediate, advanced and fluent with the corresponding indexes $\{0, 1, 2, 3, 4, 5, 6\}$. The learning level of a new learner in the system is marked as 0 and it is then modified when this learner does practice tests or examinations.

We propose a method to determine automatically the learning level for each learner in the system. To infer context, the hard level attribute of questions and the study results data are examined. The hard level of a question is manually pre-defined by teachers and it has value range $[1, h], h = 10$ in our application. The larger the hard level is, the more difficult a question is.

For each the uth learner, a score vector when the context inferring step is occurred, $s_u = (z_1, z_2, ..., z_j, ..z_h), j = 1..h$, is constructed. The value of z_j is computed as the number of times the uth learner answers correctly questions with the hard level j, divided by the total number of times this learner answers these questions. Note that $z_j = -1$ when the uth learner does not answer any question with the hard level j. Then the value of $Score_u$ is calculated as follow:

$$Score_u = \frac{\sum_{j=1, z_j \neq -1}^{h} j z_j}{\sum_{j=1, z_j \neq -1}^{h} j} \tag{1}$$

Table 1. The $Score_u$ value table for context inferring

Range of $Score_u$	Context value
$[0, 0.3)$	0
$[0.3, 0.5)$	1
$[0.5, 0.6)$	2
$[0.6, 0.7)$	3
$[0.7, 0.8)$	4
$[0.8, 0.9)$	5
$[0.9, 1]$	6

The value of $Score_u$ indicates the study result of the uth user from doing exercises until the context inferring step is occurred. Having the value of $Score_u$, we use Table 1 to decide the context of the uth learner. This table is based on the marking scale $(0 - 10)$ and ranking in our university, we transform it to scale $(0 - 1)$ to identify the learning level.

5   Rating Acquisition

As in the vast majority of RSs, the rating 1-5 or like-dislike scale is the most commonly used and it often shows the preference of a user to an item (in a certain context). However, in StudyAdvisor, giving the rating for each question (item) in a long practice test or an examination seems to be unreasonable. Moreover, a question learners might like could be very easy and vice versa. Thus, to obtain the ratings, we propose the following expression:

$$r_{uic} = \frac{p_{uic}}{s_{uic}} \qquad (2)$$

Where r_{uic} is the probability the uth learner answers correctly the ith question in context c, s_{ui} indicates the total number of times the uth learner answers the ith question in context c and p_{ui} is the number of times this user gives the correct answers in context c.

The value of r_{uic} is in range [0, 1] and the larger this value is, the larger probability a learner can the right answer. Notice that, we use the ratings with the meaning the probability a question answered correctly in the rest of this paper. The new rating larger than the previous one can reveal that this learner has improved the competency in the better way.

6 Modeling

As usual in RSs (CARSs), an algorithm is applied to predict ratings for unrated items and then items with high predicted ratings are recommended for the active user. There are numbers of CACF algorithms; however, we use the STI algorithm [20] for this task because this algorithm is quite good at addressing the sparsity problem. This algorithm is our prior research, a model-based CACF algorithm, lies on clustering and matrix factorization (SVD) [1] techniques to generate a model for recommending.

The rating matrix gained is used to build the recommender model. The recommender model is periodically re-built and stored/updated in a database. Modeling is an off-line mode process; therefore, it does not affect the speed of real-time response to the user's interface.

7 Recommendation

StudyAdvisor provides two recommendation tasks (as shown in Fig. 4) include:

Lessons Recommendation It is true to say that if a lesson is extremely hard to understand and the probability a learner give the answer for a question is high. Thus, this task is to make learners pay attention to important lessons which they need to understand clearly. Lessons are recommended immediately when a learner finishes an examination and it offers a list of lessons which this learner should review.

Questions Recommendation This task assists learners understand the content of lessons through doing related questions or to prepare for incoming examinations. Questions are recommended as the request of learners.

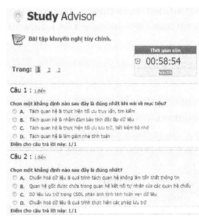

Bạn có thể học lại những nội dung sau		
Thứ tự ưu tiên	Tên bài học	Thao tác
1	Ngôn ngữ SQL	
2	Ràng buộc toàn vẹn trong một CSDL	
3	Ngôn ngữ đại số quan hệ	
4	Tối ưu hóa câu hỏi bằng đại số quan hệ	
5	Các khái niệm của một hệ CSDL	
6	Các mô hình CSDL	

(a) Lessons Recommendation   (b) Questions Recommendation

Fig. 4. Recommendation (Vietnamese GUI)

7.1 Lessons Recommendation

Given $L = \{l_0, l_1, ..., l_j, ..., l_k\}, j = 0..k$ contains lessons whose questions appear in the examination. For each lesson l_j, StudyAdvisor finds Q_j set, which includes questions belong to this lesson and predict ratings for questions in Q_j. The predicted rating for l_j is the average rating of questions in Q_j. Then such ratings are ranked in ascending order. Lessons with ratings smaller than a threshold ($\tau_l = 0.5$ in our application) are recommended to the active learner.

The rating for a lesson in this case reveals the probability a learner can answers correctly all questions belong to this lesson, that is the reason why the lessons with the smallest ratings would be on top in the recommendation list and the learner should make them priorities to review.

7.2 Questions Recommendation

Unlike RSs (CARSs) normally recommend items which users do not give their assessments, StudyAdvisor considers all questions to recommend. This is because, an extremely difficult question could be recommended many times although there are ratings for this question.

Although we consider all questions include questions done by learners, we just recommend done questions if their predicted rating values are smaller than a threshold τ_q. For instance, in StudyAdvisor, a learner answers correctly an easy question 3 times, and the predicted value is approximately 0.82 which is larger than the threshold $\tau_q = 0.6$, then we do not recommend this question.

StudyAdvisor will recommend questions base on the learner's options. There are two options for questions recommendation:

1. The first one is overall practice test which has the same structure as a real examination with questions related many lessons included. Learner is required to provide the number of questions.

2. The second one is the personalized practice test which learners can choose lessons they want to practice and how many questions belongs to the each chosen lesson.

Learners can provide duration and the hard level to obtain the suitable practice test. There are two types of hard level: ascending (default) and descending. The system predicts ratings for all questions related to the learner's configuration and recommends questions with highest predicted ratings in case the ascending hard level is selected and vice versa.

8   Conclusions and Future Work

In this paper, we introduce StudyAdvisor, an e-learning application which recommends learning materials for learners. We first present the method to determine contextual information implicitly. We then describe the technique to gain ratings from the study results data of learners. Finally, a CACF technique (STI), our prior research, is applied to predict ratings and generate suitable questions and lessons for active users. The different point of our work compared to examined prior work is considering the learning level as context role to apply a CARS algorithm and offer recommendations.

In the future, we intend to learn ubiquitous computing and combine it with CARS and mobile environment in e-learning field to assist learners study more actively and efficiently, every time and everywhere.

Acknowledgment. This research is funded by Vietnam National University - Ho Chi Minh City (VNU-HCM) under grant number 01/CNTT/2013/911VNUHCM-JAIST.

References

1. Ricci, F., Rokach, L., Shapira, B., Kantor, P.B.: Recommender Systems Handbook, 1st edn. Springer-Verlag New York, Inc., New York (2010)
2. Dey, A.K.: Understanding and Using Context. Personal Ubiquitous Comput. 5(1), 4–7 (2001)
3. Adomavicius, G., Tuzhilin, A.: Context-aware recommender systems. In: Proceedings of the 2008 ACM Conference on Recommender Systems, RecSys 2008, pp. 335–336. ACM, New York (2008)
4. Su, X., Khoshgoftaar, T.M.: Collaborative Filtering for Multi-class Data Using Belief Nets Algorithms. In: Proceedings of the 18th IEEE International Conference on Tools with Artificial Intelligence, ICTAI 2006, pp. 497–504. IEEE Computer Society, Washington, DC (2006)
5. Panniello, U., Gorgoglione, M.: Incorporating context into recommender systems: an empirical comparison of context-based approaches. Electronic Commerce Research 12(1), 1–30 (2012)
6. Baltrunas, L., Kaminskas, M., Ludwig, B., Moling, O., Ricci, F., Aydin, A., Lüke, K.-H., Schwaiger, R.: InCarMusic: Context-Aware Music Recommendations in a Car. In: Huemer, C., Setzer, T. (eds.) EC-Web 2011. LNBIP, vol. 85, pp. 89–100. Springer, Heidelberg (2011)

7. Baltrunas, L., Ludwig, B., Peer, S., Ricci, F.: Context relevance assessment and exploitation in mobile recommender systems. Personal Ubiquitous Comput. 16(5), 507–526 (2012)

8. Odić, A., Tkalčič, M., Tasič, J.F., Košir, A.: Relevant Context in a Movie Recommender System: Users Opinion vs. Statistical Detection. In: Adomavicius, G. (ed.) Proceedings of the 4th Workshop on Context-Aware Recommender Systems in Conjunction with the 6th ACM Conference on Recommender Systems (RecSys 2012), vol. 889 (2012)

9. Ono, C., Takishima, Y., Motomura, Y., Asoh, H.: Context-Aware Preference Model Based on a Study of Difference between Real and Supposed Situation Data. In: Houben, G.-J., McCalla, G., Pianesi, F., Zancanaro, M. (eds.) UMAP 2009. LNCS, vol. 5535, pp. 102–113. Springer, Heidelberg (2009)

10. Soonthornphisaj, N., Rojsattarat, E., Yim-Ngam, S.: Smart e-learning using recommender system. In: Huang, D.-S., Li, K., Irwin, G.W. (eds.) ICIC 2006. LNCS (LNAI), vol. 4114, pp. 518–523. Springer, Heidelberg (2006)

11. Thai-Nghe, N., Drumond, L., Krohn-Grimberghe, A., Schmidt-Thieme, L.: Recommender system for predicting student performance. Procedia Computer Science 1(2), 2811–2819 (2010)

12. Tan, H., Guo, J., Li, Y.: E-learning Recommendation System. In: Proceedings of the 2008 International Conference on Computer Science and Software Engineering, CSSE 2008, vol. 5, pp. 430–433. IEEE Computer Society, Washington, DC (2008)

13. Tang, T., McCalla, G.I.: Evaluating a Smart Recommender for an Evolving E-learning System: A Simulation-Based Study. In: Tawfik, A.Y., Goodwin, S.D. (eds.) Canadian AI 2004. LNCS (LNAI), vol. 3060, pp. 439–443. Springer, Heidelberg (2004)

14. Bobadilla, J., Serradilla, F., Hernando, A.: Collaborative filtering adapted to recommender systems of e-learning. Know.-Based Syst. 22(4), 261–265 (2009)

15. Wang, S.L., Wu, C.Y.: Application of context-aware and personalized recommendation to implement an adaptive ubiquitous learning system. Expert Syst. Appl. 38(9), 10831–10838 (2011)

16. Klašnja-Milićević, A., Vesin, B., Ivanović, M., Budimac, Z.: E-Learning personalization based on hybrid recommendation strategy and learning style identification. Computers & Education 56(3), 885–899 (2011)

17. Salehi, M., Kmalabadi, I.N.: A Hybrid Attribute-based Recommender System for E-learning Material Recommendation. IERI Procedia 2, 565–570 (2012)

18. Salehi, M., Kmalabadi, I.N., Ghoushchi, M.B.G.: A New Recommendation Approach Based on Implicit Attributes of Learning Material. IERI Procedia 2, 571–576 (2012)

19. Schmidt, A., Winterhalter, C.: User context aware delivery of e-learning material: Approach and architecture. Journal of Universal Computer Science (JUCS) 10, 28–36 (2004)

20. Do, P., Le, H., Nguyen, V.T., Dung, T.N.: A Context-Aware Collaborative Filtering Algorithm through Identifying Similar Preference Trends in Different Contextual Information. In: S. Obaidat, M. (ed.) Advanced in Computer Science and Its Applications. LNEE, vol. 279, pp. 339–344. Springer, Heidelberg (2014)

An Adaptive Fuzzy Decision Matrix Model for Software Requirements Prioritization

Philip Achimugu, Ali Selamat, Roliana Ibrahim, and Mohd Naz'ri Mahrin

Department of Software Engineering, Faculty of Computing, Universiti Teknologi Malaysia
{check4philo,nazri742002}@gmail.com, {aselamat,roliana}@utm.my

Abstract. Software elicitation process is the act of extracting and sorting requirements of a proposed system perceived to reflect the projected performance of the software under consideration for development. For software systems to be long-lived and satisfy stakeholder's expectations; there will be need to prioritize choices at the elicitation level. However, these choices could be distorted or misleading if appropriate techniques are not utilized in analyzing and prioritizing them. Consequently, if software systems are developed on vague prioritization results, the end product will not meet stakeholder's expectations. In this research, we present a scalable innovative prioritization model that is capable of comparing sets of elicited requirements by computing the weights of each criterion that makes up specified requirements. To achieve our aim, the weights assigned to each requirement by relevant stakeholders are normalized and a confidence function is computed to ascertain the ranking order of requirements. To validate the applicability of our model, we describe an empirical case scenario detailing the adaptability prowess of the proposed model.

1 Introduction

In most software development projects, there are more candidate requirements specified for implementation with limited time and resources. A meticulously selected set of requirements must therefore be considered for implementation. For software systems to be acceptable by users or stakeholders, its requirements must be well captured, analyzed and prioritized [1]. There are so many advantages of prioritizing requirements during software development process. Firstly, prioritization aids the implementation of a software system with preferential requirements of stakeholders. Secondly, due to the complexities and constraints associated with software development such as limited resources, inadequate budget and insufficient skilled programmers amongst others, requirements prioritization can help in planning software releases since not all the elicited requirements could be implemented in single release. Thirdly, it enhances software testing since test cases can be generated based on prioritized requirements rather than ambiguous or vague requirements. Finally, prioritization can eliminate breaches in agreement, trust and contracts.

J. Sobecki, V. Boonjing, and S. Chittayasothorn (eds.), *Advanced Approaches to Intelligent Information and Database Systems*, Studies in Computational Intelligence 551,
DOI: 10.1007/978-3-319-05503-9_13, © Springer International Publishing Switzerland 2014

To prioritize requirements, stakeholders will have to pair-wisely compare each to determine their relative importance by weighting scores. These scores are then computed across all relevant stakeholders to display prioritized requirements. The pair-wise comparison increases with rise in the numbers of requirements [2-5]. Therefore, prioritization can also be considered to be a multi-criteria decision making process. Existing prioritization techniques demonstrate high potentials; however, some limitations still exist. These limitations include scalability [6]; rank updates [7] and computational complexities [8]. Consequently, this research attempts to address the second problem (rank updates).

Rank updates refer to a situation where a prioritization technique is unable to reflect or compute the relative weights of stakeholders when a requirement is added or deleted from the set. When this problem persist, prioritization results may not be useful since it does not support requirements evolvability and reusability which are essential attributes of long-lived systems.

The rest of the paper is structured as follows: Section 2 discusses the multi-criteria decision making problem, Section 3 presents the proposed approach, section 4 deals with results and discussions of the proposed approach while section 5 concludes the paper.

2   Multi-criteria Decision Making (MCDM) Problems

Multi-criteria decision making process is the act of choosing preferential elements from some sets of alternatives based on pre-defined criteria. The MCDM problems can be divided into two types [9]. The first is referred to as classical MCDM problems [10], where ratings are achieved by computing the weights of criteria in crisp numbers; while the second is referred to as fuzzy multi-criteria decision-making (FMCDM) problems [11, 12], where ratings are achieved by computing the weights of criteria based on imprecision, uncertainty and vagueness, usually executed with the help of linguistic terms. The classical MCDM process deals with the determination of preferential elements using criteria weights provided by decision makers (DM). Crisp values are usually used to represent the ratings and weights. However, the limitation of the classical MCDM has to do with the inability to precisely assess weights and ratings due to (1) unquantifiable information, (2) incomplete information, (3) unobtainable information, and (4) partial ignorance [13]. To address this limitation, the fuzzy set theory was proposed in order to cater for the uncertainties of human judgments. Bellman and Zadeh [14] first introduced fuzzy set theory into MCDM as an approach to effectively deal with the inherent imprecision, vagueness and ambiguity associated with human decision making process. In fuzzy MCDM, ratings and weights are represented with fuzzy numbers. A preferential element is then calculated by aggregating all criteria weights and alternatives ratings across decision makers where elements with higher weights are considered to be the most valued.

3   Method

Basically, a fuzzy MCDM process has to do with the collection of some sets of M alternatives that are to be assessed based on C criteria. This will lead to the generation of a decision matrix which will consist of M rows and N columns. Each element is either a single numerical value or a single grade, representing the performance of alternative A on criterion C.

In the context of software development which is the focus of this research, requirements R can stand for the alternatives to be assessed based on C criteria (Figure 1).

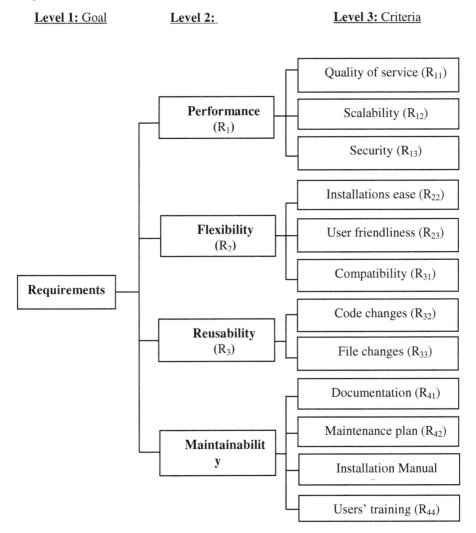

Fig. 1. Requirements and pre-defined criteria

In many cases, budget, resources and contract breeches are reasons why requirements are prioritized. Stakeholders are therefore required to find a set of requirements that maximizes the desired project objective. Figure 1 shows example of requirements and their respective criteria awaiting prioritization.

Based on linguistic variables, the weights (W) of requirements (R) across stakeholders (S) with respect to all the criteria C can be computed as follows:

$$R_1 = S_{11}(W_1) + S_{21}(W_2) + \ldots + S_{J1}(W_J)$$
$$R_2 = S_{12}(W_I) + S_{22}(W_2) + \ldots + S_{J2}(W_J)$$
$$R_3 = S_{13}(W_I) + S_{23}(W_2) + \ldots + S_{J2}(W_J)$$
$$R_4 = S_{14}(W_I) + S_{24}(W_2) + \ldots + S_{J2}(W_J)$$
$$R_I = S_{1I}(W_1) + S_{2I}(W_2) + \ldots + S_{JI}(W_J) \tag{1}$$

All the stakeholders' relative weights are elicited through questionnaire (Table 1). The valuation is done in Saaty's nine degree scale and is given as linguistic terms used for the comparison between requirements R_1, R_2, R_3 and R_4.

Table 1. Questionnaires for comparaison

Requirements	Superiority of requirements				A and B Equality
	Absolute	More	Medium	Less	
R_1					
R_2					
R_3					
R_4					

The results from each questionnaire obtained from the stakeholders are summarized in one matrix, where R_i represents the requirements; $i \in [1,\ldots, n]$ stands for the requirement number, r_{ij} are results from the comparison between the requirements. The relative weights of requirements elicited from the questionnaire across all stakeholders are computed in a decision matrix in the form shown in Equation 2.

$$D = \begin{matrix} & \begin{matrix} C_1 & C_2 & \cdots & C_n \end{matrix} \\ \begin{matrix} R_1 \\ R_2 \\ \vdots \\ R_m \end{matrix} & \begin{bmatrix} x_{11} & x_{12} & \cdots & x_{1n} \\ x_{21} & x_{22} & \cdots & x_{2n} \\ \vdots & \vdots & \ddots & \vdots \\ x_{m1} & x_{m2} & \cdots & x_{mn} \end{bmatrix} \end{matrix} \tag{2}$$

$$W = [\, w_1 \quad w_2 \quad \cdots \quad w_n\,]$$

For every requirement R, the following conditions hold:

$W_{ij} = 1$ for $i=j$; $W_{ij} = \dfrac{1}{x_{ij}}$ for $i \neq j$.

The value for the element x_{ij} is obtained as follows:

$x_{ij} = 1$, if R_i and R_j are of an <u>equal</u> importance;
$x_{ij} = 3$, if R_i is <u>less</u> important to R_j;
$x_{ij} = 5$, if R_i is <u>moderately</u> important to R_j;
$x_{ij} = 7$, if R_i is <u>more</u> important to R_j;
$x_{ij} = 9$, if R_j is <u>absolutely</u> superior to R_i.

After obtaining the relative weights of requirements from all the stakeholders, the next task is to compute the rank vector $\overline{V}$ for the given matrix A. The mean geometric values for each requirement are normalized using Equation 3.

$$
\overline{V} = \begin{pmatrix}
\overline{V_1} = V_1 / \sum_{i=1}^{n} V_i \\[2mm]
\overline{V_2} = V_2 / \sum_{i=1}^{n} V_i \\[2mm]
\cdots \\[2mm]
\overline{V_n} = V_n / \sum_{i=1}^{n} V_i
\end{pmatrix}
\tag{3}
$$

The rank vectors give the weight coefficients for every requirement. However, to calculate the relative weight of each requirement, Equation 4 is utilized.

$$
W_i = S_{+i} + \frac{\max_i S_{-i} \sum_{i=1}^{m} S_i - \min S \sum_{j=1}^{n} S_j}{S_{-i}\, n \sum_{i=1}^{m} S_{ij}}
\tag{4}
$$

Equation (5) is also used to compute the aggregated relative weights across all the stakeholders.

$$
W_i = S_{+i} + \frac{\sum_{i=1}^{m} S_{-i}}{S_{-n} \sum_{i=1}^{m} S_{-n}}
\tag{5}
$$

After computing the aggregated relative weights, the priority order of compared requirements are determined on the basis of their relative weights. The requirement with higher relative weights has higher priority (rank), and the requirement with the highest relative weight becomes most valuable requirement.

To initiate the final ranking process, the maximum d^+ and minimum d^- relative weights in each decision matrix row, is obtained using 6 and 7. Thereafter, the variance V between, the maximum and minimum weights are computed using Equation 8. The results generated from Equation 8 is divided by the total numbers of stakeholders to determine the relative ranks RR while the final rank FR is obtained by summing the relative rank weights under each category.

$$A^* \left\{ A_i \middle| \max_i W_i \right\} = \left(x_1^*, x_2^*, \cdots, x_n^* \right) \tag{6}$$

Where, x_j^* = the maximum values of the requirement entries in the aggregated decision matrix;

$$A' = \left\{ A_i \middle| \min_i W_i \right\} = \left(x_1', x_2', \cdots, x_n' \right) \tag{7}$$

Where, x_j' = the minimum values of the requirement entries in the aggregated decision matrix.

$$S_i = \left(\sum (A^*_{ij} - A^-_{ij}) \right) \tag{8}$$

Where $i = 1, 2, \cdots, m$

4   Results and Discussion

The essence of this section is to identify issues of relative disagreement and agreement on priorities between nine stakeholders. The strength of this approach has to do with the ability to identify substantial disagreement which can be isolated and subjected to more in-depth analysis so as to reach consensus or tradeoffs. The raw weights from each stakeholder matrices were normalized (by dividing the cell entry with the column total) to allow comparisons between criteria and to show the range of relative priorities. The cardinal consistency index was within the acceptable limits for all weights. Figure 2 shows the differences in priorities of the raw weights provided by the stakeholders over all the requirements while Figure 3 shows the stakeholder's preferences or relative weights in normalized form. The ranking process was conducted over four major requirements with twelve criteria. It can be seen that the largest difference lies in the reliability and flexibility requirements with a distinct divergence of priorities evident amongst many stakeholders. There is a high level of consistency amongst the separation distances between ranked requirements as seen in

Figure 4 while the final prioritized requirements are displayed in Figure 5. Meanwhile, as a reflection of the stakeholder's weights, Tables 2, 3 and 4 depict the pair wise comparison, normalized weights and priorities of specified requirements, which has led to the ranking order of $R_4 \succ R_1 \succ R_2 \succ R_3$. The geometric mean over all weights of each matrix cell was implored in addressing rank update problems. For the four main requirements, the most cherished requirement is maintainability (0.0621), followed by performance (0.0620). Flexibility was ranked third (0.03026) and the fourth was Reliability (0.0266).

Table 2. Pairwise comparison

Requirements	S_1	S_2	S_3	S_4	S_5	S_6	S_7	S_8	S_9
Performance									
Quality of service	1	0.11	0.20	0.20	1	1	1	0.20	1
Scalability	1	1	1	1	1	1	1	1	1
Security	0.33	1	1	1	0.20	0.11	0.33	1	0.33
Flexibility									
Installation ease	0.11	0.33	0.20	0.33	0.33	1	0.20	0.11	1
Use friendliness	0.14	0.20	0.33	0.20	1	1	1	0.22	1
Compatibility	0.20	0.14	0.14	0.14	1	0.20	1	0.14	0.20
Reliability									
File changes	0.14	0.11	0.20	0.14	0.20	1	1	1	1
Code changes	0.20	0.33	0.11	0.33	0.14	1	1	1	1
Maintainability									
Documentation	1	0.14	1	0.2	1	0.20	0.20	0.20	0.20
Maintenance plan	0.20	0.11	1	0.11	1	1	0.33	0.33	0.14
Installation manual	0.33	0.33	1	1	1	0.33	1	1	1
Users' training	0.14	0.20	1	1	1	1	1	1	1

Therefore, prioritizing software requirements is actually determined by subjective perceptions and weights of stakeholders based on the criteria used for the choice. Interestingly, fuzzy MCDM approach can be used to explain how stakeholders make decisions to select the best software requirements. Crisp data are inadequate to model the real life situations in MCDM; consequently, linguistic variables are implored to describe the various membership degrees of criteria used for ranking. In order to facilitate the making of subjective assessment by the stakeholders using fuzzy numbers, five sets of linguistic terms are used for assessing criteria weights and performance rating on each qualitative criterion respectively. A linguistic variable is a variable which apply words or sentences in a natural or artificial language to describe its degree of value. These kinds of expressions are used to compare each requirement based on pre-defined criteria.

Table 3. Normalized weights

Requirements	S_1	S_2	S_3	S_4	S_5	S_6	S_7	S_8	S_9
Performance									
Quality of service	0.2088	0.0275	0.0279	0.0354	0.1127	0.1131	0.1104	0.0278	0.1127
Scalability	0.2088	0.2500	0.1393	0.1770	0.1127	0.1131	0.1104	0.1389	0.1127
Security	0.0689	0.2500	0.1393	0.1770	0.0225	0.0124	0.0364	0.1389	0.0372
Flexibility									
Installation ease	0.0230	0.0825	0.0279	0.0584	0.0372	0.1131	0.0221	0.0153	0.1127
Use friendliness	0.0292	0.0500	0.0460	0.0354	0.1127	0.1131	0.1104	0.0306	0.1127
Compatibility	0.0418	0.0350	0.0195	0.0248	0.1127	0.0226	0.1104	0.0194	0.0225
Reliability									
File changes	0.0292	0.0275	0.0279	0.0248	0.0225	0.1131	0.1104	0.1389	0.1127
Code changes	0.0418	0.0825	0.0153	0.0584	0.0158	0.1131	0.1104	0.1389	0.1127
Maintainability									
Documentation	0.2088	0.0350	0.1393	0.0354	0.1127	0.0226	0.0221	0.0278	0.0225
Maintenance plan	0.0418	0.0275	0.1393	0.0195	0.1127	0.1131	0.0364	0.0458	0.0158
Installation manual	0.0689	0.0825	0.1393	0.1770	0.1127	0.0373	0.1104	0.1389	0.1127
Users' training	0.0292	0.0500	0.1393	0.1770	0.1127	0.1131	0.1104	0.1389	0.1127

Table 4. Prioritized requirements

Requirements	d^-	V	RR	FR	P
Performance					
Quality of service	0.2088	0.0275	0.1813	0.0201		
Scalability	0.2500	0.1104	0.1396	0.0155	0.0620	2
Security	0.2500	0.0124	0.2376	0.0264		
Flexibility					
Installation ease	0.1131	0.0153	0.0978	0.0109		
Use friendliness	0.1131	0.0292	0.0839	0.0093	0.0306	3
Compatibility	0.1127	0.0194	0.0933	0.0104		
Reliability					
File changes	0.1389	0.0225	0.1164	0.0129	0.0266	4
Code changes	0.1389	0.0153	0.1236	0.0137		
Maintainability					
Documentation	0.2088	0.0221	0.1869	0.0207		
Maintenance plan	0.1393	0.0158	0.1235	0.0137		
Installation manual	0.1393	0.0373	0.1020	0.0113	0.0621	1
Users' training	0.1770	0.0292	0.1478	0.0164		

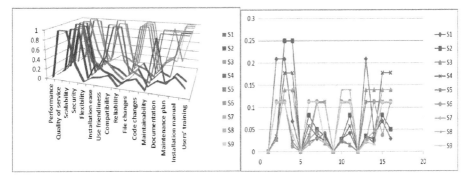

Fig. 2. Preferences of stakeholders   **Fig. 3.** Normalized stakeholder's preferences

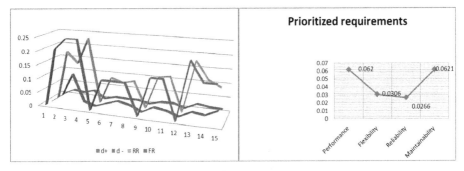

Fig. 4. Separation distances between ranked   **Fig. 5.** Final rank of requirements
requirements

5   Conclusion/Future Work

Determining the weights of stakeholder's requirements was achieved by synthesizing the priorities over all levels obtained by varying numbers of requirements. In conclusion, this research implored the fuzzy decision matrix to solve prioritization problems. However, the performance analysis of the results demonstrated promising capabilities. The approach was empirically evaluated. Four user's requirements with twelve respective criteria were identified for the design and implementation of the proposed approach. Furthermore, fuzzy numbers were used in establishing the pair-wise comparisons of requirements and their respective criteria through linguistic scales. By using this model, the subjective judgments can be quantified to make comparisons more efficient and reduce assessment biasness in pair wise comparison processes. For future work, it will be expedient to develop relevant algorithms that will aid prioritization processes, implement them and carry out evaluation with real-life project.

Acknowledgement. The Universiti Teknologi Malaysia (UTM) under research grant 03H02 and Ministry of Science, Technology & Innovations Malaysia, under research grant 4S062 are hereby acknowledged for some of the facilities utilized during the course of this research work.

References

1. Babar, M., Ramzan, M., Ghayyur, S.: Challenges and future trends in software requirements prioritization. In: 2011 International Conference on Computer Networks and Information Technology (ICCNIT), pp. 319–324. IEEE (2011)
2. Beg, R., Abbas, Q., Verma, R.P.: An approach for requirement prioritization using b-tree. In: First International Conference on Emerging Trends in Engineering and Technology, ICETET 2008, pp. 1216–1221. IEEE (2008)
3. Lima, D.C., Freitas, F., Campos, G., Souza, J.: A fuzzy approach to requirements prioritization. In: Cohen, M.B., Ó Cinnéide, M. (eds.) SSBSE 2011. LNCS, vol. 6956, pp. 64–69. Springer, Heidelberg (2011)
4. Aasem, M., Ramzan, M., Jaffar, A.: Analysis and optimization of software requirements prioritization techniques. In: 2010 International Conference on Information and Emerging Technologies (ICIET), pp. 1–6. IEEE (2010)
5. Kukreja, N., Payyavula, S., Boehm, B., Padmanabhuni, S.: Value-based requirements prioritization: usage experiences. Procedia Computer Science 16, 806–813 (2013)
6. Tonella, P., Susi, A., Palma, F.: Interactive requirements prioritization using a genetic algorithm. Information and Software Technology, Information and Software Technology 55, 173–187 (2012)
7. Perini, A., Susi, A., Avesani, P.: A Machine Learning Approach to Software Requirements Prioritization. IEEE Transactions on Software Engineering 39(4), 445–460 (2013)
8. Babar, M.I., Ramzan, M., Ghayyur, S.A.K.: Challenges and future trends in software requirements prioritization. In: 2011 International Conference on Computer Networks and Information Technology (ICCNIT), pp. 319–324. IEEE (2011)
9. Wang, Y.J., Lee, H.S.: Generalizing TOPSIS for fuzzy multiple-criteria group decision-making. Computers and Mathematics with Applications 53, 1762–1772 (2007)
10. Feng, C.M., Wang, R.T.: Performance evaluation for airlines including the consideration of financial ratios. Journal of Air Transport Management 6, 133–142 (2000)
11. Hsu, H.M., Chen, C.T.: Aggregation of fuzzy opinions under group decision making. Fuzzy Sets and Systems 79, 279–285 (1996)
12. Wang, Y.J., Lee, H.S.: Generalizing TOPSIS for fuzzy multiple-criteria group decision-making. Computers and Mathematics with Applications 53, 1762–1772 (2007)
13. Yeh, C.H., Deng, H.: An Algorithm for Fuzzy Multi-Criteria Decision Making. In: IEEE International Conference on Intelligent Processing Systems, pp. 1564–1568 (1997)
14. Bellman, R.E., Zadeh, L.A.: Decision-Making in a Fuzzy Environment. Management Science 17, 141–164 (1970)

An Ontology Design for Recommendation Filling-in for a Crime Scene Investigation Report of the Forensic Science Police Center 4 Thailand

Boonyarin Onnoom, Sirapat Chiewchanwattana,
Khamron Sunat, and Nutcharee Wichiennit

Advanced Smart Computing Laboratory
Department of Computer Science KhonKaen University, KhonKaen, Thailand
{boonyarin_onnoom,nutcharee.w}@kkumail.com, sunkra@kku.ac.th,
khamron_sunat@yahoo.com

Abstract. The Forensic Science Police Center 4 in Thailand wants to have a system that can help the officers to produce a report of a crime scene investigation. An obstacle is the sentence that is needed to be typed is a long sentence. However, the format of the report can be structured. Nouns or phases of 50 criminal cases were extracted and analyzed using noun-phase analysis. Ontology was constructed from the analysis. The ontology has five topics that are Indoor scene, Outdoor scene, Clue at scene, Evidence, Lost asset and Criminal gate. Each topic comprises of several categories. The system was also developed to test the coverage of the ontology. As measured by Precision, Recall and F-measure which is 74.21%, 95.71% and 83.86% respectively.

Keywords: recommendation filling, ontology, noun phrase analysis, crime scene investigation.

1 Introduction

Crime scene investigation is a critical step in the scientific investigation of a case. Crime scene investigators and crime-scene specialists are responsible for identifying, securing, collecting, and preserving evidence, which is submitted to the crime laboratory. An important thing to do at the scene is recording the detail of scene description. The Forensic Science Police Center 4 of Thailand currently uses the paper for taking note of initial data regarding the crime scene investigation and they will generate a full report later. It took a lot of time and there were many redundancies. Afterward, technology is playing an increasingly large role in organizations [1]. Currently electronic devices are downsizing to a smaller size. A tablet can now perform equivalently to a notebook. It can be used for taking a note, taking a picture, recording voice, and connecting to the internet, etc. There are many advantages of using a tablet, such as easy to carry and use, saving paper resource and reduce document storage. Nevertheless, the recorded data is unstructured that means the data and image cannot be easily queried. So, it should have a standard form for fill out data to create a structured form.

J. Sobecki, V. Boonjing, and S. Chittayasothorn (eds.), *Advanced Approaches to Intelligent Information and Database Systems*, Studies in Computational Intelligence 551,
DOI: 10.1007/978-3-319-05503-9_14, © Springer International Publishing Switzerland 2014

In crime investigation form for record data, a user must fill in data of several topics and each topic has continuity and a sequential list. For example, when a user fill in the outside scene in first topic, suppose that, that is home, the data in second topic about the inside scene is a room of the home. And the data in the next topic should also relate to the home. If there is a framework for suggestion the data of next topic regarding the previous data, the filling-in should be faster, more accurate and more convenient.

There are several literatures that used relation from text to produce ontology for the system to support several agencies. Maira Gatti [2] proposed an approach to use the user tagging information to learn the domain ontology and then used the tag ontology to re-rank the recommended tags. This approach can improve accuracy of the tag recommendation. Takumi Kato [3] developed a system to provide herbal recommendation based on ontology and applied them to an application. The research showed that ontology can describe the domain of interest and can link the relationship of things within that domain. Vi-sit Boonchom [4] proposed the Automatic Thai Legal Ontology Building (ATOB) algorithm which is an algorithm for constructing legal ontologies automatically. ATOB consists of three main modules: automatic seed ontology building, legal document retrieval and the legal ontology expansion module. ATOB can use Thai legal terminology to generate and expand ontology. And the precision, recall, F-measure and diversity values are 0.90, 0.91, 0.90 and 0.39, respectively. This suggests that the performance of the ATOB algorithm is better than the Baseline method. Rung-Ching Chen [5] developed a Diabetes Medication Recommendation system for hospital specialist by using ontology. In this system, researchers use Protégé to build patient ontology knowledge and drug knowledge, SWRL was used to build association rules and used JESS to make inference of system. This system can analyze the symptoms of diabetes and select the most appropriate drug.

Therefore, the researcher realized that development of a recommended clause framework for record data in crime investigation report which the information required in each topic is recommended by analysis of previous data. And then the clause recommendations which query based on ontology of crime scene investigation are represented in the list instead of type the full sentence. When apply this clause recommendation system into the crime scene investigation application on the tablet for record preliminary data at the scene, it will allow users to work faster and more accurate. So, we have developed a recommended clause framework for record crime scene investigation data based on ontology to provide recommended clause to users whose system is running on a mobile device. We create ontology structure specifically for property crime scene investigation with structural analysis of the clause that must be completed and uses data entered previously for query meet user's needs. Moreover, our approach can define pattern of recommended clause to support the conversion of data into a report.

This paper is structured as follows: Section 2 describes the architecture of the proposed system and explains how each part works and interacts. An ontology development process is described in Section 3. Section 4 explains about query the ontology to recommended clause and Section 5 contains the experimental results and discussion. Finally, we conclude this paper in Section 6.

2   System Overview

The system architecture illustrated in Figure 1 represents process in the system. Firstly, a user can access to the system via the application on mobile deviceby choosing topic that a user want to fill which was described in Section 5. Then client side request recommended clause to server side. A web service in server side receives recommended clause request, and then execute query crime scene investigation ontology using SPARQL query language [6] and return to web service. All recommended clause is being send to client side and presented on the application by explained in Section 4. When a user selects clause, system will send clause to server side and store keywords which are specified in web service to database. A small database on a mobile device was used to store keyword for limit recommended clause to exactly as growing demand. When a user chooses next topic that a user want to fill, client side request keyword from database for scope recommended clause of this topic, and then client side send keywords to server side and request recommended clause of this topic, While execute query crime scene investigation ontology, a keyword is used as a conditions for query. Next, all recommended clause sent to client side and presented on the application again. In each query, the system will use rules to increase accuracy which rules are used by inference engine. The inference engine infers the facts and priority of each recommended clause.

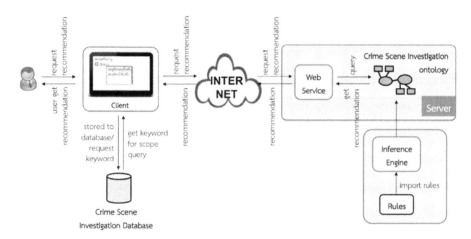

Fig.1. System architecture

We implement application on client side and web service on server side using Java programming language, which application is running on mobile device with Android operating system [7]. We have used Jena for building recommendation system. Jena is a framework to make Java technology and ontology and inference engine compatible [8].

3   Crime Scene Investigation Ontology Construction

In this section, we are presenting a process for building ontology from original data as text. We analyzed the cases of 50 documents from the crime scene investigation report about properties of Forensic Science Police Center 4, each case contains 5 topics. These topics included "สถานที่เกิดเหตุภายนอก" (Outside scene), "สถานที่เกิดเหตุภายใน" (Inside scene), "ทางเข้าออกคนร้าย" (Criminal gate), "ร่องรอยที่ปรากฏ" (Clue), "วัตถุพยานและตำแหน่งที่ตรวจพบ" (Evidence and its position), and "ทรัพย์สินที่สูญหาย" (Lost asset). When users go to the scene, he fills out to application. Then report is generated from the application. Text in the report is a long sentence which the user has filled in the form. For example, topic "Gate of criminals", standard form that a report required is "ตรวจพบกลอนประตูห้องครัวถูกงัดโดยวัตถุแข็งเสียสภาพสามารถเปิดออกได้" (Found the bolt of the kitchen door was pried broken by a solid object, made the door opened.) or "ตรวจพบมุ้งลวดประตูเหล็กดัดถูกกรีดโดยวัตถุแข็งขาดเป็นช่องสามารถใช้มือสอดเข้าไปได้" (Found the screen-door torn apart, made a hole to fit a hand in.), etc.

We will create ontology using the syntactical structure of original data. The detailed process can be described as follows:

3.1   Noun Phrase Analysis

Noun phrase analysis is component in Natural Language Processing, it used for relation extraction from noun phrase [9]. We are using Word Formation which as one of Noun phrase analysis solution. Due to Term (What used to mean a specification of a conceptualization) is important to build ontology because it instead of subject, conception, idea and process in ontology. We must use this method to extract terms out. Noun phrases can be generated ontology shown in Table 1.

Table 1. Example Instance

Pattern	Example
NP = [cnn|ctn+cnn] + [cnn|ppn]	อาคาร(cnn) คอนกรีต(cnn)
NP = [cnn|ctn+cnn|ppn|NP]+adj	บ้าน(cnn) 2 ชั้น(adj)
NP = [cnn|ctn+cnn|ppn|NP]+Prep VP = vi|[vt+NP]	ประตู(cnn) ถูกงัด (vi)เป็นช่อง(cnn)
NP = [cnn|ctn+cnn|ppn|NP]+Prep PP = prep+NP	หน้าต่าง(cnn) ด้านหลัง(prep) ห้องแต่งตัว(cnn)
NP = pref+VP VP = vi|[vt+NP]	การ (prep) โจรกรรม (vi)

Annotation: cnn = common noun, ctn = collective noun, ppn = proper noun, adj = adjective, vi = intransitive verb, vt = transitive verb, prep = preposition, pref = prefix, VP = Verb Phrase, PP = Prepositional Phrase.

The original data analysis finds Semantically Related in sentence. When separate sentence into word or phrase, each word or phrase will serve as subject or verb or object which based on situation and a word to have its category. For example, "คนร้ายงัดแงะประตู" (criminals tamper the door.) can extract to "คนร้าย" (criminal), "งัดแงะ" (temper), "ประตู" (the door) and "คนร้ายงัดแงะหน้าต่าง" (criminals tamper the window.) can extract to "คนร้าย" (criminal), "งัดแงะ" (temper) and "ประตู" (the window). The phrase "คนร้าย" is Noun and Subject and it is in category "person" (identified by WordNet

[10]). The phrase "วัดแงะ" is Verb. The phrase "ประตู" and "หน้าต่าง" is Noun and Object and it is in category "Object". Each word or phrase will serve as the subject or verb or object which based on situation, the example report shown in Figure 2 and can be divided to instance data as Table 2.

ผลการตรวจ จากการตรวจสถานที่เกิดเหตุ ปรากฏรายละเอียดดังนี้

1.   ทางเข้าออกคนร้าย ตรวจพบ

    1.1   กลอนประตูห้องครัว ถูกงัดโดยวัตถุแข็งเสียสภาพสามารถเปิดออกได้

    1.2   มุ้งลวดประตูเหล็กดัดถูกกรีดโดยวัตถุแข็งมีคมขาดเป็นช่องสามารถใช้มือสอดเข้าไปได้

Fig. 2. Example report of crime scene investigation

Table 2. Example Instance extract from sentence in reports

case	furniture part	furniture	preposition	clue	material	effect
242	กลอน	ประตู	ห้องครัว	ถูกงัด	วัตถุแข็ง	เสียสภาพสามารถเปิดออกได้

We extract noun phrase the following. (See Figure 3)

Sentence: กลอนประตูห้องครัวถูกงัดโดยวัตถุแข็งเสียสภาพสามารถเปิดออกได้

    (The latch of the door in kitchen was yank by solid object until it broken and can open.)

กลอน	ประตู	ห้องครัว	ถูกงัด	โดย	วัตถุแข็ง	เสียสภาพสามารถเปิดออกได้
latch	door	kitchen	was yank	by	solid object	broken and can open
cnn	cnn	cnn	cnn*	prep	cnn	cnn*

Fig. 3. Example sentence is separated by Noun Phrase Analysis approach

We provide a noun phrase for class of ontology. A phrase of verb is relation between classes and two adjacent words. Likewise, two adjacent words is relation class. (* star we decide that it is Class.)

3.2   Crime Scene investigation Ontology

We are building crime scene investigation ontology after extract phrase and define phrase type by taking those words to create a term or terms of the relations in ontology. The ontology was developed from information gathered by domain experts and assigned to the ontology expert in concepts, relationships and definitions. This methodology is finish by using steps defined in [11]. The steps are:

Step 1: *Determine domain and scope of the ontology*
Step 2: *Consider to reusing existing ontology*
Step 3: *Enumerate the important terms in the ontology*
Step 4: *Define the classes and hierarchy*

Step 5: *Define the properties of classes*
Step 6: *Define relation of classes*
Step 7: *Create instance of class*

First step is determining scope. We desire to build crime scene investigation ontology based on case study of the crime scene investigation report about properties for Forensic Science Police Center 4 in Thailand. We use OWL [12] language to create ontology and use SPARQL language to query recommended clause. Next step, considered reuse existing ontology. Due to desired ontology is a specialized ontology, so there is no such ontology exists. But we provide WordNet ontology for support about type and category of word. Third step, enumerated the important terms in the ontology. We define phrase of class or relation of class. The word, type of word, meaning of word and sample data must be clearly defined to reduce redundancy. Then we define the classes and the class hierarchy. The data divided into phrases from a sentence, and each phrase has specifically categories. One category is one class. Due to some topics in form use the words in the same category, so each topic has a unique relationship with another topic. We design ontology using a top-down approach that is identified root class first, and then identified subclass. Next step is defining the properties of classes. We define ID of instance in each class is English word and Thai name property which has type of Literal. Sixth step is defining relation between classes, this step is important. Each topic sentence requires recommended clause, we have to consider category of phrase in sentence and then create the relation between topic class and phrase class. Finally, we create instance of class for query recommended clause. The instance has English word as ID and Thai names property as meaning of the word. There is also a grouping of related instances to make the query more accurate. The word for query is formal or frequently words to use.

When the building ontology is complete, Figure 4 demonstrates the structure of ontology.

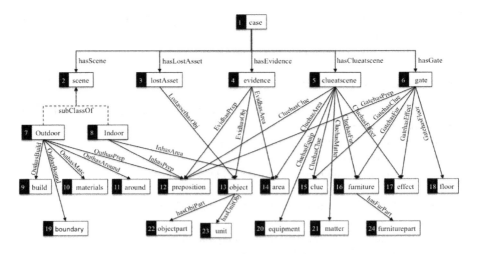

Fig.4. Crime Scene Investigation ontology structure

4   Recommendation Filling Application

In Section IV presents the result of query when require recommended clause. For example, topic "ร่องรอยที่ปรากฏ" (Clue at scene) must be query relation between *Case* class and *Clueatscene* class which relationship is named *hasClueatscene* and query class that is associated with *Clueatscene* class. We consider that *Clueatscene* class consists of the following relation: *GateHasPrep, GatehasEffect, GatehasFur, GatehasEffect* and *GatehasFloor* which correlated with *Preposition* class, *Clue* class, *Furniture* class, *Effect* class and *Floor* class respectively. Moreover, *Furniture* class correlated with *FurniturePart* class and relation between 2 classes represented as *hasFurPart*. We present sample instance in Table 3.

Table 3. Example Instance

Class	Properties	
	ID	Name in Thai
Preposition	front	ด้านหน้า
	left	ด้านซ้าย
Clue	tamper	งัดแงะ
	ransacked	รื้อค้น
Furniture Part	hasp	สายยู
	jamb	วงกบ
furniture	door	ประตู
	window	หน้าต่าง
Floor	floor	ชั้นล่าง
	second floor	ชั้นสอง
Effect	undetected	ตรวจไม่พบร่องรอยรื้อค้นในบริเวณส่วนต่าง ๆ ของสถานที่เกิดเหตุ
	can Turnout	เสียสภาพสามารถเปิดออกได้

Example query as follow:

Query 1

```
prefixthiv:  <Thieveontology.owl#>
SELECT ?furniture  WHERE {
?clueatscene thiv:CluehasFurniture ?furniture }
```

Query 1 is query RDF language SPARQL language which consist of two parts : SELECT statement   and WHERE statement. SELECT statementis desired results from a query and WHERE statement is condition of query. WHERE statement has "?clueatscene" is Subject and "?furniture" is Object. There is relation between subject and object is *CluehasFurniture* relation. After the execution of **Query 1,** the system show instance in *Furniture* class which has *CluehasFurniture* relation with *Clueatscene* class.

Query 2

```
prefixthiv:   <Thieveontology.owl#>
SELECT ?furniturepartthai WHERE {
?furniturepart thiv:haspartfurthai ?furniturepartthai.
FILTER regex(?furniturepartthai, "keyword") }
```

From **Query 2**, WHERE statement has "?furniturepart" is subject and "?furniturepart-thai" is object and relation between subject and object is *thiv:haspartfurthai*. The desired result is "?furniturepartthai". The meaning of **Query 2** is "return all furniture-partthai that is range of furniturepart, and only query furniturepartthai that contain 'keyword'."

Then, query Furniture, Preposition, Floor, Clue and Effect same as **Query 2** and match all. The result of query is show in Figure 5.

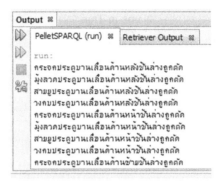

Fig. 5. Example recommended clause

5   System Evaluation

From the test filling in the information of the real situation by investigating the property crimes scene 30 scenarios and experts determine the recommended sentence from the system, we evaluated precision and recall of each topic, the experimental results as follows.

Table 4. Precision and recall for each topics.

Topic	precision	recall	F-measure
Outdoor scene	95.56%	96.67%	96.00%
Indoor scene	80.00%	100.00%	87.00%
Clue at scene	67.95%	96.15%	76.15%
Criminal gate	53.33%	90.00%	76.29%
average	**74.21%**	**95.71%**	**83.86%**

The evaluation has found that the system can be recommended as well which have a precision **74.21%**, recall **95.71%** and F-measure **83.86%**. Limitations of the system included a lot of lexicon and requires several recommended sentences to guide each time, result in a precision decreases. But the system has a high recall because the answer is always the same as the answer of experts. And because in Evidence topic and Lost Asset topic have a lot of lexicon and need to suggested many sentence result in a low precision and recall, so we ignore those topics.

6   Conclusion

In this paper, we proposed the recommendation filling system using semantic relation of ontology developed to use in the recommendation filling which each topic, a user can select clause that is frequently used and has relation with the previous clause. One of the technical problems remaining to be solved is words or phrase used to recommended clause is a term used only in the agency. A major next step is show the most accurate sentence is expected to be in the first order of the introduction for quick selection and link the words from WordNet conjunction with own ontology to provide comprehensive sentence.

References

1. Top Technologies Trends for 2014. IEEE Computer Society (2013)
2. Djuana, E., Xu, Y., Li, Y., Josang, A.: Ontology Learning from User Tagging for Tag Recommendation Making. Presented at the Proceedings of the 2011 IEEE/WIC/ACM International Conferences on Web Intelligence and Intelligent Agent Technology, vol. 3 (2011)
3. Kato, T., Maneerat, N., Varakulsiripunth, R., Izumi, S., Takahashi, H., Suganuma, T., Takahashi, K., Kato, Y., Shiratori, N.: Provision of Thai herbal recommendation based on an ontology. 2010 3rd Conference on Presented at the Human System Interactions (HSI) (2010)
4. Boonchom, V.-S., Soonthornphisaj, N.: ATOB algorithm: an automatic ontology construction for Thai legal sentences retrieval. Journal of Information Science 38(1), 37–51 (2012)
5. Chen, R.C., Huang, Y.H., Bau, C.T., Chen, S.M.: A recommendation system based on domain ontology and swrl for anti-diabetic drugs selection. Expert Systems with Applications 39(4), 3995–4006 (2012)
6. SPARQL query language for RDF. W3C Consortium,
   http://www.w3.org/TR/rdf-sparql-query/
7. Meier, R.: Professional Android 4 Application Development. Wrox Ed. (2012)
8. Zhang, W., Duan, L.G.: Reasoning and Realization Based on Ontology Model and Jena. In: IEEE Fifth International Conference on Theories and Applications (BIC-TA), Taiyuan, pp. 1057–1060 (2010)
9. Pengphon, N., Kawtrakul, A., Suktarachan, M.: Word formation approach to noun phrase analysis for Thai. In: Proceedings of SNLP2002, Thailand (2002)
10. Miller, A.G.: WordNet: A lexical database for English. Communications of the ACM 38(11), 39–41 (1995)
11. Gasevic, D., Djuric, D., Devedzic, V.: Model Driven Engineering and Ontology Development, 2nd edn. Springer, Heidelberg (2009)
12. Horridge, M., Bechhofer, S.: The OWL API: A Java API for OWL ontologies. Semantic Web Journal 2(1), 11–21 (2011)

Inter-annotator Agreement
in Coreference Annotation of Polish*

Mateusz Kopeć and Maciej Ogrodniczuk

Institute of Computer Science, Polish Academy of Sciences

Abstract. This paper discusses different methods of estimating the inter-annotator agreement in manual annotation of Polish coreference and proposes a new BLANC-based annotation agreement metric. The commonly used agreement indicators are calculated for mention detection, semantic head annotation, near-identity markup and coreference resolution.

1 Introduction

A substantial annotation of Polish coreference has been recently carried out in the course of creation of the Polish Coreference Corpus (PCC) — a large manually annotated resource of general Polish coreference. The annotation procedure consisted of marking up the following entities:

- mentions — all nominal groups constituting reference to discourse-world objects
- mention semantic heads — the most relevant word of the group in terms of meaning; typically equal to syntactic head, but different for numerals or elective expressions (cf. one_{synh} of the $girls_{semh}$)
- identity clusters — groups of mentions having the same referent
- near-identity links — associations between a pair of semi-identical mentions, carrying some of their properties (cf. *prewar Warsaw* and *Warsaw today*[1])
- dominating expressions — a mention in a cluster which carries the richest semantics or describes the referent with most precision.

For each of the above-mentioned subtasks inter-annotator agreement can be evaluated to show difficulty of each assignments individually. In this paper we present conclusions stemming from the investigation of 210 texts (60,674 segments) from 14 domains (15 texts per domain).

---

* The work reported here was carried out within the *Computer-based methods for coreference resolution in Polish texts (CORE)* project financed by the Polish National Science Centre (contract number 6505/B/T02/2011/40). The paper was also co-founded by the European Union from resources of the European Social Fund, Project PO KL "Information technologies: Research and their interdisciplinary applications".

[1] See [1] for general introduction to the concept and section 4 of [2] for details of Polish annotation of near-identity in PCC.

J. Sobecki, V. Boonjing, and S. Chittayasothorn (eds.), *Advanced Approaches to Intelligent Information and Database Systems*, Studies in Computational Intelligence 551,
DOI: 10.1007/978-3-319-05503-9_15, © Springer International Publishing Switzerland 2014

2   The Annotation Process

The texts for annotation, 250-350 token each, had been randomly selected from
the National Corpus of Polish [3]. The samples were automatically processed to
detect mention candidates with a newly implemented software based on existing
language processing tools for Polish: Spejd shallow parser [4], Pantera tagger
and Nerf named entity recognizer [5]. Baseline coreference resolution tool Ruler
[6] was used for initial mention clustering.

Each pre-processed text from the sample selected for the experiment has been
annotated independently by two annotators (hence: A and B) from the team of
eight linguists co-operating with the project. The annotation was performed in
a customized MMAX2 tool [7]. The annotators were instructed (with detailed
guidelines[2]) to correct the pre-annotation results by removing existing markup,
changing it or adding new entities or associations. As a result of the process, 420
annotated texts were produced with a total of 41,006 mentions, 4,410 clusters
and 1,009 near-identity links.

It is worth pointing out that the quantity of the annotated content is relatively large and surpasses the previous attempts of evaluation of inter-annotator
agreement in coreference annotation, while the number of annotators per text is
minimal. For example, the authors of AnCora-CO-Es corpus [9] evaluated agreement of 8 annotators processing only 2 texts (approx. 1100 segments altogether).

3   Mentions

According to [10], estimation of the inter-annotator agreement including the
chance-based factor for the task of marking up mentions (which can be nested,
discontinuous and overlapping) has not been yet investigated. Since it is difficult
to estimate the probability of a random markup of a mention, we present the
observed agreement only.

In our sample of 210 texts the annotation of the annotator A contained 20,420
mentions while for annotator B — 20,560 mentions. 17,530 mentions were shared
which means they had exactly the same borders (including the inner borders
for discontinuous mentions). Regarding annotation A as gold and B as system,
precision is 85.26%, recall = 85.85% which gives $F_1 = 85.55\%$.

When comparing mention heads only, the annotation A contained 19,394 mentions (after excluding mentions having the same heads), annotation B — 19,522
mentions; 18,317 mentions were shared. For this setting precision = 93.83%,
recall = 94.47%, $F_1 = 94.14\%$.

4   Semantic Heads

Agreement in annotation of mention heads can only be investigated for shared
mentions. For 17,363 mentions out of 17,530 the same heads were marked, which
gives the observed agreement: $p_{A_O} \approx 0.9905$.

---

[2] For the PCC annotation schema and strategies see [8].

The chance agreement (p_{A_E}) was calculated in the following way. For each mention its head was selected by pointing out one segment from all mention tokens. We assumed that the probability of choosing the same head by chance for given mention was equal to the inverse of its token count. Chance agreement was therefore calculated as an average of chance agreement probabilities for individual mentions and yielded: $p_{A_E} \approx 0.6832$. This value is high due to a high count of one-token mentions, having the chance agreement equal to 1.

Having computed both the above-mentioned values the S inter-agreement measure [11] (as the chance probability distribution is uniform) could be calculated, yielding satisfactory result:

$$S = \frac{p_{A_O} - p_{A_E}}{1 - p_{A_E}} \approx 0.9700$$

5   Near-Identity

As in the case of semantic heads, the agreement of near-identity linking was investigated only for mentions present in both annotations.

For each mention pair in a text the annotator could decide on their linking. Combined agreement results for each mention pair are presented in Table 1.

Table 1. Near-identity link agreement in all texts

		Annotation B	
		Near-identical	Non-near-identical
Annotation A	Near-identical	67	306
	Non-near-identical	367	741,584

For this table the Cohen κ [12] could be calculated, but this approach would be wrong since annotators cannot add a near-identity link between mentions in two different texts. Therefore we calculated κ for each text separately and then averaged it. This time we assumed per-text-and-annotator probability distribution (details about κ calculation are presented in section 7.1).

When text did not contain any links, agreement value of 1 was assumed. When one annotator did not mark any link while the second one did, the agreement was assumed as 0 (with predicted value equal to the observed one).

Applying this procedure to all texts we have calculated the average $\kappa \approx 0.2220$. The result is low which can be interpreted as a difficulty in linking mentions with near-identical relation. The notion seems vague — in 128 cases mention pairs were marked as near-identical by one annotator and at the same time as purely-identical (i.e. were clustered) by the other annotator.

6   Dominating Expressions

As with previous cases, this type of agreement concerns only mentions annotated by both annotators.

Dominating expressions were marked for non-singleton clusters only. The number of common mentions with a marked dominating expression was 6,162, with 4,115 mentions sharing the same dominating expression ($\approx 66.78\%$). When only one representative of each cluster is investigated (which makes sense since each element of a cluster has the same dominating expression) 1,146 out of 1,818 cluster representatives has the same dominating expression in both annotations ($\approx 63.04\%$).

Chance agreement analysis is not carried out since apart from choosing a cluster element as the dominating expression the annotators could also enter arbitrary text value, which makes good chance agreement estimations impossible.

7 Coreference

According to [10], coreference resolution is a very specific task, dealing with clustering and not classification, atypically for the whole field of computational linguistics. Moreover, selection of the best evaluation method for the new resource is difficult since there is no consensus in the scientific world about 'the best' metric. In this extensive section we analyse the most popular ones to present their properties and results for our annotation.

7.1 Cohen's κ

In [13] Passonneau describes two inter-annotator agreement metrics: Cohen's κ [12] and original Krippendorff α [14].

$$\kappa = \frac{p_{A_O} - p_{A_E}}{1 - p_{A_E}} \qquad \alpha = 1 - \frac{p_{D_O}}{p_{D_E}}$$

κ is defined as the difference of observed agreement (p_{A_O}) and chance agreement (p_{A_E}) while α involves the observed non-agreement (p_{D_O}) and chance non-agreement (p_{D_E}). Passonneau shows that $\alpha = \kappa$, so in the further part of the text we concentrate on the procedure of calculating κ.

Table 2. A sample coincidence matrix

		Annotator A		
		Label X	Label Y	Σ
Annotator B	Label X	47	14	61
	Label Y	10	29	39
	Σ	57	43	100

α and κ can be calculated when coincidence matrix is available for multiple annotators' decisions (with data how frequent each of the annotators were choosing each label). For example, when it is defined as in Table 2, we can calculate:

$$p_{A_O} = \frac{47 + 29}{100} = 0.76$$

$$p_{A_E} = \frac{57}{100} * \frac{61}{100} + \frac{43}{100} * \frac{39}{100} = 0.5154$$

Observed agreement shows diagonally in the matrix. Expected agreement is based on the probability of selection of each label by each annotator. For instance, when annotator A selected label X in 57% decisions and annotator B in 61% decisions, the chance that they accidentally chose the same label A is $(57/100) * (61/100)$. The sum of probabilities for all labels gives the expected agreement. Finally for Table 2 we have:

$$\kappa = \frac{p_{A_O} - p_{A_E}}{1 - p_{A_E}} = 0.5$$

For the coreference annotation agreement assessment, crucial decision is to choose how to represent coreference annotation in coincidence matrix similar to Table 2. We present some approaches in the following sections. For the reason described in the near-identity section, we suggest calculating the agreement for each text separately and then averaging it.

7.2 MUC-Based Metrics

In [13] the coincidence table for agreement is calculated similarly to MUC metrics. The matrix similar to Table 7.2 is created, where $Link+$ denotes annotation of association between mentions (minimal) and $Link-$ — no association. The details of calculation of MUC metrics can be found in [15].

Unfortunately, because of certain properties of MUC metrics (e.g. not taking singletons into account) it was not widely accepted standard and is usually used as a supporting metrics only.

Table 3. Coincidence matrix for MUC metrics

		Annotator A		
		$Link+$	$Link-$	Σ
Annotator B	$Link+$	47	14	61
	$Link-$	10	29	39
	Σ	57	43	100

7.3 Weighted Krippendorff α

Passonneau in [16] presented a different approach, making use of the weighted version of Krippendorff α. To calculate it, annotator's decision for a given mention should be understood as assignment of a set of mentions from the same cluster (apart from this mention).

For instance, if for five mentions with labels $1, 2, 3, 4, 5$ annotator A created clusters: $\{1, 3\}, \{2, 4, 5\}$, and annotator B clusters: $\{1\}, \{2, 4\}, \{3, 5\}$, their annotation will be represented as in Table 4. It shows that e.g. according to the annotator A the mention 1 is clustered only with 3 and it is a singleton according to the annotator B.

Table 4. Representation of a sample cluster annotation

Mention number	1	2	3	4	5
Annotator A	{3}	{4, 5}	{1}	{2, 5}	{2, 4}
Annotator B	{}	{4}	{5}	{2}	{3}

Passonneau's idea was to punish differences between annotators subject to the degree of difference between the clusters assigned to the same mention. To apply this technique, Krippendorff α was used with weights assigned to each error, according to the selected distance metrics. Let's define distance between e_1 and e_2 as $\delta(e_1, e_2)$. For the first mention in our example the weight of an error is $\delta(\{3\}, \{\})$ — and analogically for all other mentions. Let's mark the set of all clusters used by the annotators as E (in our example it contains 9 elements which is the number of rows and columns in the coincidence matrix).

The α equation for two annotators is the following:

$$p_{D_O} = \frac{1}{n} \sum_{e_1 \in E} \sum_{e_2 \in E} o_{e_1 e_2} \delta(e_1, e_2)$$

$$p_{D_E} = \frac{1}{n(n-1)} \sum_{e_1 \in E} \sum_{e_2 \in E} n_{e_1} n_{e_2} \delta(e_1, e_2)$$

$$\alpha = 1 - \frac{p_{D_O}}{p_{D_E}} = 1 - (n-1) \frac{\sum_{e_1, e_2 \in E} o_{e_1 e_2} \delta(e_1, e_2)}{\sum_{e_1, e_2 \in E} n_{e_1} n_{e_2} \delta(e_1, e_2)}$$

where $o_{e_1 e_2}$ is he number of mentions assigned by one of the annotators to e_1 cluster, and by the second one — to e_2 cluster, n_{e_1} is the number of all assignment of e_1 label and analogically, n_{e_2} is the number of assignment of e_2 label. In our example $o_{\{3\}\{2,4\}} = 1$, $o_{\{3\}\{2,5\}} = 0$, and $n_{\{3\}} = 2$, $n_{\{2,5\}} = 1$.

Passoneau in [16] defines $\delta(e_1, e_2)$ function in the following manner (with the result calculated with first matching rule counting from the top):

- $\delta(e_1, e_2) = 0$, when $e_1 = e_2$
- $\delta(e_1, e_2) = 0.33$, when $e_1 \subset e_2 \vee e_2 \subset e_1$,
- $\delta(e_1, e_2) = 0.67$, when $e_1 \cap e_2 \neq \emptyset$,
- $\delta(e_1, e_2) = 1$, when $e_1 \cap e_2 = \emptyset$.

In [17] she proposes another variant of this metric for the same task — MASI (Measuring Agreement on Set-valued Items). MASI is calculated as the product of the previous δ metrics and Jaccard coefficient [18]:

$$MASI(e_1, e_2) = \delta(e_1, e_2) * \frac{|e_1 \cap e_2|}{|e_1 \cup e_2|}$$

We calculated the agreement for the Polish Coreference Corpus following the procedure described in [16] (weighted Krippendorff α) to achieve 79.08% and 59.54% according to [17] (MASI).

7.4   Recasens Approach

In [9] Recasens describes the study of agreement between 8 annotators for 2 texts from the AnCora-CO-Es corpus (approx. 1,100 tokens in total). Assuming identity of mentions presented to annotators, the work were organised to test two aspects:

1. Annotator agreement concerning assignment of a mention to a cluster of a certain type. For each mention annotators could mark it as
   - non-coreference
   - discourse deixis
   - predicative
   - identity

   which made it a fairly standard classification task, investigated with weighted α (0.85 for the first text, 0.89 for the second one).
2. Annotator agreement concerning clustering of each mention from *predicative* or *identity* categories. Labels were cluster numbers, so it was a classification task again, investigated with κ (0.98 for text one, 1 for text two).

We have calculated the agreement for PCC by investigating, for each mention marked by both annotators, whether the mention is clustered or not. This binary decision is presented in Table 5 and examined with Cohen's κ (this time it can be calculated for the whole corpus at once, without averaging).

Table 5. Inter-coder decision agreement on singleton/cluster element for all texts in PCC

| | | Annotation B | |
		Clustered	Singleton
Annotation A	Clustered	6,238	975
	Singleton	1,223	9,094

According to the data in Table 5 the observed agreement (p_{A_O}) is:

$$p_{A_O} = \frac{6238 + 9094}{6238 + 9094 + 975 + 1223} \approx 0.8746$$

while the predicted agreement (p_{A_E}) is:

$$p_{A_E} = \frac{6238 + 1223}{17530} * \frac{6238 + 975}{17530} + \frac{9094 + 1223}{17530} * \frac{9094 + 975}{17530} \approx 0.5132$$

which makes:

$$\kappa = \frac{p_{A_O} - p_{A_E}}{1 - p_{A_E}} \approx 0.7424$$

7.5 BLANC-Type Agreement

Statistics of coreferential and non-coreferential links for mentions (as in BLANC metrics) marked by both individual annotations are listed in Table 6.

Table 6. Agreement of BLANC links in all texts

		Annotation B	
		Coreferential	Non-coreferential
Annotation A	Coreferential	16,638	3,448
	Non-coreferential	3,353	718,822

Cohen's kappa could be calculated for this data, but (again) it would not take into account that annotators cannot (even by chance) cluster mentions coming from two different texts. This means that κ should be calculated for each text separately and then averaged.

Application of such procedure to all texts in PCC and grouping results for different text types is shown in Table 7. The data can be interpreted by taking into account several factors such as:

- discrepancy in the speaker and recipient's conceptual systems, resulting in difficulty in interpretation of academic books by a non-expert annotator,
- higher readability of fiction than academic or spoken texts, boosting the agreement value.

Table 7. κ values for individual text domains

Text type	κ
Academic writing	0.699
Instructive writing and textbooks	0.727
Internet non-interactive (static pages, Wikipedia)	0.730
Dailies	0.740
Quasi-spoken (parliamentary transcripts)	0.746
Internet interactive (blogs, forums, usenet)	0.764
Spoken — conversational	0.765
Other periodical	0.772
Spoken from the media	0.785
Non-fiction	0.795
Unclassified written	0.807
Journalistic books	0.817
Misc. written (legal, ads, manuals, letters)	0.826
Fiction	0.871
Any	0.775

8   Conclusions

We have presented several approaches of calculating the inter-annotator agreement in coreference annotation of Polish and its results for four coreference-related tasks. We have investigated two typical tasks: mention detection and coreference resolution as well as two less common ones: semantic head annotation and near-identity markup.

The results of the analysis confirm the assumption that coreference is more of a semantic and conceptual phenomenon which cannot reach scores as high as those achieved in lower-level linguistic tasks such as segmentation or morphosyntactic annotation. The average coreference agreement result of 0.775 seems to show the upper limit of coreference resolution capabilities, currently being reached by the state-of-the art tools for Polish (cf. e.g. [6]). Results of near-identity annotation prove the difficulty of its reliable annotation in the current understanding of this phenomenon which should be verified in the further coreference annotation projects.

References

1. Recasens, M., Hovy, E., Martí, M.A.: A Typology of Near-Identity Relations for Coreference (NIDENT). In: Proceedings of the Seventh International Conference on Language Resources and Evaluation (LREC 2010), pp. 149–156 (2010)
2. Ogrodniczuk, M., Głowińska, K., Kopeć, M., Savary, A., Zawisławska, M.: Interesting Linguistic Features in Coreference Annotation of an Inflectional Language. In: Sun, M., Zhang, M., Lin, D., Wang, H. (eds.) CCL and NLP-NABD 2013. LNCS, vol. 8202, pp. 97–108. Springer, Heidelberg (2013)
3. Przepiórkowski, A., Bańko, M., Górski, R.L., Lewandowska-Tomaszczyk, B. (eds.): Narodowy Korpus Języka Polskiego. Wydawnictwo Naukowe PWN, Warsaw (2012) (Eng.: National Corpus of Polish)
4. Przepiórkowski, A., Buczyński, A.: Spejd: Shallow Parsing and Disambiguation Engine. In: Vetulani, Z. (ed.) Proceedings of the 3rd Language & Technology Conference, Poznań, Poland, pp. 340–344 (2007)
5. Waszczuk, J., Głowińska, K., Savary, A., Przepiórkowski, A., Lenart, M.: Annotation Tools for Syntax and Named Entities in the National Corpus of Polish. International Journal of Data Mining, Modelling and Management 5(2), 103–122 (2013)
6. Ogrodniczuk, M., Kopeć, M.: End-to-end coreference resolution baseline system for Polish. In: Vetulani, Z. (ed.) Proceedings of the Fifth Language & Technology Conference: Human Language Technologies as a Challenge for Computer Science and Linguistics, Poznań, Poland, pp. 167–171 (2011)
7. Müller, C., Strube, M.: Multi-level annotation of linguistic data with MMAX2. In: Braun, S., Kohn, K., Mukherjee, J. (eds.) Corpus Technology and Language Pedagogy: New Resources, New Tools, New Methods, pp. 197–214. Peter Lang, Frankfurt a.M, Germany (2006)
8. Ogrodniczuk, M., Zawisławska, M., Głowińska, K., Savary, A.: Coreference Annotation Schema for an Inflectional Language. In: Gelbukh, A. (ed.) CICLing 2013, Part I. LNCS, vol. 7816, pp. 394–407. Springer, Heidelberg (2013)

9. Recasens, M.: Coreference: Theory, Annotation, Resolution and Evaluation. PhD thesis, University of Barcelona (2010)
10. Artstein, R., Poesio, M.: Inter-coder agreement for computational linguistics. Computational Linguistics 34(4), 555–596 (2008)
11. Bennet, E.M., Alpert, R., Goldstein, A.C.: Communications through limited response questioning. Public Opinion Quarterly 18, 303–308 (1954)
12. Cohen, J.: A Coefficient of Agreement for Nominal Scales. Educational and Psychological Measurement 20(1), 37–46 (1960)
13. Passonneau, R.J.: Applying reliability metrics to co-reference annotation. CoRR cmp-lg/9706011 (1997)
14. Krippendorff, K.H.: Content Analysis: An Introduction to Its Methodology, 2nd edn. Sage Publications, Inc. (December 2003)
15. Vilain, M., Burger, J., Aberdeen, J., Connolly, D., Hirschman, L.: A model-theoretic coreference scoring scheme. In: Proceedings of the 6th Conference on Message Understanding, MUC6 1995, pp. 45–52. Association for Computational Linguistics, Stroudsburg (1995)
16. Passonneau, R.J.: Computing reliability for coreference annotation. In: LREC. European Language Resources Association (2004)
17. Passonneau, R., Habash, N., Rambow, O.: Inter-annotator agreement on a multilingual semantic annotation task. In: Proceedings of LREC (2006)
18. Jaccard, P.: Nouvelles recherches sur la distribution florale. Bulletin de la Sociète Vaudense des Sciences Naturelles 44, 223–270 (1908)

Revenue Evaluation Based on Rough Set Reasoning

Khu Phi Nguyen[1], Sy Tien Bui[1], and Hong Tuyet Tu[2]

[1] University of Information Technology, Vietnam National University-HCMC
[2] University of Technical Education-HCMC,
Linh Trung, Thu Duc District, HCM City, Vietnam
khunp@uit.edu.vn

Abstract. Budget revenue evaluation is an important issue in planing economic policies, especially in making profit by financial investment. It is dealt with this paper, a method of rating turnover efficiency using the methodology of rough set-a technique for data mining, and information entropy- a measurement of the average amount of information contained in an information system. The proposed method, firstly reduces superfluous data sources, and then extracts important decision rules from data set. By using this method, organizations of economic management and corporations management can evaluate of economic policies, business strategies and determine what revenue sources are efficient and need be invested. Through testing of collected dataset, it is shown that the proposed method is feasible and need be developed in practice.

Keywords: Rough set, information entropy, mutual information, decision rules, ACO algorithm, financial evaluation.

1 Introduction

A prediction model of budget revenues plays a very important role, which will influence economic and budgetary policies. All recent works focus around short-term forecasting for the annual operating budget but in lack of attention and coordination among agencies, [1]. In general, budget consists of main groups, income taxes and non-tax revenues. This paper is dealt with rating budget revenues from the major sources, including Nontax revenue, Capital revenue, Grant revenue and Indirect tax, Direct tax, Personal Income tax, Corporate Income tax, Value-Added tax, Export-Import tax. Total revenues from these sources are balanced with the rates of budget deficit with respect to gross domestic product.

To analyze the efficiency of these budget revenues and their impact to the rates of budget deficit, methods based on rough set theory along with information entropy and ant colony optimization algorithm are applied. The initial expectation of this analyzing is to deliver a systematic approach to budget revenues forecasting and planning, as well as building financial strategies, [2]. Yet, there are still limitations of predicting methods in this paper due to the data sources are quite short, only twelve years from 1999 to 2010, in which nearly the third of them is used for verifying. More data and efficient techniques will be found and developed in future research to improve gradually this result.

J. Sobecki, V. Boonjing, and S. Chittayasothorn (eds.), *Advanced Approaches to Intelligent Information and Database Systems*, Studies in Computational Intelligence 551,
DOI: 10.1007/978-3-319-05503-9_16, © Springer International Publishing Switzerland 2014

In this paper, some basic concepts of rough set, information entropy and finding reducts are firstly recalled in Section 2. After designing algorithm for solving the problem, a case study related to the problem is illustrated in Section 3. Finally, a summarization of this paper is presented.

2   Methodology

Since the late of 1980s, the rough set approach seems to be basically important to artificial intelligence and cognitive sciences [3]. Some others are in the fields of knowledge discovery [4], knowledge acquisition, machine learning [5], decision analysis, expert systems, inductive reasoning and pattern recognition. Recently, the rough set theory has been successfully applied in many real-life problems in medicine [6], pharmacology, engineering, banking, financial and market analysis, application in linguistics, environment, etc. Rough set method has become a powerful tool for data mining, knowledge discovery in databases, especially in analyzing indiscernible, inconsistent, incomplete information systems. The main advantage of rough set theory is that it is easy to understand and does not require any preliminary or additional information about data-like probability in statistics, or possibility in fuzzy set, [7].

2.1   Preliminary Concepts of Rough Set

Let Ω be a non-empty finite set of objects called universe, A be a finite set of attributes, then IS = (Ω, A) is an information system. An observation of such a system is represented by an information table or IT which the first column illustrates objects, other columns are attributes of each object. Thus, an IT can be considered as an outcome of IS. Each $a \in$ A known as a condition attribute corresponds with a map $I_a: \Omega \rightarrow V_a$, from Ω into the value set V_a of a, the so-called information function of a. For $B \subseteq$ A, let $R_B \subseteq \Omega \times \Omega$ be a binary relation, defined as follows:

$$(x, y) \in R_B \Leftrightarrow I_a(x) = I_a(y) \quad \forall a \in B, x, y \in \Omega \tag{1}$$

In this case, R_B is an equivalence relation, called B-indiscernibility relation. Equivalence class of $\omega \in \Omega$ with respect to B is denoted by $[\omega]_B$. Set of all these equivalence classes is a partition of Ω, called the quotient space and denoted by Ω/R_B. Let $X \subseteq \Omega$, approximations to X are constructed using $B \subseteq A$. B-lower and B-upper approximation of X, denoted by $B_L X$, $B_U X$ respectively are defined in [8]:

$$B_L X = \{ \omega \mid [\omega]_B \subseteq X \}, \quad B_U X = \{ \omega \mid [\omega]_B \cap X \neq \varnothing \} \tag{2}$$

It is obtained $B_L X \subseteq X \subseteq B_U X$, the difference $B_B X = B_U X - B_L X$ is the B-boundary region of X. If the boundary region $B_B X$ is non-empty then X is rough, otherwise crisp.

A decision system is an information system (Ω, A) with a set of decision attributes D, denoted by DS = (Ω, A$\cup$D). In simple case D = {d} where d$\notin$A, the function I_d: $\Omega \rightarrow V_d$ defines a set $D(\Omega) = \{ i \in V_d /\ I_d(\omega) = i, \omega \in \Omega \}$ and $r(d) = card(D(\Omega)) = |\ D(\Omega)\ |$ is the rank of d in D.

Consider B $\subseteq$ A, the union of $B_L X$ for $X \in \Omega/R_D$ is the B-positive region of Ω/R_D with respect to D. For D ={d}, Ω/R_d consists of elements: $\Omega_i = \{ \omega \in \Omega /\ d(\omega) = v_{d,i} \}$, i = 1, 2,.. , r(d) called the i-th decision class of Ω. The B-positive region with respect to d, simply denoted by $B_+(d)$ as follows:

$$B_+(d) = \bigcup_{i=1}^{r(d)} B_L(\Omega_i) \tag{3}$$

A given decision system $(\Omega, A\cup\{d\})$ is consistent if $A_+(d) = \Omega$, otherwise inconsistent. The dependency between attributes is an important issue. Let B and C be attribute sets, it can be asserted that B depends on C if all values of attributes from B are uniquely determined by C, its degree of dependency is measured by:

$$\gamma_C(B) = |C_+(B)| / |\Omega| \tag{4}$$

Obviously, $0 \leq \gamma_C(B) \leq 1$. If $\gamma_C(B) = 0$ then B does not depend on C; $\gamma_C(B) = 1$, B depends totally on C; and if $0 < \gamma_C(B) < 1$; B depends partially on C. A reducts of decision system DS = $(\Omega, A\cup D)$ is a set of attributes of A that preserves partitioning of Ω with A, i.e. $\Omega/R_A = \Omega/R_B$ and this is equivalent to $\gamma_B(D) = \gamma_A(D)$. An information system may have many attribute reducts, denoted by Red, precisely:

$$Red = \{ B\subseteq A|\ \gamma_B(D) = \gamma_A(D), \forall C\subseteq A: \gamma_C(D) \neq \gamma_A(D)\} \tag{5}$$

A set of reducts with a minimal cardinality is called minimal reducts. The intersection of all reducts is called score of DS. Searching a minimal reduct has been a subject of much research and many algorithms have paid attention to this problem. The QuickReduct algorithm attempts to find a minimal reduct. It starts with an empty set and adds in turn, step by step, those attributes that result in the greatest increase in dependency, until this produces its maximum possible value for the dataset.

An other algorithm with the same complexity is the EBR or entropy-based reduction, whose procedures are similar to the QuickReduct algorithm except the dependency condition is replaced with the entropy condition of attributes, [9]. In general, finding a minimal reducts is NP-hard problem, just like the travelling sale-man problem. Nowadays, usual algorithms involve heuristic or random search strategies in an attempt to avoid this prohibitive complexity, e.g. the attribute reducts task is reformulated in Ant Colony Optimization reasoning or ACO algorithm [2, 9]. Besides this, BCA or Bee Colony Algorithm is also proposed.

2.2   Attribute Reduction and ACO-Reducts

In the attribute reduction problem of a decision system, the relevant attributes contain more important information than the irrelevant attributes. So, the task for attribute reduction is to determine what attributes contain as much information as possible. For this purpose, the measurements of entropy and mutual information in information theory can be used as a measure of information table for attribute selection, [10].

Let B be any subset of attributes taken from a decision system $(\Omega, A \cup D)$, the quotient space Ω/R_B with respect to R_B can be considered as an outcome space. Assume that $m = |\Omega/R_B|$ and $\Omega/R_B = \{ X_1, X_2,.. , X_m \}$, in which X_i, $i = 1,2,.. , m$, is considered as an event whose occurrence frequency is $p(X_i) = | X_i |/|\Omega|$ from the view point of equally likely events. In this case, information entropy of B is estimated by:

$$H(B) = -\sum_{i=1}^{m} p(X_i) \log_2 p(X_i) \tag{6}$$

For $\Omega/R_A = \{ P_1, P_2,... , P_n \}$ and $\Omega/R_D = \{ Q_1, Q_2,.. ,Q_m \}$, conditional information entropy of D given A is evaluated by:

$$H(D|A) = -\sum_{j=1}^{n} p(P_j) \sum_{i=1}^{m} P(Q_i | P_j) \log_2 P(Q_i | P_j) \tag{7}$$

Based on the above estimations, mutual information between A, D is determined by I(A,D) and the importance of an attribute $a \in A$ defined in [0, 1] as follows:

$$I(A,D) = H(D) - H(D|A) \tag{8}$$

$$Impt(a) = | H(A) - H(A-\{a\}) | \tag{9}$$

In case of Impt(a) > 0 then a is a necessary, otherwise a is redundant attribute. Using (8), a subset B of attributes A is a reduct set of attributes, or a reduct, respect to D if $I(B,D) = I(A,D)$ and $\forall b \in B$, $I(B-\{b\},D) < I(A,D)$. From this, an attribute $a \in A$ is a core attribute if a cannot be eliminated from DS unless decreasing I(A,D) strictly:

$$I(A-\{a\}, D) < I(A, D) \tag{10}$$

Set of the core attributes is the core of DS, denoted by Core(C), which is a subset of all possible reducts. Besides, significance of a given attribute $a \in A$ with respect to $B \subseteq A$ and D is determined as follows:

$$Sgnf(a; B,D) = I(B \cup \{a\}, D) - I(B,D) \tag{11}$$

Based on recommendations in [2,9], ACO algorithm is used to find possible reducts. To construct a reduct of DS, each ant in the ant colony starts from a selected attribute in Core(C) and find out some other attributes in the reduct. Assume that a given ant is currently at an attribute q, the ant should search a next unselected attribute p in C – (Core $\cup\{q\}$) with a probability determined by (12), [2].

At q and time step t, let τ_{qp} be the pheromone of choosing attribute p. The τ_{qp} is updated by $\rho\tau_{qp} + \Delta\tau_{qp}$ from the previous time step, in which $\rho\in(0, 1)$: evaporation factor of pheromone, $\Delta\tau_{qp}$: amount of pheromone laid between q and p. Information of p with respect to q determined by $\eta_{qp} = $ Sgnf(p; Core$\cup\{q\}$, D). If Φ_k is the set of unselected attributes of the ith ant at t, then:

$$p_{qp}^i(t) = \frac{\tau_{qp}^\alpha \eta_{qp}^\beta(t)}{\sum_{s\in\Phi_k} \tau_{qs}^\alpha \eta_{qs}^\beta(t)} \tag{12}$$

Where α and β are positive parameters which determine the relative importance of the pheromone trail and heuristic information, [12]. The following ACO-Reduct algorithm demonstrates how to get such a reduct.

```
Algorithm ACO-Reduct
Input: a DS=(Ω,A∪D),imax
Output: minimal-reduct Rm,Lm=|Rm|
  i:=0, Rm:=A, Lm:=|A|
  calculate I(A,D) // using (8)
  find core=Core(A) // using (9)
  while i<imax
  for each ant p
    Rp=core, Lp=|core|
      select ap∈A-Rp, Lp=Lp+1
      repeat
        for every sp∈A-Rp
        compute η(sp,ap)// using (11)
        endfor
        select next bp∈A-Rp // using (12)
      until ( I(Rp,D)=I(A,D) or Lp≥Lm )
      if ( I(Rp,D)=I(A,D) and Lp<Lm )
        Rm=Rp, Lm=Lp
      endif
  endfor
  update pheromone
  i=i+1
endwhile.
```

2.3   Decision Rule Extraction

Let $a \in B \subseteq A$, $v \in V_a$, $\omega \in \Omega$, proposition $I_a(\omega) = v$, briefly $I_a = v$ or somewhere (a, v), takes a specific Boolean value. The notation $\phi := I_a = v$ is used to define a logic variable ϕ whose value is the one of $I_a = v$, in other words, ϕ is true if there is an $\omega \in \Omega$ so that $I_a(\omega) = v$ or false otherwise.

The set of logic variables on B and logical operations form a set of logic expressions with respect to B, denoted by $\Lambda(B)$ and called decision language from B. A decision rule in DS is a Boolean expression: $\phi \rightarrow \psi$, read " if ϕ then ψ " in which $\phi \in \Lambda(A)$ and $\psi \in \Lambda(D)$ are referred to as condition and decision of the rule, respectively.

For each $\psi \in \Lambda(\{d\})$ and $v_d \in V_d$, let $[\psi] = \{\omega \in \Omega / I_d(\omega) = v_d, d \in D\}$. It is assumed that objects in $[\psi]$ and objects in the complement of $[\psi]$ are numbered with consecutive integral subscripts i and j, respectively. A decision matrix $M = (m_{ij})$ of a decision table with respect to $[\psi]$ is determined as a matrix whose entries are sets of attribute-value pairs, [11]:

$$m_{ij} = \{ (a, I_a(\omega_i)) / I_a(\omega_i) \neq I_a(\omega_j), a \in A \} \tag{13}$$

Where $I_a(\omega_i)$ denotes the common value of the attribute a on corresponding objects. The set m_{ij} consists of all attribute-value pairs (attribute, value) in which values are not identical between ω_i and ω_j. The set of all minimal value reducts of the collection of attribute value pairs corresponding to row i, and so, the set of all minimal decision rules for that row can be obtained by forming the following Boolean expression, called a decision function:

$$B_i = \wedge_j (\vee m_{ij}) \tag{14}$$

Where $\wedge$ and $\vee$ are respectively generalized conjunction and disjunction operators. B_i is constructed out of row i of the decision matrix by taking conjunctions of disjunctions of pairs in each set m_{ij}.

Finally, the decision rules are obtained by turning each decision function into disjunctive normal form and using the absorption law of Boolean algebra to simplify it and obtain the prime implicants of the decision function correspond to the minimal decision rules.

3   A Case Study

Data source is provided continuously in each quater of years from 1999 to 2010 by DFIS-Vietnam. This dataset contains a lot of features, but only some main ones are choosen to be processed by a computer program named Anodisys.

3.1   Dataset

In Table 1, conditional attributes A_i - in trillion VND, i = 1,2,.. ,9, discretized using the equal-width discretization, and a decision attribute D mean as follows: .

Table 1. DFIS discretized dataset for revenue evaluation in quaters of eigth years

A_1	A_2	A_3	A_4	A_5	A_6	A_7	A_8	A_9	D
±1.072	±0.055	±0.273	±3.266	±1.269	±0.438	±1.925	±2.197	±0.278	
2.61	0.12	0.52	1.92	2.59	0.66	4.41	4.70	2.68	S
2.61	0.12	0.52	1.92	2.59	0.66	4.41	4.70	3.23	S
2.61	0.01	0.52	1.92	2.59	0.66	4.41	4.70	3.23	S
2.61	0.12	1.07	1.92	2.59	0.66	4.41	4.70	3.79	S
2.61	0.23	0.52	1.92	2.59	0.66	8.26	4.70	3.79	S
2.61	0.12	0.52	1.92	2.59	0.66	8.26	4.70	3.23	A
2.61	0.12	0.52	1.92	2.59	0.66	8.26	4.70	3.23	G
2.61	0.23	0.52	1.92	2.59	0.66	8.26	4.70	3.79	B
2.61	0.56	0.52	5.18	5.13	0.66	8.26	4.70	3.23	L
2.61	0.23	0.52	5.18	2.59	0.66	8.26	4.70	3.23	A
2.61	0.23	0.52	1.92	2.59	0.66	8.26	4.70	3.79	A
2.61	0.12	0.52	1.92	5.13	0.66	12.11	9.10	3.79	G
11.18	0.56	1.61	5.18	7.67	0.66	15.96	9.10	3.79	B
2.61	0.12	0.52	1.92	5.13	0.66	12.11	9.10	4.34	S
4.76	0.23	0.52	5.18	7.67	0.66	12.11	13.49	3.23	A
17.61	0.78	1.07	8.45	17.82	0.66	23.66	9.10	3.23	A
2.61	0.23	0.52	5.18	7.67	0.66	15.96	13.49	3.23	G
4.76	0.23	0.52	5.18	7.67	1.54	15.96	13.49	4.90	S
2.61	0.23	0.52	5.18	7.67	1.54	15.96	13.49	4.34	S
4.76	0.56	0.52	5.18	7.67	1.54	19.81	13.49	5.45	A
19.76	0.89	5.42	5.18	17.82	1.54	39.06	17.89	7.67	G
21.90	1.00	3.25	14.98	25.43	2.42	39.06	17.89	7.12	L
4.76	0.78	0.52	11.72	10.20	2.42	27.51	22.28	5.45	E
19.76	0.23	5.42	14.98	20.35	4.17	39.06	13.49	6.56	B
4.76	0.12	1.07	14.98	7.67	2.42	23.66	22.28	5.45	A
9.04	0.23	1.61	18.25	10.20	3.29	23.66	22.28	6.01	A
6.90	0.34	1.61	18.25	12.74	2.42	27.51	26.67	6.01	A
15.47	0.78	2.16	34.58	25.43	5.92	35.21	31.07	6.01	B
6.90	0.12	1.07	21.51	10.20	4.17	31.36	31.07	6.01	S
9.04	0.34	1.07	14.98	15.28	5.92	35.21	35.46	6.01	A
9.04	0.34	1.61	21.51	17.82	6.80	39.06	44.25	7.12	G
13.33	0.78	1.07	28.05	25.43	8.55	39.06	44.25	7.12	B

A_1 :   Nontax revenue, A_6 :   Personal Income tax,
A_2 :   Capital revennue, A_7 :   Corporate Income tax,
A_3 :   Grant revenue, A_8 :   Value-Added tax,

A_4 :	Indirect tax,	A_9 :	Export-Import tax,
A_5 :	Direct tax,	D :	GDP-balanced level.

Here, GDP-balanced levels D can be classified into six types, E: excellent, G: good, S: satisfactory, A: avarage, L: low and B: bad. Dataset for processing lasted in eight years are illustrated in Table 1.

3.2   Characteristic Numbers from Dataset

Significance of attribute defined by (11) is applied to evaluate significance of condition attributes A_i listed in Table 1:

$$Sgnf(A_1,C,D) = Sgnf(A_2,C,D) = Sgnf(A_4,C,D) = Sgnf(A_5,C,D) = Sgnf(A_8,C,D) = 0.$$

$$Sgnf(A_3,C,D) = Sgnf(A_6,C,D) = Sgnf(A_7,C,D) = 0.08609, Sgnf(A_9,C,D) = 0.0625.$$

These show that A_3, A_6, A_7 are the most significant attributes. Similarly, from (9):

$$Impt(A_1) = Impt(A_4) = Impt(A_5) = Impt(A_8) = 0.0.$$

$$Impt(A_2) = 0.0625, Impt(A_3) = Impt(A_6) = Impt(A_7) = 0.08609, Impt(A_9) = 0.125.$$

Using ACO-Reduct coded in a computer program, named Anodisys, a reduct of the above information table consists of A_2, A_3, A_6, A_7, A_9 and then A_1, A_4, A_5, A_8 redundant.

3.3   Decision Rules

Applying the scheme for extracting decision rules presented in 3.3 and programmed in Anodisys, the following rules are found:

R1: If (A_7 in 2.75±1.925) and (A_9 in 5.45±0.278) then D = E.

R2: If (A_2 in 0.89±0.055) or (A_6 in 6.80±0.438) then D = G.

R3: If (A_3 in 0.52±0.273 and A_6 in 0.66±0.438) or
       (A_6 in 1.54±0.438 and A_7 in 15.96±1.925) or
       (A_7 in 31.36±1.925) or (A_7 in 4.41±1.925 and A_9 in 3788±0.278) then D = S.

R4: If (A_2 in 0.23±0.055 and A_3 in 0.52±0.273) or (A_7 in 23.66±1.925) or
       (A_2 in 0.34±0.055 and A_9 in 6.01±0.278) or (A_7 in 19.81±1.925) then D = A.

R5: If (A_2 in 1.00±0.055) or (A_2 in 0.56±0.055 and A_7 in 8.26±1.925) then D = L.

R6: If (A_3 in 2.16±0.273) or (A_6 in 8.55±0.438) or (A_9 in 6.56±0.278) or
       (A_2 in 0.56±0.055 and A_9 in 3.79±0.278) then D = B.

Notes: Here, a logic expression, e.g., A_7 in 5.45±0.278 means that the attribute A_7 takes its value between A_7 in 5.45-0.278 = 5.172 and 5.45+0.278 = 0.728.

3.4   Verification

The extracted rules are verified using the testing data set. This set consists of sixteen data objects of five reductive attributes. Then the extracted decision rules are applied to assign values to decision attitude D.

Table 2. Verification of prediction results using test dataset

Ω	A2 ±0.055	A3 ±0.273	A6 ±0.438	A7 ±1.925	A9 ±0.278	D	Test
1	0.01	2.16	3.29	35.21	4.34	E	Error
2	0.12	0.52	0.66	8.26	2.68	S	Good
3	0.89	1.07	1.54	23.66	4.34	G	Good
4	0.34	1.61	3.29	35.21	5.45	G	Error
5	0.12	0.52	0.66	4.41	3.79	S	Good
6	0.23	0.52	0.66	8.26	3.79	S	Good
7	0.23	0.52	0.66	8.26	3.23	S	Good
8	0.12	0.52	0.66	19.81	4.90	S	Good
9	0.45	1.07	0.66	12.11	4.34	A	Error
10	0.23	0.52	0.66	12.11	3.79	A	Good
11	0.34	1.07	1.54	19.81	4.34	A	Good
12	0.34	0.52	1.54	19.81	4.90	A	Good
13	0.45	1.07	1.54	23.66	5.45	A	Good
14	0.56	1.07	0.66	8.26	3.79	L	Good
15	0.45	0.52	0.66	19.81	6.56	B	Good
16	0.78	2.16	1.54	19.81	4.90	B	Good

The accuracy of different approaches is given by using error types. The 1st type error refers to the situation when matched data is classified as unmatched one, and the 2nd type error refers to unmatched data classified into matched data. The result listed in Table 2 shows that there are only three Error-s of the 2nd type in 16 verification cases or 18.75% errors, i.e., the best overall prediction accuracy level at 81.25%. These results are accepted in comparison with conclusions in [1].

4   Conclusion

A novel approach for analysing of revenue efficiency is proposed based on the rough set reasoning, the method of attribute reduction using information entropy, mutual information and ACO algorithm. Some decision rules are extracted using the method of discernible matrix and prime implicants of disjunctive normal form. This approach focuses on the classification object and makes the most of the significance of condition attributes and classification objects, and then generates decision rules with a pay attention in discretization of condition attribute values.

Data source was provided continuously in each quater of years from 1999 to 2010 by DFIS-Vietnam. This dataset contains a lot of features, but only some main ones are choosen to be processed by a computer program named Anodisys. Experiments on real data in twelve years provoded by DFIS shown that the proposed methods are feasible and in accordance with a greater than 80% real testing data.

The direction of further research is that the increase in revenues will have to depend on how detailed the data are. Moreover, as for the macro approach, the forecasting of budget revenue is based on the forecasting of macro-economic indicators, e.g. GDP and its components.

Acknowledgments. This work is funded by Vietnam National University VNU-HCMC under grant number C2013-26-03. The authors are grateful to DFIS, Dept. of Financial Informatics and Statistics, in support of experimental data for this research.

References

1. Thang, V.N.: Developing Prediction Model on National Budget Revenue in Vietnam. UNDP Report, in Project: Strengthen Decision and Supervision Capacity on Budget of Elective Institutions in Vietnam, Committe of Finance and Budget - Vietnam National Parliament. Hanoi, Vietnam (2012)
2. Chen, Y., Miao, D., Wang, R.: A rough set approach to feature selection based on ant colony optimization. Pattern Recognition Letters 31, 226–233 (2010), doi:10.1016/j.patrec.2009.10.013
3. Ahna, B.S., Chob, S.S., Kim, C.Y.: The Integrated Methodology of Rough Set Theory and Artificial Neural Network for Business Failure Prediction. Expert Systems with Applications, Pergamon 18, P II, 65–74 (2000)
4. Sun, S., Zheng, R., Wu, Q., Li, T.: VPRS-Based Knowledge Discovery Approach in Incomplete Information System. Journal of Computers 5(1), 110–116 (2010)
5. Das, V., Pathak, V., Sharma, S., Sreevathsan, Srikanth, M.V.V.N.S., Gireesh Kumar, T.: Network Intrusion Detection System Based on Machine Learning Algorithms. Intl' Journal of Computer Science and Information Technology 2(6), 138–151 (2010), doi:10.5121/ijcsit.2010.2613
6. Chen, Y., Wang, S., Chan, C.-C.: Application of Rough Sets to Patient Satisfaction Analysis. In: Proc. of the 11th Intl' DSI and the 16th APDSI Joint Meeting, Taipei, Taiwan (2011)
7. Hua, Q., Zhangb, L., Chenc, D., Pedryczd, W., Yua, D.: Gaussian kernel based fuzzy rough sets: Model. Uncertainty measures and Applications. Intl' Jour. of Approximate Reasoning (2010), doi:10.1016/j.ijar.2010.01.004
8. Pawlak, Z., Skowron, A.: Rough sets: Some extensions. Information Sciences (177), 28–40 (2007)
9. Jensen, R.: Combining rough and fuzzy sets for feature selection. Dr. Thesis, pp. 112-120. School of Informatics, University of Edinburgh (2005)

10. Gray, R.M.: Entropy and Information Theory, Stanford University, pp. 17–47. Springer, New York (2009)
11. Nguyen, K.P., Tu, H.T.: Data mining based on Rough set theory. A chapter in the book: Knowledge Discovery in Databases. Academy Publisher Inc., Quastisky Building, P.O. Box 4389, USA (2013)
12. Dorigo, M., Birattari, M., Stutzle, T.: Ant Colony Optimization. IEEE Computational Intelligence Magazine, 1556-603X/06/$20.00©2006IEEE, 28–39 (2006)

Part III

Database Systems Applications, Data Models and Analysis

Measuring Population-Based Completeness for Single Nucleotide Polymorphism (SNP) Databases

Nurul A. Emran[1], Suzanne Embury[2], and Paolo Missier[3]

[1] Computing Intellingence Technologies (CIT) Lab, Centre of Advanced Computing Technology (C-ACT), Universiti Teknikal Malaysia Melaka (UTeM), Hang Tuah Jaya, 76100, Melaka, Malaysia
[2] The University of Manchester
[3] The University of Newcastle
nurulakmar@utem.edu.my, embury@cs.man.ac.uk, paolo.missier@ncl.ac.uk

Abstract. Completeness of data sets is an important aspect of data quality as observed in biological domain such as Single Nucleotide Polymorphism (SNP). In order to decide on the acceptability of the data sets of concerned, biologists need to measure the completeness of the data sets. One type of data completeness measure is population-based completeness (PBC) that has been identified as relevant to deal with data completeness problem in this domain. In this paper, the implementation of PBC measurement will be presented as a system prototype involving real SNP data sets. The result of the analysis on the practical problems encountered during the implementation of PBC will also be presented.

Keywords: population-based completeness (PBC), Single Nucleotide Polymorphism (SNP), data completeness measurement.

1 Introduction

In genetics study, it is important for biologists to understand the relationship between the genes and the 'phenotype' (i.e., specific traits and behavioural characteristics) of an organism [1]. One form of phenotype is susceptibility towards certain diseases (e.g., cancer or asthma). In order to test the hypothesis of the relationship between genes and diseases, biologists use 'genetic information' taken from a DNA sequence. A DNA sequence is made up of four types of simple units called nucleotides, which are represented by the "letter" A (adenine), C (cytosine), T (thymine), and G (guanine), which carries the genetic information of an organism [1]. In particular, biologists are interested in a variation that occurs in the DNA sequences of the organisms under study that they hypothesise to be the cause of the diseases. This variation called SNP (pronounced *snip*), that occurs in a single nucleotide of a gene, is observed in a sufficiently large proportion of an organism to determine the specific phenotype of interest (e.g., the biological function that causes susceptibility towards diseases and physical appearance [2]) [1]. The underlying assumption behind many SNP studies is that the variation

J. Sobecki, V. Boonjing, and S. Chittayasothorn (eds.), *Advanced Approaches to Intelligent Information and Database Systems*, Studies in Computational Intelligence 551,
DOI: 10.1007/978-3-319-05503-9_17, © Springer International Publishing Switzerland 2014

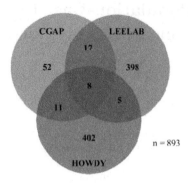

Fig. 1. Completeness differences in three human SNP databases involving 74 genes (taken from [8])

that contributes to an increased risk for a particular disease (or other observed phenotype) should occur at higher rates in the population which exhibits the disease compared to the population which does not [1]. For human genes for example, it has been explicitly stated that the variation must occur in at least 1% of the human population in order to say whether there is a link between SNPs and a phenotype under study [1,3]. Because of this requirement, many forms of studies involving SNPs are sensitive to completeness, as, if some of the SNPs under study are missing, a conclusion about its correlation to a specific trait under study cannot be supported.

Every SNP which is discovered by biologists has some information that is stored about it. The information includes the scientific submitter's SNP id, name of organism, known genes in the DNA sequence region, sex of organism, tissue type and host (laboratory name) [4]. Our initial exploration of the SNP domain [5] revealed that in the analysis where SNPs are used, not all the information stored about them is of interest. For example, in association studies to link genes and diseases, biologists are only interested in SNPs that occur in certain chromosomal locations [6] or species [7]. Therefore, we say that not all information about SNPs is sensitive to completeness. For example, given a SNP data set, its completeness as regards to the chromosomes or species will be of concern, instead of other information that is stored about them.

The availability of multiple sources for SNPs (such as dbSNP by NCBI)[9] and Ensembl by the European Bioinformatics Institute and The Wellcome Trust Sanger Institute [10])) increases the possibility of supporting many forms of studies involving SNPs, but there is a concern among SNPs users community regarding the completeness of SNPs data sources. A study by Marsh et al. compared three well-known human SNP databases for 74 human genes: CGAP-GAI [1], LEELAB [2] and HOWDY in terms of their SNP coverage [8]. Their work discovered a small overlap between the databases, as illustrated by the Venn

---

[1] CGAP-GAI - http://gai.nci.nih.gov/cgap-gai/
[2] LEELAB - http://www.bioinformatics.ucla.edu/snp/

diagram in Figure 1. Based on the discovery, we learnt that our confidence in the analyses performed over only a single database (or a small selection) should be reduced, because this would result in incomplete sets of SNPs being covered.

The lesson to be learnt from our observation of the SNP domain is that a concern for SNPs data set completeness is specific to some sets of individuals of interest that are present within the SNP data set itself (e.g., mouse species or certain chromosome). Data completeness requirements in the SNP domain hinted at the applicability of the population-based completeness (PBC) notion [11] in supporting questions regarding the completeness of SNPs data sets. For example, to answer the question of how complete a SNP data set is as regards to a specific chromosome, we need a reference population for chromosomes to tell us whether some chromosomes are missing from the SNP data set. Without a system that could help to answer this question, the individuals of the desired reference population must be gathered manually, from the source(s) that can provide them. This is a non-trivial task, especially if the number of individuals of the reference population is large or they must be gathered from multiple sources. If the reference population is to be used frequently, we need to help to answer PBC measurement requests that could ease the non-triviality of manual gathering of reference populations.

To show how PBC conceptual ideas can be put into practice, in the next section we will describe an implementation of a SNP PBC system prototype. The rest of this paper is organised as follows. Section 3 provide details on the framework used to implement PBC measurement processor; Section 4 covers findings of the problems discovered from the implementation. Finally Section 5 concludes the paper.

2   A SNP PBC System Prototype

One factor that motivates the notion of the universe as presented in our early work on data completeness is the absence of a single, good approximation of the true populations (for further details please refer [11]). The universe is therefore proposed to support PBC measurement in such situations in which the individuals of a reference population are integrated from a range of sources. As the SNP domain faces the same situation, where it is hard to find a single SNP source that fully overlaps with other SNP sources, we need to gather the individuals of reference populations that are needed to answer PBC measurement requests.

We developed a SNP PBC system prototype based on a PBC reference architecture for a fully materialised universe (one type of reference architecture as published in [12]) to determine the problems that are actually encountered in this architecture. We implemented and configured most of the components of PBC within the architecture and omitted some components for certain reasons. In no particular order, the description of how we configured these components will be given in the following sections.

SNP Universe. Before the SNP universe can be configured, we must identify the reference populations to be included in the universe based on PBC measurement requests in the SNP domain. We considered in the implementation the PBC measurement requests hinted at the SNP literature, which are:

- to measure the completeness of a SNP data set in regards to chromosomal locations [13],
- to measure the completeness of a SNP data set in regards to chromosomal locations where certain genes are located [6], and
- to measure the completeness of a SNP data set in regards to certain species [7].

If we focus on SNPs for mouse species, we can reduce these requests to the first and second requests only. To answer these requests, we must provide a SNP's chromosome reference population that consists of information about its chromosomal locations for mouse species. We also need the identifier for this reference population. The schema of the chromosome population table where information about SNP's chromosome reference population is stored should be at least in the form of $\langle I, A \rangle$, where I is the set of identifier attributes and A is the set of attributes used to retrieve the reference sub-population. But if there is a request that puts some conditions on the CSs from which the genes are gathered, then we must add a *source* attribute within the schema, as outlined in the reference template of the basic configuration of PBC.

For this SNP system prototype, we designed the schema of chromosome population table, chromosome, as: $\langle$chromosomeLocation, geneName,$\rangle$, where chromosomeLocation is the identifier for the chromosome reference population and the geneName is the information about the gene that lies within the chromosome region. We created the SNP universe (and the population table) using MySQL[3].

SNP Contributing Sources. In this SNP system prototype, we selected two primary sources of mouse SNPs as the CSs of a SNP chromosome reference population, namely Ensembl [10] and dbSNP [9], for their easy public access, and the amount of SNPs covered by these sources that, together, we suspect could make a good approximation of the universe. We configured these SNPs CSs based on CSs configuration outlined in the reference template of the basic configuration of a PBC system, where information about the CSs name, their URI, and population maps must be provided. Both Ensembl and dbSNP have public databases that are downloadable in a variety of formats (i.e., text (tab-delimited or FASTA[4]) and SQL dumps).

Population Map and PBC Components Configuration. PBC components configuration information consists of information about the reference pop-

---

[3] http://mysql.com
[4] Text-based format for representing either nucleotide sequences or peptide sequences, in which base pairs or amino acids are represented using single-letter codes.

ulations in the universe and the CSs for PBC elements variables and type definitions). For a SNP PBC system prototype, the configuration information stored for the universe is: UP ={(`chromosome`, ⟨`chromosomeLocation,geneName`⟩)}, where UP is the set of the universe's population tables and their schemas, and `chromosome` is the name of the population table for the chromosome reference population. Configuration information for the CSs is shown in Table 1.

Table 1. Configuration Information Stored for CSs

| CS name | URI | Population Map | |
		population table Name	Query Against CS
Ensembl	`ftp://ftp.ensembl.org/pub/release-63/mysql/mus_musculus_variation_63_37/`	chromosome	`SELECT chromosome_position_id,` `associated_gene` `FROM variation v, variation_annotation va` `WHERE v.variation_id=va.variation_id;`
dbSNP	`ftp://ftp.ncbi.nih.gov/snp/database/organism_data/mouse_10090/`	chromosome	`SELECT loc_snp_id,gene_name,` `FROM subSNP;`

ETL Pipeline. We devise The Clover ETL framework (http://cloveretl.berlios. de/) in order to implement Extract, Transform and Load components as included in the PBC reference architecture (please see the details in [12]). In particular, for our SNP system prototype, we use the CloverETL community Edition[5] which is a java-based, open source ETL tool deployed on a Mac OS X operating system.

We downloaded the data sets that consisted of the individuals of the chromosome reference population from Ensembl and dbSNP in the form of SQL dumps. We executed the SQL dumps and stored these raw data sets in temporary tables for each CS in which those tables matched with the schema of the universe (relational schema). These tables are:

- For Ensembl:
  - `variation(variation_id,chromosome_position_id,source_id,name,` `validation_status,ancestral_allelle,flipped,class_attrib_id,somatic),`
  - `variation_annotation(variation_annotation_id,variation_id,` `phenotype_id,study_id, associated_gene,associated_variant_risk_allele,` `variation_names,risk_allele_freq_in_controls,p_value).`
- For dbSNP:
  - `subSNP(subsnp_id,variation_id,known_snp_handle,snp_loc_id,batch_id,` `chr_id,gene_name,create_time, sequence_len,samplesize,` `ancestral_allelle,validation_status).`

At this point, format and semantic differences had not been addressed yet. The tasks we just described should be performed by the wrappers, but for the SNP PBC system prototype, we performed them manually as the number of CSs is small. We used the Eclipse plug-ins provided by The Clover ETL to perform the subsequent Setup tasks after the 'raw' data sets had been downloaded and stored as temporary tables.

---

[5] http://www.cloveretl.com/products/community-edition

Basically, the functions provided by Clover are described as transformation graphs in which the components (such as Copy, Sort, Reformat and Filter) involved interact. To implement the Extract component (of PBC reference architecture), we used the Copy component of Clover to execute the queries in the population map, which takes the raw, temporary tables as the input. The result of this step is the relations that match with the `chromosome` table in the universe in terms of the table's structure (i.e. a table with two columns), as follows:

- Ensembl: `ensemblTemp(chromosome_position_id,associated_gene)`,
- dbSNP: `dbSNPTemp(snp_loc_id,gene_name)`.

To implement the Transform component, we used the Reformat component of Clover, for which we resolved format differences among the columns of the extracted table. For example, Ensembl uses 'integer type' for `chromosome_position_id`, while dbSNP uses 'varchar' type for `loc_snp_id` that both map with the universe's `chromosomeLocation` attribute. The Transform component takes `ensemblTemp` and `dbSNPTemp` as the inputs and produces the transformed tables for each CSs that conforms to the schema of `chromosome` table. These tables are:

- Ensembl: `ensemblTransformed(chromosomeLocation,geneName)`,
- dbSNP: `dbSNPTransformed(chromosomeLocation,geneName)`.

We add a function to integrate these transformed tables within the Reformat component of Clover using a simple SQL command with union operator as follows:

```
SELECT chromosomeLocation, geneName FROM ensemblTransformed UNION
SELECT chromosomeLocation, geneName FROM dbSNPTransformed;
```

The result of the query, which was the `integratedChromosome` table, was exported into a temporary text file called `integratedChromosome.csv`. We loaded the integrated chromosomes from `integratedChromosome.csv` into the `Chromosome` table in the SNP universe using the following SQL command:

```
LOAD DATA INFILE 'integratedChromosome6' INTO TABLE chromosome;
```

About 42,505,315 tuples were loaded into the `chromosome` table, out of which Ensembl and dbSNP contributed 15,514,284 and 26,991,031 tuples respectively. Even though the SNP universe only consists of a single table, about 40.5 gigabytes of space has been used to store the chromosome reference population alone. Storage space will be an issue for the SNP domain if we wish to include more reference populations for mice (or other species) that individually require a large storage space.

---

[6] For brevity, we omit the full path of the file location.

3   PBC Measurement Processor

Having the universe readily loaded with the reference population, the question that arises is how can PBC measurement be implemented (i.e. is there any existing technological framework that we can use to implement it)? We found that one DQ framework that is specifically designed for implementing quality measures, called Quality View (QV) [14], could support us in answering the question just raised. QV, which is independent of our SNP PBC system prototype, is a software component proposed based on the DQ framework provided by the Qurator project[7]. Basically, the default behaviour of QV is to: 1) take a data set under measure i.e. D={a,b,c,d} (where D is the data set under measure), 2) gather the 'evidence' of the quality of each element in the data set, 3) classify each element in the data set according to its evidence (i.e. a and b (in D) are good), and 4) output a version of the data set, which is 'transformed' based on the classification. For example, a common transformation is filtering out the elements of the data set that do not meet some predefined DQ acceptance standard. For example, D'={a,b}, where D' is the version of D which was transformed based on exclusion of 'rejected' elements. The evidence is the term used for the result of some forms of domain-specific DQ measure performed on the elements of the data set under measure.

For the SNP PBC system prototype, we implemented all three layers of QV to answer the PBC measurement requests posted earlier in Section 2 (page 175). We call the QV for our SNP system QVsnp. Based on the PBC measurement requests, two data sets are of interest: the data set of the chromosomal region (denoted as D_1), and the data set of the chromosomal region of specific genes or sub-population of chromosome (denoted as D_2). For D_2, we used genes Gm11767-001, Gm12390-001 and Gm12401-001 to artificially represent the genes of interest. Therefore, the data set under measure used for the SNP PBC system prototype is $D = \{D_1, D_2\}$. In addition to D, we need to provide the parameter(s) as the inputs to QVsnp. As both D_1 and D_2 require the same chromosome reference population, then the only parameter needed is the name of the reference population, which is equivalent to the name of the population table stored in the SNP universe, chromosome. D_1 and D_2 consist of the chromosome locations (the identifiers) of the chromosome population that was taken from a SNP source called Celera[8]. As illustrated in Figure 2, data flow a1 is a PBC measurement request which consists of the inputs just stated, provided to the QEF component of QVsnp. Based on the inputs, QEF issued the following query against the SNP universe that stores the reference population for D_1, RP_1:

```
SELECT chromosomeLocation FROM chromosome;
```

Then, to retrieve the reference population for D_2, RP_2, the following query was issued by QEF:

---

[7] www.qurator.org

[8] Mouse SNPs taken from Celera was downloaded from mouse phenome database, http://phenome.jax.org/db/q?rtn=snps/download

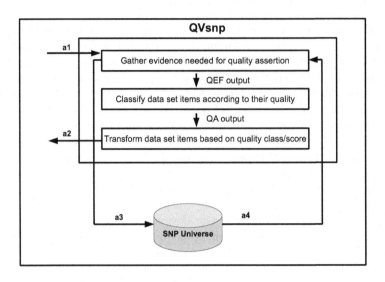

Fig. 2. SNP Quality View

SELECT chromosomeLocation FROM chromosome WHERE geneName IN
('Gm11767-001', 'Gm12390-001', 'Gm12401-001');

These queries were denoted as data flow a3 in Figure 2. The SNP universe
returned the results of the queries (data flow a4), and based on the result of
the queries, QEF instantiated the PBC measurement formula to compute PBC
measurement results as follows:

- $Completeness(D_1, RP_1) = \frac{|(D_1 \cap RP_1)|}{|RP_1|}$,
- $Completeness(D_2, RP_2) = \frac{|(D_2 \cap RP_2)|}{|RP_2|}$.

The results of PBC measurement (in the form of percentages) were issued to
the second layer of QVsnp (data flow QEF output), in which the QA component
will perform a classification upon D_1 and D_2. We added a decision procedure
that classifies the results based on a predefined acceptance threshold. We defined
PBC measurement results that are between 50 and 100 per cent as being within
the 'acceptable' class while the results which are below 50 per cent are defined
as being within the 'rejected' class. The PBC measurement results and their
classification (data flow QA output) were issued to the third layer of QVsnp,
where the transformation components were invoked. We set the transformation
component to transform D into D', which consists of data sets under measure
whose PBC measurement results are within the acceptable class, together with
the percentage of their PBC measurement results. The result of the transformation (data flow a2) was issued to the person requesting PBC measurement.

4 Findings of PBC Implementation Problems

Based on the implementation of the SNP PBC system prototype, we encountered several problems. In addition to the format differences, we faced a storage space constraint problem during the setup of the SNP PBC system prototype. It is difficult to say whether duplicates and errors problem is present without a proper data cleaning procedure being implemented for the SNP PBC system prototype. However, as far as we know, dbSNP exercises strict data submission regulation so that any detected errors or duplicates will be removed from the dbSNP [15]. For certain, we did not experience any CSs accessibility problem for both selected CSs that are public databases. Even though we have not encountered the data freshness issue during the implementation of the SNP PBC system, we anticipate this problem occurring as it is inevitable that modifications will be made to SNP CSs. This is based on an observation made on DbSNP and Ensembl that announce any modifications that occur to their databases and the tools that they provide in their websites on a periodic basis[9]. For PBC measurement, we encountered a format differences problem, in which the format of the chromosome location for the populations under measure extracted from Celera is different from the format of the chromosome location used for our chromosome reference population in the SNP universe. Celera uses a Mega base pair (Mgp) unit format for its chromosome location, while the chromosome location in our SNP universe is formatted in a base pair (bp) unit. For example, for SNP that occurs in chromosome 11 at location 89916077 (in bp unit) is stated as 89.916077 by Celera. We resolved the format problem by converting the format of chromosome locations in D_1 and D_2 that conform to the format of the chromosome location in the SNP universe in order to enable successful PBC measurement provided by the SNP PBC system prototype.

5 Conclusion

In this paper, we looked at how the conceptual PBC reference architecture can be put into practice, and showed how PBC measurement requests raised in the SNP domain were answered. By revealing the problems encountered in PBC system implementation in SNP domain, we draw the principal lesson which is: a PBC system is implementable in practice with proper handling of the problems that are barriers to answering PBC measurement requests and barriers to answering PBC measurement requests accurately. Although we observed PBC measurement requests in the SNP domain, it seems plausible that the PBC system will be of value in other domains as well, where the need for PBC is present.

Acknowledgments. The authors would like to thank the financial assistance provided by the Universiti Teknikal Malaysia, Melaka (UTeM) during the course of this research.

---

[9] dbSNP announcement site: `http://www.ncbi.nlm.nih.gov/projects/SNP/`, and Ensemble release news: `http://www.ensembl.info/blog/category/releases/`

References

1. Brookes, A.J.: The essence of SNPs. Gene 234, 177–186 (1999)
2. Syvïen, A.: Accessing genetic variation: genotyping single nucleotide polymorphisms. Nature Reviews Genetics 2, 930–942 (2001)
3. Human Genome Project Information: SNP fact sheet (2011),
   http://www.ornl.gov/sci/techresources/Human_Genome/faq/snps.shtml
   (Online; accessed July 22, 2011)
4. Information, N.C.F.B.: Submission of SNPs to dbSNP (2006), http://
   www.ncbi.nlm.nih.gov/projects/SNP/how_to_submit.html#DATA_ELEMENTS
   (Online; accessed July 23, 2011)
5. Emran, N., Embury, S., Missier, P.: Model-driven component generation for families of completeness. In: 6th International Workshop on Quality in Databases and Management of Uncertain Data, Very Large Databases (VLDB) (2008)
6. Halperin, E., Kimmel, G., Shamir, R.: Tag SNP selection in genotype data for maximizing SNP prediction accuracy. Bioinformatics 21, 195–203 (2005)
7. Frazer, K.A., Eskin, E., Kang, H.M., Bogue, M.A., Hinds, D.A., Beilharz, E.J., Gupta, R.V., Montgomery, J., Morenzoni, M.M., Nilsen, G.B., Pethiyagoda, C.L., Stuve, L., Johnson, F., Daly, M., Wade, C., Cox, D.: A sequence-based variation map of 8.27 million snps in inbred mouse strains. Nature 448, 1050–1053 (2007)
8. Marsh, S., Kwok, P., Mcleod, L.H.: SNP database and pharmacogenetics: great start, but a long way to go. Human Mutation 20, 174–179 (2002)
9. Sherry, S.T., Ward, M.H., Baker, J., Phan, E.M., Smigielski, E.M., Sirotkin, K.: dbSNP: the NCBI database of genetic variation. Nucleic Acids Research 29, 308–311 (2001)
10. Hubbard, T.J.P., Aken, B.L., Ayling, S., Ballester, B., Beal, K., Bragin, E., Brent, S., Chen, Y., Clapham, P., Clarke, L., Coates, G., Fairley, S., Fitzgerald, S., Fernandez-Banet, J., Gordon, L., Graf, S., Haider, S., Hammond, M., Holland, R., Howe, K., Jenkinson, A., Johnson, N., Kahari, A., Keefe, D., Keenan, S., Kinsella, R., Kokocinski, F., Kulesha, E., Lawson, D., Longden, I., Megy, K., Meidl, P., Overduin, B., Parker, A., Pritchard, B., Rios, D., Schuster, M., Slater, G., Smedley, D., Spooner, W., Spudich, G., Trevanion, S., Vilella, A., Vogel, J., White, S., Wilder, S., Zadissa, A., Birney, E., Cunningham, F., Curwen, V., Durbin, R., Fernandez-Suarez, X.M., Herrero, J., Kasprzyk, A., Proctor, G., Smith, J., Searle, S., Flicek, P.: Ensembl 2009. Nucleic Acids Research 37, D690–D697 (2009)
11. Emran, N.A., Embury, S.M., Missier, P., Isa, M.N.M., Muda, A.K.: Measuring data completeness for microbial genomics database. In: Selamat, A., Nguyen, N.T., Haron, H. (eds.) ACIIDS 2013, Part I. LNCS, vol. 7802, pp. 186–195. Springer, Heidelberg (2013)
12. Emran, N.A., Embury, S.M., Missier, P., Ahmad, N.: Reference architectures to measure data completeness across integrated databases. In: Selamat, A., Nguyen, N.T., Haron, H. (eds.) ACIIDS 2013, Part I. LNCS, vol. 7802, pp. 216–225. Springer, Heidelberg (2013)
13. Tiffin, N., Andrade-Navarro, M.A., Perez-Iratxeta, C.: Linking genes to diseases: it's all in the data. Genome Medicine 1, 1–7 (2009)
14. Missier, P., Embury, S., Greenwood, R., Preece, A., Jin, B.: Quality views: capturing and exploiting the user perspective on data quality. In: Proceedings of the 32nd international conference on Very Large Databases (VLDB), pp. 977–988. ACM Press (2006)
15. Information, N.C.F.B.: Submission of SNPs to dbSNP (2006),
    http://www.ncbi.nlm.nih.gov/projects/SNP/how_to_submit.html#Withdrawn
    (Online; accessed July 26, 2011)

SPARQL Processing over the Linked Open Data with Automatic Endpoint Detection

Gergő Gombos and Attila Kiss

Eötvös Loránd University, Budapest, Hungary
{ggombos,kiss}@inf.elte.hu

Abstract. The LOD Cloud is a collection of the available datasets of the Semantic Web. The individual datasets typically store information about one specific area. These datasets can be queried using the SPARQL query language. Since the SPARQL 1.1 standard, it is possible to query several remote datasets with one query, and to combine the received data. This can be done with the SERVICE clause. SERVICE clauses generate join operations, and each join operation increases the load of the endpoint, that is why some endpoints do not allow at all to run queries using the SERVICE clause on their web interface. Some endpoints use SPARQL 1.0 and they do not know the SERVICE clause. There are endpoints which estimate the execution time of the query, and if that reaches a limit, then the endpoint refuses to execute it. Our aim is to create a system, which is able to execute queries over the LOD Cloud, without explicitly mentioning the necessary endpoints in the SERVICE clauses. Instead, the proposed system automatically recognizes the endpoints based on conditions of the query.

Keywords: Semantic Web, Linked Data, LOD Cloud, SPARQL.

1 Introduction

The LOD Cloud is a collection of the available datasets of the Semantic Web [6]. The individual datasets typically store information about one specific area. These datasets can be queried using the SPARQL query language [7]. The WHERE clause of the SPARQL query contains triple patterns which must be satisfied by the results. Since the SPARQL 1.1 [1] standard, it is possible to query several remote datasets with one query, and to combine the received data. This can be done with the SERVICE clause. Each SERVICE clause tries to access the defined remote endpoint where the corresponding triples can be found. SERVICE clauses generate join operations, and each join operation increases the load of the endpoint, that is why some endpoints do not allow at all to run queries using the SERVICE clause on their web interface. There are endpoints which estimate the execution time of the query, and if that reaches a limit, then the endpoint refuses to execute it. Our aim is to create a system, which is able to execute queries over the LOD Cloud, without explicitly mentioning the necessary endpoints in SERVICE clauses. Instead, the proposed system automatically recognizes the endpoints based on conditions of the query.

---

[1] http://www.w3.org/TR/sparql11-federated-query

J. Sobecki, V. Boonjing, and S. Chittayasothorn (eds.), *Advanced Approaches to Intelligent Information and Database Systems*, Studies in Computational Intelligence 551,
DOI: 10.1007/978-3-319-05503-9_18, © Springer International Publishing Switzerland 2014

In this paper we present a solution to join data from multiple endpoints using a single query. In order we can query multiple endpoints, we need to know the URL of the SPARQL endpoints. Our system helps this problem with automatic endpoint detection. We need the following informations about each endpoint: the URL of the endpoint to be accessed, and the namespaces used in the dataset. The system needs an initial step. In this step we add endpoints with its namespaces to the system. After that, the system tests the endpoint, if it is able to use the VALUES clause. This clause is neccessary to the join. After this initial configuration, the system is able to automatically determine the endpoints on which the requests must be evaluated by analyzing the namespaces in the queries. This helps to simplify the SPARQL queries, since the SERVICE clause has not any URL, and it also helps to use the LOD Cloud, because we do not need to search the sparql endpoint, or do not need to know the URL of the endpoints. Some namespaces can be found on more than one endpoints, therefore the evaluation of a triple pattern is not always unambigous. In this paper we show a method to resolve conflicts. If a triple pattern is in conflict, we resolve that by examining the environment of the variables.

The system is designed to allow queries not to specify which endpoints they would like to use. Determination of the endpoints is automatic. The basis of this is that we store the namespaces used by each endpoint. This way we can determine which endpoint should satisfy which triple pattern. Most of the endpoints have their own prefix, which makes the endpoint resolution unambigous. However, there are some namespaces, which can be found in virtually all endpoints, such as rdf, rdfs or owl. If the namespaces in a triple pattern are not unique, then we can determine which should be used by investigating the variables in the pattern. If a triple pattern has a variable that appears in another triple pattern which corresponds unambiguously to one endpoint based on its namespaces, then the triple pattern in question will be sent to that endpoint, too. There may be cases when a variable is used to join two endpoints. In this case, we start by sending the triple pattern to the endpoint, which has fewer possible values for this variable. After that, we send the values to the other endpoint. Nowadays, the number of endpoints is increasing, so the problem of endpoint resolution is becoming increasingly necessary. Our system provides an opportunity to automatically select the endpoint needed for the query.

2   Related Work

A semantic search system called LODatio has been presented in [1] which aims to support users to find relevant sources in Linked Open Data Cloud. LODatio searches information with a schema-level index. The system provides some functionality such as ranked result lists, result snippets and estimation of the total result set size. Our query planner uses the information of the result set size for making the plan.

In a survey [10] the authors presented a state of the art in this topic. They presented the relevant federation systems and techniques, that we describe these too.

The ADERIS [11] stores the predicates of the data sources. This can be used to choose the appropriate endpoint for subqueries. It has a setup phase where the predicate information must be set. Our approach is similar, but we uses the prefixes for describing the datasets. The prefixes are usually smaller than the number of the predicates.

The FedX [12] is a Sesame framework. It can use the local repository and SPARQL endpoints. It resolves the source selection problem with the SPARQL ASK query. The FedSearch [13] based on the FedX framework. This query engine provides full-text search in a federation of SPARQL endpoints. FedX has a query optimization technique with which we do not deal in this paper.

The DARQ [15] is an extension of JENA ARQ. It decomposes the query to subqueries, and it rewrites the query with cost-based optimization. It uses a service description for endpoint selection. The service descriptions describe the data available from a data source. It is based on the predicates too.

The LODStats [2] system calculates several statistics about the datasets. Our system uses statistics during the query evaluation process. We calculate the size of the intermediate result set as well.

In the paper [3], the authors describe a problem that many endpoints are not available. The available endpoints which can use SPARQL 1.1 standard was tested by the authors. Our solution is available to use the SPARQL 1.0 and SPARQL 1.1 standard.

Several join techniques (e.g. Remote Join, Mediator Join, Semi-join, Bind Join) have been described in the [4]. The Semi-join technique that was defined in the paper was used by our system. This means that we use the intermediate result set in the filter of the next subquery.

3   Endpoint Description

The endpoint resolution framework is based on the stored information about the endpoints, which plays a key role in the preparation of the query plan. The endpoints are usually described with some features of its dataset. In the previous paper [5] they offered the property because it occurs few times comparing to the size of the dataset, and this property is rare as a variable in the usual queries. Our system uses the namespaces of the datasets for endpoint selection. Usually the number of namespaces used in a dataset is relatively small, and they can be usually available. Adding an endpoint to the framework is currently done manually, however, we plan to automate this process as well.

There are numerous namespaces, which occur in more than one endpoint, such as rdf (e.g. in rdf:type). These properties would be the cause of the conflicts. The endpoint selection is based on the namespaces and variables appearing in the triple patterns. First we check the namespaces in the pattern. If they unambiguously correspond to an endpoint, then the resolution is finished. If the endpoint is ambiguous – because more than one endpoint store information in these namespaces – we have to resolve the conflict. The details about resolving conflicts are described in the next section.

Another stored information about the endpoints is the usable standard of the SPARQL. If the endpoint only uses the 1.0 standard, then we can not use the VAL-UES clause. We query these endpoints with FILTER and BIND clause.

4   Conflict Resolving

Our system selects the endpoint which can evaluate a given triple pattern. The selection is based on the namespace of the non-variable values of the triple pattern. In case of some namespaces such as *http://dbpedia.org/ontology* (dbpo) and *http://data. linkedmdb.org* (imdb) the proper endpoint can be easily chosen. However, there are some often used namespaces such as *http://www.w3.org/1999/02/22-rdf-syntax-ns#*

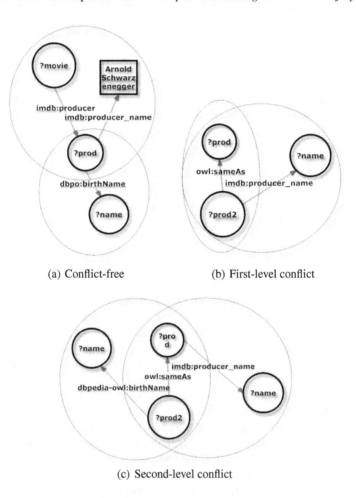

(a) Conflict-free   (b) First-level conflict

(c) Second-level conflict

Fig. 1. Conflicted triple patterns

(rdf) and *http://www.w3.org/2002/07/owl#* (owl) when the choice is ambiguous because multiple endpoints store data with the given namespace. If at least one of the variables of the triple pattern occurs in another one, then the system tries to infer which endpoint is the proper choice based on the other triple pattern. If the variables are contained by exactly one other triple and the evaluation of that triple is decidable, then the triple pattern will be sent to the same endpoint, otherwise it is a second-level conflict which needs special conflict resolution.

If a query consists of only the following line and no other triple pattern is given that can restrict the list of the potential endpoints, then the system can not resolve the conflict.

Example 1. ?s rdf:type ?p

In this case, the query are sent to all possible endpoints, as it is described in [4], and the union of the results is computed.

Figure 1 shows examples of various conflicted situations. On the 1(a) there is a conflict-free situation, because the triple patterns ?movie imdb:producer ?prod, ?prod imdb:producer_name 'Arnold Swarzenegger', ?prod dbpo:birtName ?name are evaluated on the Linked Movie DataBase[2] and the DBpedia[3] endpoint respectly to the namespaces. On the 1(b) a first-level conflict is shown. The endpoint which evaluates the triple ?prod owl:sameAs ?name is ambiguous, but it can be infered based on the ?prod2 imdb:producer_name ?name pattern. Lastly on the 1(c) there is a second-level conflict because no single endpoint can be chosen. In this case, the system sends the triple to all the potential endpoints and computes the union of the results.

5 Query Processing

In this section we present the algorithm which split the original query into subqueries that can be evaluated on the proper endpoint with respect to the used namespaces. The algorithm first identifies and manages the conflict-free patterns and marks the other ones. In next step, it tries to resolve the first-level conflict based on the common variables. In case of second-level conflicts the triples will be sent to all the previously defined endpoints. The system works well for the simple query without SERVICE clause. In this case the readServiceTriplePattern reads all triples from WHERE clause, and searches the endpoint. The system resolves the endpoint of all SERVICE clauses. After that, the query runs as in the second case. The last of query types is when the query has join and it has not SERVICE clauses. If we use this type of query, the query rewriting is simple. The system resolves the necessary endpoints, and it collects the corresponding triples to each one.

The Prog 1 shows the endpoint resolving mechanism. The w stores the triples, that we want to check. The e stores the triples of each endpoint, and the v stores triples of each variable. The *conf_w* stores the triples that have conflict. At the end of the program the e will store the triples of each endpoint. These triples can be evaluated on the endpoints.

---

[2] http://data.linkedmdb.org/sparql
[3] http://dbpedia.org/sparql

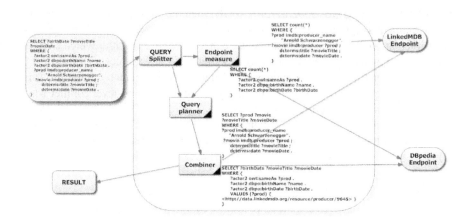

Fig. 2. Running of the system

After all triples have been assigned to the proper endpoint, the system calculates the order of the request. In order to calculate the best order, it tries to estimate the number of the rows for every request. The query plan is made by using these numbers and the SPARQL standard of the endpoints. Those subqueries are running first that have the fewest rows. Its result defines additional conditions for the next subquery which decreases the number of the possible values on the second endpoint. The result of the second subquery adds extra condition to the next, and so on. The addition conditions are built in the queries using VALUES keyword.

Example 2.
```
SELECT ?movieTitle ?movieDate ?birthDate
WHERE {
      ?actor2 owl:sameAs ?prod .
      ?actor2 dbpo:birthDate ?birthDate .
      ?prod imdb:producer_name "Arnold Schwarzenegger".
      ?movie imdb:producer ?prod ;
            dcterms:title ?movieTitle ;
            dcterms:date ?movieDate .
}
```

Example 2 shows an example, where we query films that are produced by Arnold Schwarzenegger and we would like to know the title and the date of the film and the birth date of Schwarzenegger. Data related to films are from the Linked Movie DataBase [9] and the information about their producers are from the DBpedia [8]. On the left side of the Figure 2 the query is shown. It can be seen that the query does not contain the SERVICE keyword, however, the system is able to split the query into subqueries based on the namespaces and the variables of it.

Prog. 1. Endpoint resolver.

```
1  List<Cond> w = readServiceTriplePatterns();
2  Map<Endpoint,List<TriplePattern>> e = {};
3  Map<Variable,List<TriplePattern>> v = {};
4  List<TriplePattern> conf_w = ();
5
6  foreach c in w:
7     // goodEndpoints return the endpoints that can serve a condation
8     if size(goodEndpoints(c)) = 1:
9        // push the triple pattern to all the list of the endpoint
10       push(e{goodEndpoints(c)}, c)
11       // push the endpoints to all the list of the variable
12       pushAll(v{variables(c)}, goodEndpoints(c))
13    //Line conflicted
14    else:
15       ///collect the conflicted lines
16       pushAll(conf_w,c);
17    end if
18 end foreach
19
20 foreach c in conf_w:
21    // check all variables in the line
22    // canResolveConflit check that we can resolve the conflict with variable
23    if canResolveConflit(c,v):
24       //resolvedEndpoints return the resolved endpoints
25       push(e{resolvedEndpoints(c)}, c)
26       // push the endpoints to all the list of the variable
27       push(v{variables(c)}, resolvedEndpoints(c))
28    else:
29       // push the triple to all endpoints
30       pushAll(e{goodEndpoints(c)}, c)
31    end if
32 end foreach
```

Table 1. Decided endpoint

Line Triple	Endpoint(s)
1 ?actor2 owl:sameAs ?prod	LinkedMDB,**DBpedia**
2 ?actor2 dbpo:birthDate ?birthDate	**DBpedia**
3 ?prod imdb:producer_name "Arnold Schwarzenegger"	**LinkedMDB**
4 ?movie imdb:producer ?prod	**LinkedMDB**
5 ?movie dcterms:title ?movieTitle	DBpedia, **LinkedMDB**
6 ?movie dcterms:date ?movieDate	DBpedia, **LinkedMDB**

Table 1 shows the selected endpoints for each triple of the where clause. Some triples can be sent to more endpoints. The system decides using the namespaces, and the previously calculated endpoints of the variables. The first two lines will be sent to the DBPedia endpoint, and line 3-6 will be sent to the LinkedMDB endpoint. The endpoint measure component estimates the number of the results. Based on these results the query planner creates the plan which determines the order of the evaluation of the subqueries. According to the plan the Combiner executes the subqueries and computes the result of the original query from the responds.

6 Evaluation

The FedBench [14] is used for benchmarking the federated queries. Our paper does not focus on the optimization of the SPARQL query. Our system uses the live endpoints, like DBpedia, LinkedMDB or FactForge [16]. We evaluated our solution with the Arnold Schwarznegger's movie query (q1), that we mentioned before, and the query1 of the FedBench benchmark (q2). The q1 query uses two datasets, the DBpedia and the LinkedMDB. Each dataset has SPARQL endpoint. We tried to run the query with several different techniques. We ran the query on Web Interface of the endpoint and we tried it from program code with Jena[4] framework. The result is shown on table 2. The first row shows the results when we used the Web Interface. The query used one SERVICE clause when we ran on the DBpedia or the LinkedMDB endpoint. It used two SERVICE clauses when we ran it on external endpoint, this time we used the endpoint of the sparql.org. Finally the query has not any SERVICE clauses when we ran it with our system. The results show the DBpedia throwed an exception when we ran the query. It happened on Web Interface and from the program code too. The LinkedMDB did not throw an exception, but did not give a correct answer. The external endpoint (sparql.org) had a time out when it tried to ran it our query in the Web Interface. This endpoint also throwed an exception, when we tried to ran it from program code. Finally our solutions completed the query. It checked the possible rows from the endpoints. The first subquery was sent to the LinkedMDB and it generated VALUES clause from the result. After that another subquery ran with this condition on the DBpedia endpoint, and the endpoint gave the correct answer. The q2 query showed the same. The results are shown on the table 3.

Table 2. Query run over endpoints

	DBpedia	LinkedMDB	FactForge(NYTimes)	Sparql.org	Our system
q1 Web Browser	Exception	0 row	-	Time out	-
q1 Code	Exception	0 row	-	Exception	2 row
q2 Web Browser	Exception	-	1 row	Time out	-
q2 Code	Exception	-	1 row	Exception	1 row

Table 3. The results of the q1 with our system

?movieTitle	?movieDate	?birthDate
Last Action Hero	1993-06-18	1947-07-30 (xsd:date)
The 6th Day	2000-11-17	1947-07-30 (xsd:date)

7 Conclusion

The SPARQL 1.1 introduced the SERVICE keyword that enables the user to connect to remote endpoints from the LOD Cloud. However, due to the costs of this operation most

---

[4] http://jena.apache.org/

of the public endpoints do not permit to use this. For example, some endpoints estimate the runtime of the query and they drop the query if its runtime is more than a predefined limit. The runtime of SERVICE operation is usually estimated to be more than the limit. In this paper we presented a system that allows to query multiple endpoints from LOD Cloud. The system automatically infers the proper endpoints if neccessary. The endpoint is determined based on the namespaces of datasets of the endpoints. The namespaces of the triple patterns which are contained in the query determine the possible endpoints that can answer the query. The system can use SPARQL 1.0 and 1.1 standard too.

8   Future Work

At this time, endpoints are added manually to the system which can result in situations where the namespaces of the endpoints are not consistent. To avoid that situations, we would like to make the assignments automatically with SPARQL queries.

This system does not use any optimalization. Our aim is the optimalization of the system, and we compare with the available federation systems.

Another aim is a smart solution for a query writing. If we write a SPARQL query, we need to know the structure of the dataset. If we do not know, then it is difficult. We would like to make a recommendation system for this problem. This system can help to use the LOD Cloud.

Acknowledgments. This work was partially supported by the European Union and the European Social Fund through project FuturICT.hu (grant no.: TAMOP-4.2.2.C-11/1/KONV-2012-0013) and the Hungarian and Vietnamese TET (grant agreement no. TT 10-1-2011-0645). We are grateful to Tamás Matuszka, Balázs Pinczel and Gábor Rácz for helpful discussion and comments.

References

1. Gottron, T., Scherp, A., Krayer, B., Peters, A.: LODatio: Using a Schema-level Index to Support Users in Finding Relevant Sources of Linked Data. In: K-CAP, pp. 105–108 (2013)
2. Ermilov, I., Martin, M., Lehmann, J., Auer, S.: Linked Open Data Statistics: Collection and Exploitation. In: Klinov, P., Mouromtsev, D. (eds.) KESW 2013. CCIS, vol. 394, pp. 242–249. Springer, Heidelberg (2013)
3. Buil-Aranda, C., Hogan, A., Umbrich, J., Vandenbussche, P.-Y.: SPARQL web-querying infrastructure: Ready for action? In: Alani, H., Kagal, L., Fokoue, A., Groth, P., Biemann, C., Parreira, J.X., Aroyo, L., Noy, N., Welty, C., Janowicz, K. (eds.) ISWC 2013, Part II. LNCS, vol. 8219, pp. 277–293. Springer, Heidelberg (2013)
4. Görlitz, O., Staab, S.: Federated data management and query optimization for linked open data. In: Vakali, A., Jain, L.C. (eds.) New Directions in Web Data Management 1. SCI, vol. 331, pp. 109–137. Springer, Heidelberg (2011)
5. Husain, M.F., McGlothlin, J., Khan, L., Thuraisingham, B.: Scalable Complex Query Processing Over Large Semantic Web Data Using Cloud. In: 2011 IEEE International Conference on Cloud Computing (CLOUD). IEEE (2011)

6. Berners-Lee, T., Hendler, J., Lassila, O.: The semantic web. Scientific American 284(5), 28–37 (2001)
7. Prud'Hommeaux, E., Seaborne, A.: SPARQL query language for RDF. W3C Recommendation 15 (2008), http://www.w3.org/TR/rdf-sparql-query/
8. Auer, S., Bizer, C., Kobilarov, G., Lehmann, J., Cyganiak, R., Ives, Z.G.: DBpedia: A nucleus for a web of open data. In: Aberer, K., Choi, K.-S., Noy, N., Allemang, D., Lee, K.-I., Nixon, L.J.B., Golbeck, J., Mika, P., Maynard, D., Mizoguchi, R., Schreiber, G., Cudré-Mauroux, P. (eds.) ASWC 2007 and ISWC 2007. LNCS, vol. 4825, pp. 722–735. Springer, Heidelberg (2007)
9. Hassanzadeh, O., Consens, M.P.: Linked Movie Data Base. In: Workshop on Linked Data on the Web (LDOW 2009), Madrid, Spain (2009)
10. Rakhmawati, N.A., Umbrich, J., Karnstedt, M., Hasnain, A., Hausenblas, M.: Querying over Federated SPARQL Endpoints-A State of the Art Survey. arXiv preprint arXiv:1306.1723 (2013)
11. Lynden, S., Kojima, I., Matono, A., Tanimura, Y.: ADERIS: An adaptive query processor for joining federated SPARQL endpoints. In: Meersman, R., Dillon, T., Herrero, P., Kumar, A., Reichert, M., Qing, L., Ooi, B.-C., Damiani, E., Schmidt, D.C., White, J., Hauswirth, M., Hitzler, P., Mohania, M. (eds.) OTM 2011, Part II. LNCS, vol. 7045, pp. 808–817. Springer, Heidelberg (2011)
12. Schwarte, A., Haase, P., Hose, K., Schenkel, R., Schmidt, M.: FedX: a federation layer for distributed query processing on linked open data. In: Antoniou, G., Grobelnik, M., Simperl, E., Parsia, B., Plexousakis, D., De Leenheer, P., Pan, J. (eds.) ESWC 2011, Part II. LNCS, vol. 6644, pp. 481–486. Springer, Heidelberg (2011)
13. Nikolov, A., Schwarte, A., Hütter, C.: FedSearch: Efficiently Combining Structured Queries and Full-Text Search in a SPARQL Federation. In: Alani, H., et al. (eds.) ISWC 2013, Part I. LNCS, vol. 8218, pp. 427–443. Springer, Heidelberg (2013)
14. Schmidt, M., Görlitz, O., Haase, P., Ladwig, G., Schwarte, A., Tran, T.: FedBench: A benchmark suite for federated semantic data query processing. In: Aroyo, L., Welty, C., Alani, H., Taylor, J., Bernstein, A., Kagal, L., Noy, N., Blomqvist, E. (eds.) ISWC 2011, Part I. LNCS, vol. 7031, pp. 585–600. Springer, Heidelberg (2011)
15. Quilitz, B., Leser, U.: Querying distributed RDF data sources with SPARQL. In: Bechhofer, S., Hauswirth, M., Hoffmann, J., Koubarakis, M. (eds.) ESWC 2008. LNCS, vol. 5021, pp. 524–538. Springer, Heidelberg (2008)
16. Bishop, B., Kiryakov, A., Ognyanov, D., Peikov, I., Tashev, Z.: Velkov, "Factforge: A fast track to the web of data. Semantic Web 2(2), 157–166 (2011)

Application of Ontology and Rough Set Theory to Information Sharing in Multi-resolution Combat M&S

Dariusz Pierzchała

Military University of Technology, Faculty of Cybernetics,
Gen. Sylwestra Kaliskiego Str. 2, 00-908 Warsaw, Poland
dpierzchala@wat.edu.pl

Abstract. Military decision support and simulation training tools are mostly complex and large-scale IT systems and therefore multi-resolution distributed simulation models have been playing a leading role. The paper considers an approach which combines a graph theory, HLA simulation standard, a special ontology and rough set formalisms into a synergistic software. The first issue is the way to enhance HLA object model by an ontology. The subsequent problem is construction of a software plugin to explicit handle shared information. Furthermore, the Rough Set Theory provides the solid foundation for the construction of classifiers as well as generation of decision rules from dataset. The proposed approach might be perceived in terms of distributed computational intelligence and ontology-based information sharing.

Keywords: distributed multi-resolution simulation, graphs, ontology, rough set.

1 Introduction

The cutting-edge military systems are very complex, that's why no single, unified simulator can satisfy users' needs and meets multi-resolution modelling (MRM) assumptions. A concept "resolution" is typically defined as a conceptual level at which an object (entity) is simulated. It is strictly related to modelling constraints and to the question what criteria define the dimensions and the degree of detail. A simple object (like a vehicle) is a disaggregated high-resolution entity (HRE) while a group of objects states aggregate low-resolution entity (LRE). The significant problem of MRM is related to uncertain data utilised to fill out a state vector of a HRE/LRE object as well as to state transformation and consistency management functions. Several projects have proposed a mechanism Multiple Resolution Entity (MRE) for handling consistency of object at multiple resolution levels concurrently and mutually. In [5] the Attribute Dependency Graph is proposed for maintaining consistency and then cost of operations' aggregation/disaggregation is estimated. Some questions regarding specification of multi-resolution modelling space in simulations are taken up by [6]. Next work [7] deals with formal modelling of MRM supported by both state mapping functions and consistency correction methods. In turn, the author [8] analyses multi-resolution HLA federation supporting CAX exercises. Nonetheless, despite the huge

J. Sobecki, V. Boonjing, and S. Chittayasothorn (eds.), *Advanced Approaches to Intelligent*
Information and Database Systems, Studies in Computational Intelligence 551,
DOI: 10.1007/978-3-319-05503-9_19, © Springer International Publishing Switzerland 2014

effort and hundreds of ideas, only few of results are practically mature for large-scale simulations. Therefore, a fully justified step is regarding expanding standards based on outcomes originating from non-simulation researches: e.g. knowledge-based.

Ontology is a contemporary tool to precisely define vocabulary (terminology, data types) related to objects or attribute sets throughout both lexical and conceptual point of view. The advantage of ontology is shared knowledge about a domain, semantic interoperability as well as reusability of legacy models. Authors of [9] emphasise that successful combining data from numerous different applications into a coherent complex system might be supported by semantics. According to [10] physical aspects of communication and interoperation in HLA need to be enhanced in a semantic way. And, some research on military domain ontology are reported in [11].

The Rough Set Theory (RST) facilitates discovering patterns hidden in data based information. The novelty originated from the distinct concepts for recognize and manipulation of uncertain data-based information: reasoning with incomplete information, reducing redundant both objects and attributes, classifying new cases.

Taking all these details into account, the combined approach based on HLA-ontology-rough sets has been firmly recommended by this paper.

2   HLA-Based Federated Multi-resolution Architecture

A distribution of conflict model components seems to be very natural as it results from the real system objects' dispersion. Integrated multi-simulation environments for mission planning and training for both staff and field personnel on a different command levels are usually realized using well-known standards. The two primary COTS standards in M&S are: DIS-Distributed Interactive Simulation (IEEE 1278) and HLA-High Level Architecture (IEEE 1516 family, NATO STANAG 4603). The second one was defined for the most general purposes: to meet a need of interoperability, to support reusable distributed computer simulations composed of different software federates (theoretically regardless of computing platforms) and to maximize the flexibility for all kinds of local time management mechanisms. It offers all federates possibilities to use any combination of: event-driven, time-stepped, discrete and continuous simulation. Simulation events (interactions) and objects' states (attributes) are defined by Federation Object Model (FOM). Related information is propagated via HLA Runtime Infrastructure (RTI) services in a form of: updating/reflecting attribute's value and sending/receiving interactions. The above crucial concepts states the M&S paradigm, nonetheless they are a root of problems in clear communication between objects implemented inside separated simulators on different levels of resolution. Thus, despite significant success of case studies, such M&S interoperability standards seem to be insufficient to obtain fully integrated multi-resolution simulation and there is still a huge lack of gateways between different simulation models.

With regarding to [2], we consider a distributed simulation environment (DSE) represented by both hardware and software architecture models. Software architecture can be formally defined as a triplet:

$$S = < S^S, S^O, S^I > \tag{1}$$

The three complementary parts (S^S - communication and synchronization part, S^O - object data part, S^I - interaction data part) are to be subsequently defined. Communication and synchronization structure is modelled as a theoretical network:

$$S^S = << W^S, \Gamma^S>, FW^S, \emptyset >, \text{ relying on Berge graph } G^S = < W^S, \Gamma^S >, \text{ where:} \quad (2)$$

- $W^S = W^{FED} \cup W^f$ is a set of vertices, in which:
  - $W^{FED}, /W^{FED}/ = 1$ – set of synchronization components' tags,
  - $W^f, W^f \neq \emptyset$ – set of federates' tags;
- $\Gamma^S : W^S \rightarrow 2^{WS}$ is assignment function of component to component;
- and $\forall (x \in W^f, y \in W^{FED}) \, y \in \Gamma^S(x) \Rightarrow x \in \Gamma^S(y)$.

Object data part can be defined using a network:

$$S^O = << W^S, \Gamma^O>, FW^O, FL^O>, \text{ where for example:} \quad (3)$$

- $f^O_1 \in FW^O : < G^O_{SOM}, F^O_{SOM}, \emptyset > $ – represents FOM's object data interface described by several functions:
  - $f^O_{SOM\,1}(\bullet, id^O): W^O_{SOM}(\bullet) \rightarrow W^A_{SOM}(\bullet, id^O)=\{a_i = <a\_name, a\_type>\}$ – a set of attributes of object id^O;
  - $f^O_{SOM\,2}(\bullet, id^O): W^O_{SOM}(\bullet) \rightarrow \{P(ublish), S(ubscribe), PS, N(on)\}$ – a declaration of operations regarding transmission of object data;
  - $f^O_{SOM\,3}(\bullet, id^O): W^O_{SOM}(\bullet) \rightarrow W^O_{SOM}(\bullet)$ – a declaration of aggregation operations regarding selected object attributes;
  - $f^O_{SOM\,4}(\bullet, id^O): W^O_{SOM}(\bullet) \rightarrow W^O_{SOM}(\bullet)$ – a declaration of disaggregation operations regarding selected object attributes;
  - $f^O_{SOM\,5}(\bullet, id^O) = \text{card}(W^A_{SOM}(\bullet, id^O)$ – a resolution level of object id^O;
- $f^O_2(\bullet, t): W^f \, x \, T \rightarrow W^{OP}_{SOM}$, where:
  - $W^{OP}_{SOM} \subset W^O_{SOM}$ – is an subset of object identifiers in a moment t, which attributes' updates are published for;
- and $\forall id^O \in W^{OP}_{SOM}(\bullet) \, (f^O_{SOM\,2}(\bullet, id^O)=P \lor f^O_{SOM\,2}(\bullet, id^O)=PS)$ – is a publication constraint: the object id^O has to be declared as "publish" or "publish and subscribe" in order to publish any relevant data.

The interaction data part $S^I = << W^S, \Gamma^I>, FW^I, \emptyset >$ is being defined analogously so we can lossless omit those description. Finally, having a complete model of HLA-based environment we might proceed to project of DSE software architecture.

A case study simulation environment is composed of three COTS simulators: *Zlocien Au2* (constructive stochastic land combat models, platoon-to-brigade levels), *VBS2* (virtual first-person simulation) and another one for maritime operations – each of them represents both different level of resolution and aggregation. What is more, each simulator operates on the own simulation data model (SOM) and thus for the whole federation a multi-resolution integrated FOM should be established.

The most basic question is how to realize data integration in order to provide interoperable information exchange within federation? Considering COTS products that are mostly closed to internal modification, there is possible one way: to adopt simulators using a dedicated software gateway. Where such a tool ought to be inserted?

From the point of view of *Zlocien Au2* (created at Military University of Technology with contribution of the paper's author, hence it remains open) a plugin extension for each external simulator are to be constructed and merged into its interface. Nevertheless, even if any modifications were prohibited in the federates a gateway could be situated as the subsequent federate on the path between the RTI and simulators and then might subscribe incoming data / publish converted. According to HLA Rules information assigned to resolution levels have to be shared by federates using RTI.

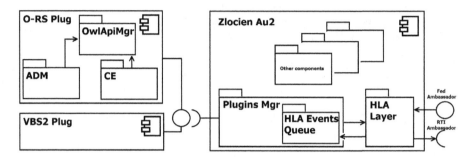

Fig. 1. Concept of software gateway with plugins to handle FOM-based data

Considering a construction of the proposed gateway (see Fig. 1.) we should mention *"HLA Layer"* component responsible for two-way communication with RTI and *"Plugins Mgr"* as a central part of the gateway. Every data package the RTI sends to *Zlocien Au2* is stored inside a cache *"HLA Events Queue"* and then appropriate plugin gets it to service (aggregation, disaggregation, conversion, mapping, propagation). Information is stamped with a time of occurrence, thus *Zlocien Au2* receives them in a proper order and at the correct simulation time.

A detailed explanation of *VBS2* plugin [1] was focused on a mechanism to mapping attributes of *Zlocien Au2* objects to the attributes of *VBS2* objects. Although that solution allows to compare and map two different models of the same real entities, the crucial limitation concerns the same resolution level. This work proposes a relaxation of that constraint in *"O-RS Plug (Ontology-rough set-based plugin)"* by improvement HLA based models and methods with ontology reasoning and rough set classifiers.

3   A Short View at Modelling the Semantics of Conflict Domain

A military conflict domain, like any others, requires unified semantics that ensures clear information sharing across multi-resolution components or even systems. HLA Federation Object Model is reported as a standard template for structured information. Nonetheless, the vocabulary (terminology, data types, …) related to object classes or attribute sets is not well-defined yet. Following the work [4], let ontology be simply defined as a domain model for the purpose of describing shared knowledge about specific modelling area. The proposed formal meaning is the following tuple:

$$O = < C, R_c, H_c, R_{Rel}, A^O >, \text{ where:} \tag{4}$$

- $R_c = \{r_c: r_c \in C \times C\}$ – is structural relations;
- $H_c = \{h_c: h_c \in (C\backslash\{c_0\}) \times C\}$ – is taxonomical relationships providing information on super/sub- concepts;
- R_{Rel} – is able to describe inverse relationships, possible relationship taxonomies along with symmetry, transitivity, reflexivity relationship characteristics;
- and $A^O = <L(G^A), I^L>$ – is the axiom definitions (including logical statements of model restrictions).

The ontology-based approach gives us new possibilities to unify meaning of concept and relationship from the military conflict domain. It provides the inference abilities to verify the instance data correctness according to constraints specified by model consistency. The last but not least useful function is merging and mapping objects of varying resolutions into proper representation based on semantic pattern recognition. In order to provide software with capability to parse models explicitly and to interpret information based on semantics, ontologies are encoded with formal languages. In many works the Web Ontology Language (OWL) has been adapted – for example to code the DeMO (Discrete Event Modeling Ontology) which is a general level ontology for discrete and event simulation (DES) [12]. Despite formal description of states, activities, events and processes it omits any specific domains. The other project [13] is based on detailed model of military forces and tasks in the context of both "course of action" and "Center of Gravity" analyses. Furthermore, the work [10] refers a

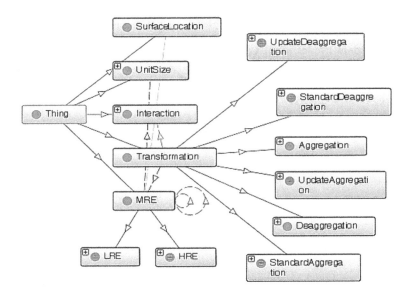

Fig. 2. A chosen part of the MRE ontology (generated with Protégé editor)

research on two-level military ontology architecture. A very reusable approach is introduced in [4] – the UBOM ontology has been adaptively generated with respect to JC3IEDM data schemas. Unfortunately, the results mentioned above almost completely avoid multi-resolution requirements. Therefore, taking into consideration the status of the work, we propose an approach to enhance FOM model by ontology as well as construction of software plugin to explicit handle shared information.

The MRE-based object should be relevant for each level of resolution – e.g. a military unit has multi-variant representation (inter alia type, size, coordinates, commanding level, formation and task type, superior, subordinate, combat potential, radio-communications, resources and armament status, etc.). The ontology contains a detailed description hierarchy of models and restrictions for both models relationships and transitions, and then helps changing a resolution (see Fig. 2.). Finally, standard FOM structure has been enriched towards ontology-based multi-resolution model in accordance with the FOM-OWL scheme:

- *FOM object types*   →   *OWL Class Hierarchy*
- *Multi-resolution rules*   →   *OWL Restrictions*
- *Multi-resolution structures*   →   *OWL Object Property*
- *Rules for transformation: methods*   →   *OWL Class Hierarchy*
- *Rules for transformation: context of use*   →   *OWL Class Restrictions.*

replying to the questions: What is the proper level for objects? What is the actual resolution? How to change the resolution level?

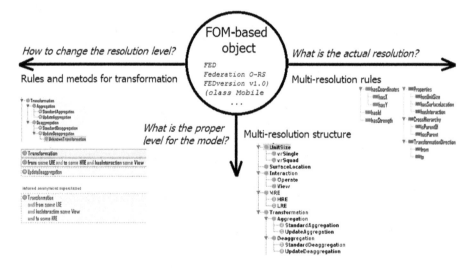

Fig. 3. Expanded definition of FOM-based object with MRE ontology

"O-RS Plug" (see Fig. 1.) implements FOM-OWL schema based on several Java object classes – the most significant are as follows:

- *ADM (Aggregate/Disaggregate Module)* – classes for transformation between resolution levels - their use depends strictly on the rules stored in the ontology;
- *CE (Consistency Enforcer)* – classes for propagation of changes to object's attributes on the other resolution levels;

- *HermiT* – classes responsible for reasoning on ontologies and models consistency;
- *OwlApiMgr* – a software envelope for *OWL Java API* enabling creating and manipulating OWL-based ontologies.

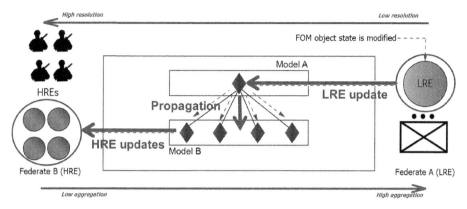

Fig. 4. A sample of an update for MRE object attributes [3]

The plugin's work is triggered with the update received from any external federate. Its classes (*ADM, CE*) perform transformation to proper resolution level(s) and propagate changes to objects. Methods implemented for aggregation and especially disaggregation object attributes, as well as for identification necessary objects are crucial for this proposition. The following program code of Java class is a skeleton of sample reasoning procedure "*recognizeMRE*" [3]:

```
public OWLClass recognizeMRE(UnitSize us){
OWLClass mre = factory.getOWLClass("MRE", pm);
OWLClass unknown = factory.getOWLClass("Unknown1", pm);
OWLClass uSize = factory.getOWLClass(us.toString(), pm);
// Object properties
OWLObjectProperty hasUnitSize =
factory.getOWLObjectProperty("hasUnitSize", pm);
// Restrictions for object
OWLClassExpression property =
  factory.getOWLObjectSomeValuesFrom(hasUnitSize, uSize);
OWLSubClassOfAxiom axiom1 =
  factory.getOWLSubClassOfAxiom(unknown, mre);
OWLSubClassOfAxiom axiom2 =
  factory.getOWLSubClassOfAxiom(unknown, property);
manager.addAxiom(ontology, axiom1);
manager.addAxiom(ontology, axiom2);
reasoner.flush();
reasoner.precomputeInferences();
Set<OWLClass> set =
  reasoner.getSuperClasses(unknown, true).getFlattened();
```

```
OWLClass ret = null;
for (OWLClass cl : set) {
  ret = cl.asOWLClass(); }
manager.removeAxiom(ontology, axiom1);
manager.removeAxiom(ontology, axiom2);
return ret; }
```

The typical problem is when the incoming updates represent lacking or imperfect knowledge. Thus, in order to achieve both efficient and accurate software implementation we propose to rely on rough set classifiers.

4   Applying Rough Set Theory in Classification and Reasoning

A large number of data, their inconsistency and incompleteness cause a lot of problems in practical applications of pattern recognition and data-mining, especially in the area of decision support and intelligent control. To date, a number of methods based on Rough Set Theory (Z. Pawlak, 1982) [15] have attempted to solve those issues. The back bone of rough set tools is indiscernibility relation, approximations of a set, reducts and classification of the subject domain into number of disjoint categories. Assume like in [14], [16], [17] that:

$$DS = (U, A = C \cup \{d\}, f) \text{ is a decision system where:} \tag{5}$$

- $U, C, \{d\}$ – are non-empty finite set of objects (called universe), sets of attributes (called conditions) and decision attributes respectively;
- and f – is a decision function.

When any $a \in A$ represents continuous data a preceding discretization process is required. The key concept in that theory is *the indiscernibility relation* intended to express inability to distinguish some objects and to induce a partitioning of the universe:

$$IND_{IS}(B) = \{(x, y) \in U^2 \mid \forall a \in B, a(x) = a(y)\}, A \subseteq B \tag{6}$$

The *B-lower (B")* and the *B-upper (B*)* approximations of a set:

$$B'' = \cup \{Y \in U/B : Y \subseteq X\} \tag{7}$$

$$B^* = \cup \{Y \in U/B : Y \cap X \neq \emptyset\} \tag{8}$$

are also key points on which *B-boundary region of X* is based:

$$BN_B(X) = B^* - B''. \tag{9}$$

BN_B consists of objects that cannot be decisively classified into X in B. And concluding, if a boundary region is non-empty a set is said to be *rough*. Accuracy of approximation is calculated as a imprecision coefficient:

$$\alpha B(X) = card(B''(X)) / card(B^*(X)) \tag{10}$$

where *card( )* denotes the cardinality of nonempty set, thus:

- $0 \leq \alpha B(X) \leq 1$ and
- if $\alpha B(X) = 1$, X is *crisp set* regarding the attributes B;
- if $\alpha B(X) < 1$, X is *rough set* with respect to B.

The next step is to find reducts that are understood as minimal subsets of attributes to preserve set approximation without losing any information from the system. It should ensure current existing classifications of objects. There is frequently more than one such subsets of attributes and in consequence finding a minimal reduct among all candidates is *NP-hard* problem [18]. Fortunately, there exist a number of classification algorithms (e.g. neural network, Bayes classifiers, heuristics genetic algorithms) that compute a lot of reducts in acceptable time.

Table 1. Sample observations in a decision table

Update No	Conditional Attributes			Decision Attribute
	in_movement (c^1)	radiocommunication (c^2)	potential (c^3)	active_or_destroyed (d)
1	movement	poor	high	active
2	at_position	good	medium	active
...	...	...	...	...
n	at_position	good	very_low	destroyed

As a final point, decision rules might be induced from a reduced rough set and derived reducts. We might note that the decision rule is true in the set A if the set of objects supporting the rule is the non-empty subset of the set of objects matching the decision rule. In the following case we consider a very simplified MRE - combat unit is characterized by the RST-based decision table and related rule. The indiscernibility function has a Boolean form "$(c^1 \text{ and } c^3) \text{ or } (c^2 \text{ and } c^3)$", $\alpha B(X) = 0.6$ while one of the rules has been formulated as: "*if c^1=movement and c^3=high then d = active*".

5   Summary

The purpose of the studies has been to merge a graph model, HLA standard, an ontology and RST theory into a synergistic software in which the main role plays the plugin *"O-RS Plug"*. A significant influence for a computational complexity have resolution and aggregation which are in the inverse relation. Commonly applied multi-resolution methods of aggregation and disaggregation cause inconvenience to the operations "up-down" between levels of resolution. The MRE is a kind of panacea for the problem of building models relevant for different levels of resolution.

The lessons learned concern development of an ontology which is critical to successful development of knowledge-based systems. The advantage of the solution is separating the rules, which determine when to use the method, of implementation of those methods. Despite growing role of the RST theory, application of RST for continuous data demand preceding discretization step. Moreover, finding reducts is one of the bottlenecks of the RST. Unfortunately, due to regular paper's restrictions only simple representative parts of applied software and cases are shown – particularly the example illustrating RST contribution is too simple.

The concurrent issue is a validation of models what is in practise really complex. However, a good way is to utilize expert knowledge and empirical cases. The proposed approach is being tested to domains different than military.

Acknowledgments. This work was partially supported by the research co-financed the National Centre for Research and Development and realized by Cybernetics Faculty at MUT: No O ROB 0021 01/ID 21/2 and DOBR/0069/R/ID1/2012/03.

References

1. Pierzchała, D., Dyk, M., Szydłowski, A.: Distributed Military Simulation Augmented by Computational Collective Intelligence. In: Jędrzejowicz, P., Nguyen, N.T., Hoang, K. (eds.) ICCCI 2011, Part I. LNCS, vol. 6922, pp. 399–408. Springer, Heidelberg (2011)
2. Pierzchała, D.: Designing and testing method of distributed interactive simulators. In: 15th International Conference on Systems Science, Wroclaw (2004) ISBN 83-7085-805-8
3. Pierzchala, D., Szymański, P.: Ontology-based adapter for data of multi-resolution battlefield simulation. In: XVIII Workshop of the Society For Computer Simulation, Poland (2011)
4. Pierzchała, D., Najgebauer, A., Antkiewicz, R., Chmielewski, M., Rulka, J., Wantoch-Rekowski, R., Tarapata, Z., Drozdowski, T.: Knowledge-Based Approach for Military Mission Planning and Simulation. In: Gutiérrez, C.R. (ed.) Advances in Knowledge Representation, pp. 251–272. InTech (2012) ISBN 978-953-51-0597-8
5. Natrajan, A., Reynolds, P.F., Srinivasan, S.: Guidelines for the Design of Multiresolution Simulations. US DoD DMSO (July 1997)
6. Su-Youn, H., Tag Gon, K.: Specification of multi-resolution modeling space for multiresolution system simulation. Transactions of the Society for Modeling and Simulation International 89(1), 28–40 (2012), doi:10.1177/0037549712450361
7. ShangGuan, W., Bai-gen, C., Si-Hui, L., Zhen-Guo, L., Jian, W.: Multi-resolution simulation strategy and its simulation implementation of Train Control System. IEEE (2011) 978-1-4577-0574-8/11
8. Cayirci, E.: Multi-Resolution Federations In Support Of Operational And Higher Level Combined/Joint Computer Assisted Exercises. In: IEEE Proceedings of the 2009 Winter Simulation Conference (2009) 978-1-4244-5771-7/09
9. Turnitsa, T., Tolk, A.: Federated Ontologies Supporting a Merged Worldview for Distributed Systems. In: Association for Advancements in Artificial Intelligence (AAAI) Fall Symposium, Technical Report FS-07-06, pp. 116–119. AAAI Press, Menlo Park (2007)
10. Hu, J., Zhang, H.: Ontology Based Collaborative Simulation Framework Using HLA & Web Services. IEEE, 116–119 (2008) 978-0-7695-3507-4/08
11. Jia, M., Yang, B., Zheng, D., Sun, W.: Research on Domain Ontology Construction in Military Intelligence. IEEE (2009), doi:10.1109/IITA.2009.80, 978-0-7695-3859-4/09
12. Silver, G., Hassan, O., Miller, J.: From Domain Ontologies To Modeling Ontologies To Executable Simulation Models. In: Proceedings of Winter Simulation Conference (2007)
13. Bowman, M., Lopez, A., Tecuci, G.: Ontology Development for Military Applications. Report from DARPA grants: no. F49620–97–1–0188 and no. F49620–00–1–0072 (2007)

14. Bazan, J., Nguyen, H.S., Nguyen, S.H., Synak, P., Wróblewski, J.: Rough set algorithms in classification problem. In: Polkowski, Tsumoto, Lin (eds.) Rough Set Methods and Applications. Physica-Verlag, Heidelberg (2000) ISBN:3-7908-1328-1
15. Pawlak, Z.: Rough Sets: Theoretical Aspects of Reasoning about Data. Kluwer Academic Publishers, Dordrecht (1991)
16. Rissino, S., Lambert-Torres, G.: Rough Set Theory – Fundamental Concepts, Principals, Data Extraction, and Applications. In: Ponce, Karahoca (eds.) Data Mining and Knowledge Discovery in Real Life Applications. InTech (2009) ISBN 978-3-902613-53-0
17. Skowron, A., Nguyen, H.: Rough Sets: From Rudiments to Challenges. In: Skowron, Suraj (eds.) Rough Sets And Intelligent Systems (2013) ISBN 978-3-642-30344-9
18. Skowron, A., Rauszer, C.: The discernibility matrices and functions in information systems. In: Słowiński, R. (ed.) Intelligent Decision Support. Handbook of Applications and Advances of the Rough Set Theory, pp. 311–362. Kluwer Academic Publishers, Dordrecht (1992)

Analysis on Smartphone Related Twitter Reviews by Using Opinion Mining Techniques

Jeongin Kim[1], Dongjin Choi[1], Myunggwon Hwang[2] and Pankoo Kim[3,*]

[1] Dept. of Computer Engineering, Chosun University,
375 Seoseok-dong, Dong-gu, Gwangju, Republic of Korea
{jungingim, dongjin.choi84}@gmail.com
[2] Korea Institute of Science and Technology Institute (KISTI),
245 Daehak-ro, Yuseong-gu, Daejeon, Republic of Korea
mgh@kisti.re.kr
[3] Dept. of Computer Engineering, Chosun University,
375 Seoseok-dong, Dong-gu, Gwangju, Republic of Korea
pkkim@chosun.ac.kr

Abstract. Due to the recent popularization of smartphones, it has become easy to get connected on Social Network Service (SNS) which caused the proliferation on the amount of tweets on Twitter. Many studies have been proposed to discover valuable meanings from Twitter text messages by using opinion mining. However, these researches have a side effect that it is only focused on positive and negative in the limited category. In this paper, we will attempt to examine which factors could affect the users interests or preferences by analyzing and comparing smartphone product reviews which were posted on Twitter, in multilateral categories, by using opinion mining.

Keywords: Social Network Services, Opinion Mining, Twitter.

1 Introduction

Recently, it has become possible to access to Social Network Service (SNS) without any temporal and locational restriction due to the rapid popularization of smartphones. Because of this issue, the amount of tweets on Twitter which is one of the most popular SNS around the world has been increasing consistently [1]. Tweet can be defined as a short message, containing a letter with minor sentences within 140 characters, noticing ones thought to others, to phrase the users feeling with positive and negative emotions [2]. As the amount of tweets has promptly increased, the Opinion mining technique, which rapidly analyzes their meanings and extract any valuable information, is being applied to various studies. Nevertheless, the analysis throughout the Opinion Mining technique is being studied with limited category. In this paper, we will analyze smartphone product reviews which were posted on Twitter in multilateral category and compare them, and attempt to make examine on primary factors of the product that

---

* Corresponding author.

J. Sobecki, V. Boonjing, and S. Chittayasothorn (eds.), *Advanced Approaches to Intelligent Information and Database Systems,* Studies in Computational Intelligence 551,
DOI: 10.1007/978-3-319-05503-9_20, © Springer International Publishing Switzerland 2014

influences the users interests by using Opinion mining technique. The reminder of this paper is organized as follows: explanation of existing related researches by using text mining and opinion mining will be described in Section 2. In Section 3, we will explain a method for providing more specified information to users concerning reviews for the certain products on Twitter[1] by analyzing selected sub categories of given product. And we will conduct simple experiments to justify our proposed method in Section 3. Finally, we will conclude this paper with future works in Section 4.

2   Related Works

2.1   Twitter

Social Network Service represents an online platform that facilitates social networks or relations between users for sharing interests, activities, and experiments with others to make and strength online or offline social relationship with their friends. There are several types of SNS: open and enclosed SNS, and Microblog, etc. Twitter is a type of open SNS as well as a Microblog with a different characteristic. There are various studies on Twitter such as a research for developing relationships among the users [3], since there is a big advantage that it is possible to share and spread information with others. There was couple of researches which Twitter was applied: a study for extracting topics by utilizing news Twitter and comparing those with Google Trend [4]. Moreover, there was a research which analyzed Twitter network information spread process by applying diverse indexes measuring information diffusion [5].

2.2   Text Mining

Text mining, as known as text analytics, Knowledge Discovery in Textual database (KDT), document mining, and more, is defined as the extraction of meaningful information on a large scaled text. Text mining is a technique which extracts and refines information on unstructured text data with applying Natural Language Processing; summarization, classification, clustering and feature extraction, etc, are the core study area. Recently, not only these text mining techniques have been applied to the big data analytic area but also, to the diverse research areas consistently. For example, there was a research for extracting context information from Wikipedia[2] documents to provide enlarge Knowledge bases [6]. Also, there was a research based on the text mining approach to expand WordNet[3] which is one of the most popular knowledge bases around world developed by Princeton University. The major contribution of this research is the fact that it is able to reduce semantic distances between noun types of words [7]. Moreover, there was a study

---

[1] http://twitter.com
[2] http://wikipedia.org
[3] http://wordnet.princeton.edu

for extracting interactional information from biomedical text data sets, automatically [8]. In addition, there was a research for classifying web based contents by using text mining and data mining approaches [9]. As we can see in the other researches, the text mining area has been steadily studied by many scientists.

2.3 Opinion Mining

Opinion mining is defined as the analysis on a person's emotional status when facing an issue or events. People emotions or opinions are caused by physical, psychological, sociological, and cultural experiences. Opinion mining which is based on Natural Language Processing is a technique for extracting only personal opinions by comparing the facts and ideas from the given documents or sentences [10]. The extracted result of opinion mining is made good use of referring other people's opinion about specific information what the user is interested in. Opinion mining, a technique which measures strength of the applicable subject based on users' opinions and evaluations from the given documents based on three steps. First, it extracts positive and negative words from the given documents. Second is for machine learning steps based on algorithms such as Naive Bayes, Maximum Entropy (ME) model, and Support Vector Machine (SVM) to construct relationship between evaluation factors and opinions. Third, it extracts the number of positive and negative words for creating the summary by extracting core sentences from given documents. We can conclude that diverse kinds of approaches have been studied for last century and people are focusing on the statistical approach based on Natural Language Processing [11].

3   Twitter Analysis System by Using Opinion Mining

In this Section, we propose a method for analyzing the smartphone product reviews which were posted on Twitter from September, 2012 to October, 2013 by using opinion mining approach. To provide more detailed opinions to users, we select sub categories for representing the given products to six kinds of specifications which are Display, Network, AP, Size, Camera, and Audio. Display category contains information concerning what type of display is and its size. Network category contains information related to whether given smartphone is based on Long Term Evolution (LTE) or 3G. Application Processor (AP) category is information for performance of CPU chip set. Size denotes for the size and weight of given smartphones. Camera is for the information related to the number of pixel of camera and the condition whether it has front camera or not. Figure 1 indicates the proposed method for analyzing smartphone related tweets corresponding to the six sub categories by using opinion mining technique.

First of all, we collect tweets which were containing information for Galaxy S4[4], iPhone 5[5], and Blackberry Q10[6] from September, 2012 to October, 2013 and save to the database monthly by using Twitter Open API.

---

[4] http://en.wikipedia.org/wiki/Galaxy_S4
[5] http://en.wikipedia.org/wiki/IPhone5
[6] http://en.wikipedia.org/wiki/BlackBerry_Q10

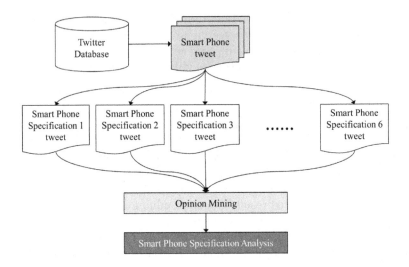

Fig. 1. The process of analysis on smartphone product review on Twitter

The extracted smartphone related tweets are classified into the six categories which are Display, Network, AP, Size, Camera, and Audio. And we go through opinion mining process. Only positive and negative values are extracted by using the opinion mining analysis program named Linguistic Inquiry and Word Count (LIWC)[7] which is a program developed by Pennebaker et al. to measure the degree of more than 80 sentiment categories, including positive and negative status. To normalize the positive and negative status resulting from LIWC, we subtract negative values from the positive values as described in the following equation 1.

$$Tweening = Positive_{(t)} - Negative_{(t)} \qquad (1)$$

where, $Positive_{(t)}$ denotes monthly positive emotion values calculated by LIWC, $Negative_{(t)}$ indicates monthly negative emotions. $Tweening$ represents to normalized Twitter opinion emotional degrees for given tweets.

We can finally examine the users interests corresponding to the specified sub categories from Galaxy S4, iPhone 5, and Blackberry Q10 related tweets.

The following Figure 2 indicates graphs after applying opinion mining process without categorizing monthly tweets via six predefined sub categories information. It was found that although the start of graph shows the highest value on iPhone 5 since it was released to the public in September 2012, but after April 2013, when Galaxy S4 was released to the public, the value of iPhone 5 was decreased. In other words, people interests for the iPhone 5 were decreased during those days. However, in Blackberry Q10's case, it was released on May 2013, but there was almost no change on the graph. The reason can be conjectured that the product could not catch smartphone user's interest.

---

[7] http://www.liwc.net

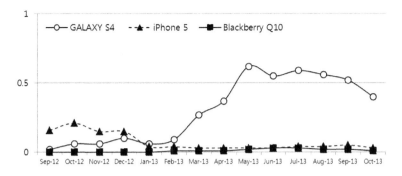

Fig. 2. Monthly opinion mining measurement value of smartphone

Table 1 below is the result of the proposed opinion mining which we considered the six sub categories for mining user opinions from tweets corresponding to Galaxy S4, iPhone 5, and Blackberry Q10 described in Figure 1. Through it, it was found that the primary issue for each smartphone may be caused by size on Galaxy S4, network on iPhone 5, and camera on Blackberry Q10. Figure 3 is a picture using disc graph to have table 1.

Table 1. Results of monthly opinion mining based on six predefined sub categories information

Galaxy S4	2013.04	2013.05	2013.06	2013.07	2013.08	2013.09	2013.10
Display	0.35	0.55	0.79	0.74	0.57	0.9	0.66
Network	0.24	0.35	0.91	0.98	1	0.79	0.68
AP	0.59	1	0.85	0.7	0.58	0.87	0.79
Size	0.37	0.83	0.89	0.84	0.95	0.91	0.84
Camera	0.27	0.72	0.79	0.79	0.98	0.78	0.74
Audio	0.45	0.56	0.98	0.98	0.74	0.84	0.76
iPhone 5	2013.04	2013.05	2013.06	2013.07	2013.08	2013.09	2013.10
Display	1	0.5	0.67	0.72	0.63	0.89	0.56
Network	0.7	0.62	0.59	0.7	0.64	0.63	0.67
AP	0	0	0	0	0	0	0
Size	0.68	0	0	0	0.57	0.69	0
Camera	0.71	0.88	0.57	1	0.48	0.76	0.71
Audio	0.87	1	0.91	0.69	0.73	0.85	0.59
Blackberry Q10	2013.04	2013.05	2013.06	2013.07	2013.08	2013.09	2013.10
Display	0	0	0	0	0	0	0
Network	0	0	0	0	0	0	0
AP	0	0	0	0	0	0	0
Size	0	0	0	0	0	0	0
Camera	0.22	0.48	1	0.85	0.77	0.74	0.36
Audio	0	0	0	0	0	0	0

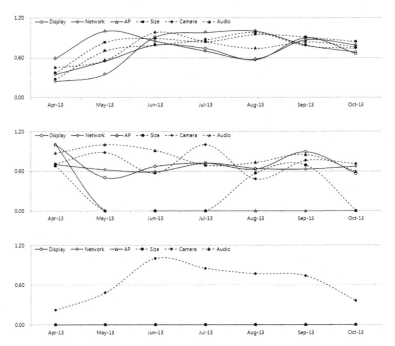

Fig. 3. Graphs for opinion mining results of the six sub-categories Galaxy S4, iPhone 5, and Blackberry Q10 based on table 1

The following Table 2 indicates the average on calculated values after applying the proposed opinion mining method onto the product reviews from Twitter when each smartphone's release date to October 2013. It was found that the specified sub-categories of Galaxy S4 catch overall interest, but iPhone 5 loses interest on Application Processor, and only Camera category has got interests for Blackberry Q10 from the users.

Table 2. Average values for the proposed opinion mining based on the six predefined categories

	Galaxy S4	iPhone 5	Blackberry Q10
Display	0.65	0.71	0.00
Network	0.71	0.65	0.00
AP	0.77	0.00	0.00
Size	0.80	0.28	0.00
Camera	0.74	0.73	0.63
Audio	0.74	0.81	0.00

Figure 4 is a picture using disc graph to have table 2. Based on this, it was easily found that Galaxy S4 catches much interest than iPhone 5 and Blackberry Q10 from the customers. According to the Figure 4, the size and AP information for the iPhone 5 is not satisfied by users when Galaxy S4 does. This is not guarantee that Galaxy S4 is better than iPhone 5. It indicates that Galaxy S4

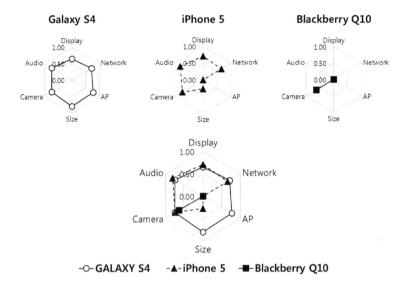

Fig. 4. Graph results values for the proposed opinion mining based on the six predefined categories

has got more interest than iPhone 5 for the size and AP related information. Moreover, Blackberry Q10 only has been interested by user for camera related information due to the fact that it has become to apply the front camera system since Blackberry Q10. Before Blackberry Q10, there were no devices have the front camera. Therefore, although Blackberry Q10 is not likely to be posted on Twitter, people might be happy with its camera performance4.

4 Conclusions and Future works

This paper has proposed the analyzing method for applying specified opinion mining technique, after categorizing smartphone product review to the six predefined sub-categories which are Display, Network, AP, Size, Camera, and Audio conditions. The classified tweets will be calculated those positive and negative score by using LIWC. And we normalized positive and negative emotions to discover hidden patterns in natural language. As a result, it was possible to measure the opinions in details which other previous researches were not dealing with. Because of this, we can conclude that it is possible to determine which kind of information is important to represent a certain device. Moreover, it is able to observe which part of specifications has advantages compared with other products. It can be easily applied to diverse review analytics such as movie opinion mining, travel opinion mining, airplane opinion mining, and more. These diverse opinion mining areas are only focused on the issue whether public has positive or negative responses to given areas. For example, we have not considered the issue that why people are not satisfied for the certain movie. Is there because of the actors? or because of the story? Our method can be applied to enhance monitoring public opinions in diverse areas. In the future, we are planning to develop our own opinion mining tools to overcome the limitation of LIWC. Also, we need to set precise criteria to represent the given product adequately. Finally, we also need to analyze whether those criteria is appropriate for opinion mining or not.

Acknowledgments. This research was supported by Basic Science Research Program through the National Research Foundation of Korea (NRF) funded by the Ministry of Education (No. 2013R1A1A2A10011667).

References

1. Marketresearch on digital media, internet marketing, `http://www.emarketer.com`
2. Kang, J., Kim, C., Yoo, S., Lee, Y., Park, H., Kim, Y.: Recommendation of Acquaintances to Follow in Twitter. In: The 38th Domestic Conference on Korean Institute of Information Scientists and Engineers, Korea, vol. 38(2), pp. 383–386 (2011)
3. Lee, G., Lee, E., Kwak, K., Park, J., Je, R., Kim, J.: A study on charactheristics of Twitter users (compared to Cyworld, Facebook). In: HCI 2011, pp. 1043–1050 (2011)
4. Kim, J., Ko, B., Jeong, H., Kim, P.: A Method for Extracting Topics in News Twitter. 7th International Journal of Software Engineering and Its Applications 7(2), 1–6 (2013)
5. Lee, J., Kim, C.: Analysis of the Information Diffusion Process on Twitter:Effects of Influentials and Hyperlinks. Korean Journal of Journalism & Communication Studies 56(3), 238–265 (2012)
6. Hwang, M., Choi, D., Kim, P.: A Context Information Extraction Method according to Subject for Semantic Text Processing. Journal of Korean Institute of Information Technology 8(11), 197–204 (2010)
7. Hwang, M., Kim, P.: An Enrichment Method on Semantic Relation Network of WordNet. Journal of Korean Institute of Information Technology 7(5), 209–215 (2009)
8. Park, K., Hwang, K.: A Bio-Text Mining System Based on Natural Language Processing. Journal of KISS: Computing Practices 17(4), 205–213 (2011)
9. Choi, Y., Park, S.: Interplay of Text Mining and Data Mining for Classifying Web Contents. Korean Journal of Cognitive Science 13(3), 33–46 (2002)
10. Pang, B., Lee, L.: Opinion Mining and Sentiment Analysis. Foundations and Trends in Information Retrieval 2(1-2), 1–135 (2008)
11. Yang, J., Myung, J., Lee, S.: A Sentiment Classification Method Using Context Information in Product Review Summarization. Journal of KISS: Database 36(4), 254–262 (2009)

Evaluation of the Factors That Affect the Mobile Commerce in Egypt and Its Impact on Customer Satisfaction

Olla Elsaadany[1] and Mona Kadry[2]

[1] Teaching Assistant at AASTMT, Egypt
olla-42@hotmail.com
[2] Vice Dean Graduate School of Business at AASTMT, Egypt
monakadry@hotmail.com

Abstract. M-Commerce is thought to be the next big phase in technology involvement following the E-Commerce era. However, its adoption and level of use is low in Egypt. This paper seeks to find out the factors influencing the Egyptian customers' behavior towards using mobile commerce in Egypt. Analyze the effective way to push the customer's intention toward M-commerce usage, and to measure their satisfaction after using the m-commerce. This study presents and tests the intention to adopt M-Commerce technology framework in Egypt through the critical antecedent factors including level of E-Commerce usage, perceived ease of use, perceived usefulness, the services cost, media, and attitude towards the M-commerce.

Keywords: Mobile commerce, Adoption level, M-commerce services, Service quality, Customer satisfaction.

1 Introduction

Recent mobile revolution offered new opportunities. The most important opportunity is the ability to buy or sell products or services using a small sized mobile device, such as a mobile phone, a Personal Digital Assistant (PDA), a smart phone, or other emerging mobile equipment. M-commerce contributes the potential to deliver the most of what the internet can offer plus the advantage of mobility. M-commerce is difficult to define and can be interpreted in a variety of ways. This is because M-commerce is a fairly new phenomenon and several definitions of it exist. M-commerce is defined as the use of wireless device to communicate, interact, and transact via high speed communication to the internet.

Despite the rapidly growing number of mobile phone users in Egypt, M-commerce is a relatively new phenomenon compared to other markets in Europe, the U.S. and Asia Pacific. The M-commerce service providers in Egypt lack a clear direction towards understanding the factors affecting the adoption of M-commerce. Therefore, the objective of this study is to explore the factors that influence the adoption of M-commerce services among mobile users in Egypt and its impact on customer satisfaction to determine the effective way to push the customers' intention toward M-commerce usage.

J. Sobecki, V. Boonjing, and S. Chittayasothorn (eds.), *Advanced Approaches to Intelligent Information and Database Systems*, Studies in Computational Intelligence 551,
DOI: 10.1007/978-3-319-05503-9_21, © Springer International Publishing Switzerland 2014

2   Factors Affecting the Adoption of M-commerce Services and Customer Satisfaction

Adoption is an individual's decision to become a regular user of a product or a service. This review explores the factors that influence the adoption of M-commerce service among working employed users. Europe leads the mobile market and the M-commerce. A number of factors have contributed to Europe's position in leading the M-commerce market. These factors include favorable pricing structure, increasing competition in greater quality of service, and declining costs of network operators (Islam, Md. A, Khan, M.A., Ramayah, T., Hossain, M. M (2011)).

Safeena, Hundewale, and Kamani (2011) considered five factors; perceived usefulness, perceived ease of use, subjective norm, consumer awareness about mobile banking and perceived risks associated with mobile banking. This study also points out that these factors have a strong and positive effect on customers to accept mobile banking system. Alsheikh and Bojei (2012) examined the factors affecting the value held by bank customers toward the use of mobile banking services. This study may enable banks to develop a marketing strategic plan based on perceived value from the customer's point of view. This paper presented a new model as an attempt to better understanding mobile banking usage based on perception value using benefit factors (performance expectancy and effort expectancy) in conjunction with sacrifice factors (cost and risk).

3   Research Model

The Technology Acceptance Model (TAM) is one of the most influential extensions of Ajzen and Fishbern's Theory of Reasoned Actions (TRA) in the information system literature. TAM used the generic Fishbein and Ajzen's TRA model Fishbein and Ajzen (1975) to the particular domain of user acceptance of computer technology. There are a lot of factors that affect success in mobile commerce and could be summarized into the component of technology acceptance model (TAM) which are; Perceived usefulness, Perceived ease of use, Attitude (toward using Mobile commerce),

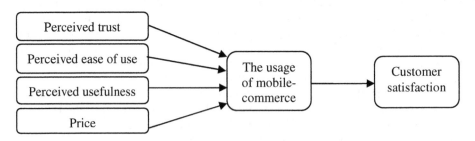

Fig. 1. Conceptual Framework; Source: own

Behavior Intention (toward using Mobile commerce) and Perceived Trust. This research examined the question of what factors lessen or increase Egyptians' intention or satisfaction towards M-commerce usage. It tried to adopt and use the component of technology acceptance model and try to put perceived cost, and media factors like main factors for decisions of customers in using M-commerce services to help in answering this question. Also how the proposed conceptual research model explains the variances of consumer's M-commerce acceptance intention or satisfaction.

4   Research Hypotheses

In terms of the micro and macro factors affecting the mobile commerce; the research develops number of hypotheses as follows.

H_1: There is a significant relationship between; Perceived usefulness, Perceived ease of use, Attitude towards the M-commerce, Media, and the Cost factors and usage of mobile commerce. And what is the effect of these factors on the usage of mobile commerce in Egypt.

H_2: There is a significant relationship between; usage of mobile commerce, the quality of the mobile services and customer satisfaction. And what is the effect of these factors on the Egyptian customer satisfaction.

In order to answer the research question data results for the questionnaire survey has been analyzed statistically to test if the hypotheses can be substantiated of the appropriate research model and hypotheses.

4.1   Samples and Demographics Statistics

In this study, the questionnaire was randomly distributed. It was administered to the customers of mobile companies (Vodafone, Mobinil, Etisalat) in Egypt this year (2013), based on all different sectors or organizations in Egypt that focus on the mobile services. Responses to the survey questions are presented for the whole sample and for sub-samples according to demographic and other researcher statistics. Most of the questions are based on a 5 likert scale, with 3 as the neutral score. The results the demographics variables of the sample are as follows: The gender distribution of the respondents is 66.5% female respondents and 33.5% male respondents. The dominant age group of the respondents was from 20 to 34 years old (77%), followed by 35 to 45 years old (13%), whereas the age 56-65 years old made up the smallest group, representing 3.9% of the respondents. With regard to respondents' individual income, the largest groups includes those with a monthly income of below 2000 and 3000-4000 the same percent (32.2%), whereas 3000-4000 is the smallest group (9.6%) the group of respondents have civil servant (28.3%), followed by student (10%l), businessman/woman (7.8%), whereas the other groups have the largest percent (53.9%). Also noticed that the dominant group of the respondents used who Vodafone services is (58.3%).

The dominant group of the respondents believe that the Perceived trust is extremely important for using new mobile services (47.4%), followed by the Cost (43.5%), and Perceived ease of use (40.9%) whereas the largest group of the respondents believe that the Perceived usefulness is important (57.4%).

4.2   Reliability of Measurement

Testing goodness of data is testing the reliability of the measures. The reliability of the factors was checked using Cronbach's alpha.

Table 1. Reliability of usage patterns

Usage Patterns dimensions	No of Items	Items Mean	Cronbach'sAlpha
Cost (expensive)	2	3.869	0.710
Perceived ease of use	5	3.895	0.907
Perceived usefulness	5	3.91	0. 946
Attitude towards using M- C	5	3.847	0.988
Media	3	3.55	0.787

Table 2. Reliability of mobile services dimensions, quality and satisfaction

Constructs	No of Items	Items Mean	Cronbach's Alpha
Existing mobile services	10	3.60	0.816
New mobile services	9	2.92	0.872
Customer Satisfaction	3	2.97	0.79
Quality Variables	12	3.843	0.938

As shown from the above tables1 and 2, all scales have Cronbach's alpha values are higher than 0.7 Thus, the results of mobile services dimensions, Perceived ease of use, Perceived usefulness, Attitude, and Media factor, quality dimensions and satisfaction variables are acceptable. The mean values of existing and new mobile services are 3.6, and 2.9 respectively, which means that most respondents use the service occasionally

4.3   Examining Relationship between Usage Patterns Factors and M-Commerce Usage

By using the chi-square test the author finds that there is a significant correlation between the using of the M-Services and the Usage Patterns, including; Perceived Usefulness (Pu), Perceived Ease Of Use (PEOU), Attitude (AT) towards the Mobile Services, and the Cost (C) of the services. That is because all possible agreement levels or combinations are reported to three categories Low Pattern, Medium Pattern, and High Pattern. Then the significant

Table 3. Chi-square tests

Factors	Chi-Square Tests	P-Value
Perceived Ease of Use	49.15	0.000
Perceived Usefulness	83.33	0.000
Attitude	33.15	0.000
Cost	18.418	0.001
Media	40.122	0.000

As can reported from the Chi-square tests there is a significant dependence between the Usage Mobile Service and the Usage patterns, as P-Values are less than the significant level (0.06). So the chi-square test shows that there is a significant relation or dependence between the Usage Mobile Service and all the Usage Patterns dimensions. The following correlation matrix examines how the analysis dimensions are correlated with each other.

Table 4. Correlation matrix

Factors	Usage M-Service	PU	PEOU	AT	C	Media
Usage M-Service	1.000					
PU	$.566^{**}$	1.000				
PEOU	$.433^{**}$	$.601^{**}$	1.000			
AT	$.402^{**}$	$.516^{**}$	$.419^{**}$	1.000		
C	$-.247^{**}$	$.430^{**}$	$.496^{**}$	$.528^{**}$	1.000	
Media	$.303^{**}$	$-.032^{**}$	$.439^{**}$	$.288^{**}$	$.178^{**}$	1
*and ** means that the correlation coefficient is significant at 0.05 and 0.001						

As shown from table 4, there is a significant strong correlation between the respondent mean Usage Mobile Services scores and the mean Usage Patterns dimensions, while the correlation coefficient take the values greater than (0.3), which means that the Mobile Services Usage will increase with the increasing of PU, PEOU, AT, and Media dimensions. The correlation between the mean Usage Mobile Services scores and the mean Cost of the mobile services is a significant negative weak correlation which take the value (-0.24). This means that the M-Services usage maybe decrease with the increasing of the service cost. This implies that, as expected Mobile Services Usage is correlated to usage patterns dimensions.

4.4   Examining Relationship between M-Commerce Usage, Services Quality and the Customer Satisfaction

In order to answer the questions whether there are significant relationships between usages of mobile commerce, service quality and the customer satisfaction, the independence using the chi-square test should be tested first. Then there is significant correlation between the usages of mobile commerce, service quality and the customer satisfaction.

Table 5. Chi-square tests

Factors	Chi-Square Tests	P-Value
M-Service Usage Levels	43.4	0.000
Quality Dimensions	43.4	0.000

As the P-Value (0.000) is less than the significant level (0.06) so this indicates that there is a significant relation or dependence between the usage M-services levels, the service quality dimensions and the customer satisfaction.

Table 6. Correlation coefficients

Spearman Correlation	Customer Satisfaction
Customer Satisfaction	1
Existing M-Services	$.367^*$
New M-Services	$-.145^*$
Quality dimension 1	$.497^{**}$
Quality dimension 2	$.650^{**}$
*, ** Correlation is significant at the 0.05 and 0.01level (2-tailed).	

As shown from table 6, there is a significant positive correlation between the Customer Satisfaction scores and the Existing M-Services, and negative weak correlation with the New M-Services scores, where the correlation coefficient take the values (0.36), (-0.14) respectively. There is a significant strong positive correlation between the Customer Satisfaction scores and the Quality dimensions, where the correlation coefficients take the values (0.49) and (0.65).

4.5   Factor Analysis

Factor analysis can identify the structure of a set of variable as well as provide a process for data reduction. A principal component for analysis associate with varimax rotation is used in the procedure. Also we determine the mobile service factors and the factors of service quality to estimate the customer satisfaction factor.

4.5.1 Usage Patterns Factor Analysis
The factor analysis underlines 4 factors where; Perceived Ease of use and Perceived Usefulness (Factor 1) contained five attributes of Perceived Usefulness, and five attributes of Perceived Ease of Use. Attitude towards using M-commerce (Factor 2) contained three attributes and Media (Factor 3) contained three attributes and Cost of the service (Factor 4) contained two attributes. The four factors explained 84.3% of the variance in the data.

Table 7. Total variance explained

Component	Initial Eigen values			Rotation Sums of Squared Loadings		
	Total	%of Variance	Cumulative %	Total	%of Variance	Cumulative %
1	12.019	48.077	48.077	8.371	33.484	33.484
2	4.659	18.634	66.712	7.965	31.859	65.343
3	2.814	11.256	77.967	2.910	11.639	76.982
4	1.586	6.343	84.310	1.832	7.328	84.310
Extraction Method: Principal Component Analysis.						

4.5.2. Mobile Services Factor Analysis

The Mobile Services could be determined by two dimensions or factors; Existing Mobile Services, and New Mobile Services.

Table 8. Total variance explained

Component	Initial Eigen values			Rotation Sums of Squared Loadings		
	Total	%of Variance	Cumulative%	Total	%of Variance	Cumulative%
1	8.236	48.448	48.448	7.374	43.377	43.377
2	3.777	22.216	70.664	4.639	27.286	70.664
Extraction Method: Principal Component Analysis.						

The factor analysis underlines 2 factors can explain the usage of mobile service variables (questions) as; Existing mobile service (Factor 1) contained eleven attributes. The second component is the New M-service (Factor 2) contained seven attributes. The two factors can explain 70.66 % of the variance in the data.

4.5.3. Customer Satisfaction Factor Analysis

Customer satisfaction variables denoted by one Factor contained three attributes. The customer satisfaction factor can explain 81.35% of the variance in the data.

Table 9. Total variance explained

Component	Initial Eigenvalues			Extraction Sums of Squared Loadings		
	Total	%of Variance	Cumulative%	Total	%of Variance	Cumulative%
1	2.441	81.350	81.350	2.441	81.350	81.350
Extraction Method: Principal Component Analysis.						

4.5.4. Quality Dimensions Factor Analysis

The quality of the mobile service denoted by the following dimensions; Reliability dimension determined by two questions Responsiveness dimension including two questions, Empathy dimension determined by three questions, Assurance dimension determined by three questions and Technical quality dimension determined by two questions. All the quality of the mobile service dimensions denoted by two factors, the first factor including Empathy, Assurance, and Technical Quality dimensions. The second factor including Reliability and Responsiveness dimensions. The two quality factors can explain 70.4 % of the variance in the data

Table 10. Total variance explained

Component	Initial Eigenvalues			Rotation Sums of Squared Loadings		
	Total	%of Variance	Cumulative%	Total	%of Variance	Cumulative %
1	7.263	60.527	60.527	4.758	39.653	39.653
2	1.194	9.950	70.477	3.699	30.825	70.477
Extraction Method: Principal Component Analysis.						

4.6   Research Regression Models

Multiple regression analysis is the study of how a dependent variable is related to two or more independent variables. They are used to test the hypotheses in this study.

4.6.1. Explaining Usage Patterns Factors Toward Using M-Services

In the first regression model, using M-commerce is entered as the dependent variable, while Perceived Usefulness (PU) and Perceived Ease of Use (PEOU) Attitude towards the mobile service (AT), Media and, Cost (C) factors which are determined by the factor analysis, and the customers characteristics variables which are; gender, age categories, Occupation, and Income levels, are simultaneously entered as independent variables. The following multiple regression will be used to test the first hypothesis denoted by H1.

Table 11a. Usage m-commerce coefficients model

Variables and Factors	Unstandardized Coefficients		Standardized Coefficients	T	Sig.
	B	Std. Error	Beta		
(Constant)	-1.759	.236		-7.451	.000
PU & PEOU	.554	.043	.547	12.765	.000
Attitude (AT)	.453	.043	.451	3.543	.000
Media	.265	.046	.363	3.566	.000
Cost (C)	-.046	.048	-.045	-.952	.342
Dependent Variable: Usage M-Services					

Table 11b. Usage m-commerce coefficients model

Variables and Factors	Unstandardized Coefficients		Standardized Coefficients	T	Sig.
	B	Std. Error	Beta		
Age 24-34	1.693	.215	.519	7.876	.000
Age 35-45	1.377	.255	.468	5.399	.000
Age 46-55	2.071	.290	.241	7.145	.000
Civil Servant	-.359	.144	-.163	-2.495	.013
Student	-.228	.147	-.069	-1.551	.122
Business Man-Woman	.081	.163	.022	.499	.618
Income 2000-3000	.393	.125	.185	3.136	.002
Income 3000-4000	.445	.171	.132	2.599	.010
Income above 4000	.429	.128	.190	3.342	.001
Adjusted R Square	0.826				
Model Sig	0.000				
Dependent Variable: Usage M-Services					

The results, presented in table 11a and table 11b, show support for H1, as AT, PU, PEOU emerged as significant predictors of Usage M-Services as the sig (P-Value) equal 0.000 which is less than the significant level Alpha (0.06), also from the value of the Beta coefficient we find that the Perceived Usefulness(PU), Perceived Ease of Use (PEOU), Attitude, and Media factors affect the mobile service by (0.54), (0.45) and (0.36) respectively, but the Cost (C) factor has insignificant effect on the Usage M-Services factor as it's P-Value equal (0.34) which is greater than the significant level Alpha (0.06). Also all the characteristics variables have a significant and positive effect on the Usage M-Services except the Occupation categories (Student and Business man-woman). The value of Adjusted R-square (0.826) indicated that 82.6 % of the variance in Usage M-Services factor is explained by the AT, PU, PEOU and Customer Age and Income.

4.6.2. Explaining Factors Affected the Customer Satisfaction
This regression model examines how the Usage M-Services factors, Quality factors and Cost factor affected the Customer Satisfaction which is entered as the dependent variable. Note that the factors are determined by the factor analysis. The following multiple regression will be used to test H2

Table 12. Customer satisfaction coefficients model

Model; Customer Satisfaction	Unstandardized Coefficients		Standardized Coefficients	t	Sig.
	B	Std. Error	Beta		
(Constant)	-.0004	.048		.000	1.000
Quality Factor1	.401	.067	.401	6.009	.000
Quality Factor2	-.123	.050	-.123	-2.449	.015
Existing M- Service	.346	.056	.346	6.142	.000
New M- Service	-.141	.055	-.141	-2.576	.011
Cost Factor	-.457	.070	-.457	-6.517	.000
Adjusted R Square	0.63				
Model Sig	0.000				
Dependent Variable: Customer Satisfaction					

The results presented in table 12, show support for Hypotheses, as the Usage M-Services two factors, Quality two factors and Cost factor affected the Customer Satisfaction significantly as the sig (P-Value) equal 0.000 which is less than the significant level Alpha (0.06), also from the value of the Standardized Beta coefficient we find that the Quality Factor 1 and the Existing M-Services factor affected the Customer Satisfaction positively which means that increasing this factors lead to increasing the customer satisfaction by (0.40), (0.34) respectively . Also we can find that the Quality Factor 2 , New M-Services factor and the Cost (C) factor have a significant negative effect on the Customer Satisfaction factor, which means that increasing the Cost, will lead to decreasing the customer satisfaction by (-0.45), and the mobile customers not satisfied about New mobile services, and quality factor2 by (-0.14), (-0.12) respectively. The value of Adjusted R-square (0.63) indicated that 63 % of the variance in Customer Satisfaction factor is explained by the Usage M-Services two factors, Quality two factors and Cost factor affected.

5   Discussions and Conclusion

To explore factors that influence the satisfaction of users to adopt mobile commerce, the simple model based on technology acceptance model (TAM) was developed and measured. The research results significantly verified the hypothesis between perceived usefulness, and perceived ease of use, service cost, media factor, and attitude towards the M-commerce and the service quality dimensions which affected the customer satisfaction.

First, this study successfully applied the TAM. Consistent with previous studies, perceived usefulness and perceived ease of use were found to be significant antecedents to use mobile commerce. Therefore, the managers should design the M-commerce interface. Also, this study found a significant direct relationship between Attitude towards the mobile services and M-commerce.

By identifying the major drivers for M-commerce adoption, the results of this paper can help managers to prioritize their M-commerce initiatives and to allocate resources accordingly. For instance, potential adopters of M-commerce are highly sensitive to the issues of quality. Accordingly, the major quality dimension may include. Also, this study shows that, recognizing both technological and quality issues are important in increasing citizen's satisfaction to use this service.

M-commerce provider should first develop attractive media mechanisms through advertising for citizens in order to attract and inform users about a new M-commerce services.

Moreover, Perceived ease of Use and Usefulness of M-commerce emerges as important issues in attracting new users and should be carefully designed in terms of users' requirements to reflect PU of this service. Without an original consideration from cost aspect, a well-designed on-line tax with significant PU will not well perform in attracting novice users.

Furthermore, training customers how to use M- commerce would make them aware of the valuable information and advantages exiting in using M-commerce and shows them how to access that information quickly and efficiently.

The Cost factor affects the customer's satisfaction significantly and it has a significant effect to use M- commerce. However perceived the usefulness, ease of use, and having a positive attitude toward using the M-commerce, are important factors, it does not mean that the customers will use this service. This might be explained within the Egyptians context. It might be explained due to one of the characteristics of Egyptians which are careful and cautious. Even when they know that M- commerce is very useful, they still need time to find more information, as well as need time to see the evaluation of the ones who are using M-commerce. They will not intend to use M-commerce until they have enough information and opinions assuring that using M-commerce is not harmful.

6   Suggestions and Future Studies

Although the findings strongly support the proposed model, they still need improvement and further investigation. As the current research focuses on M-commerce that may involve different factors in the M-commerce context, also organizational theories may be necessary to be incorporated to understand adoption behavior. Moreover, as explained above, the users in Egypt might refer to the others' opinions; therefore, the further research might study using different methodologies, such as focus groups and interviews to examine the applicability of the research model adopted in this study.

Future research should also follow a longitudinal approach and investigate the relationship between intention and actual behavior. Furthermore, in future, researches can try to put "Self Efficiency" like a main factors for decisions of customers in using M-commerce services, as when people feel they are capable of using M-commerce, they will tend to prefer and even enjoy behaviors if they feel they can. Finally, measures of perceived self-efficacy may develop and represent specific instruments that can be used to assess perceived knowledge and financial resources.

References

1. Alsheikh, L., Bojei, J.T.: Customer's Perceived Value to Use Mobile Banking Service. In: International Conference on Management, Behavioral Sciences and Economics Issues (ICMBSE 2012), Penang. Malaysia, pp. 178–182 (2012)
2. Angsana, A.T.: International Diffusion of Digital M-Technology: A Coupled-hazard Approach. Thesis. University of Minnesota, Department of Information and Decision Sciences, Carlson School of Management (2003)
3. Chew, A.A.: The Adoption of M-Commerce in the United States, Thesis, California State University, Long Beach, CA 90840 (2006)
4. Deng, Z., Lu, Y., Wei, K.K., Zhang, J.: Understanding Customer Satisfaction and Loyalty: An empirical study of mobile instant messages in China. International Journal of Information Management 30, 289–300 (2010)
5. Fishbein, M., Ajzen, I.: Belief, Attitude, Intention and Behavior: An introduction to Theory and Behavior. Addison-Wesley, Reading (1975)
6. Islam, M.A., Khan, M.A., Ramayah, T., Hossain, M.M.: The Adoption of Mobile Commerce Service among Employed Mobile Phone Users in Bangladesh: Self-efficacy as A Moderator. International Business Research 4, 80–89 (2011)
7. Lee, Kassim: Service Quality and Mobile Commerce Customer Satisfaction in Malaysia. International Journal of Network and Mobile Technologies 3, 12–23 (2012) ISSN 2229-9114 Electronic Version
8. Uddin, M.B., Akhter, B.: Customer Satisfaction in Mobile Phone Services in Bangladesh: A survey Research. Management and Marketing X, 20–36 (2012)
9. Safeena, R., Hundewale, N., Kamani, A.: Customer's Adoption of Mobile-Commerce A Study on Emerging Economy. International Journal of e-Education, e-Business, e-Management and e-Learning 1, 228–233 (2011)
10. Suki, N.M.: A structural model of customer satisfaction and trust in vendors involved in mobile commerce. Int. Journal of Business Science and Applied Management 6, 17–30 (2011)

An Agent-Based Modeling
for an Enterprise Resource Planning (ERP)

Nadjib Mesbahi, Okba Kazar,
Merouane Zoubeidi, and Saber Benharzallah

Intelligent Computer Science Laboratory, Computer Science Department,
University Mohamed Khider of Biskra, 07000 Algeria
{najib.mes,mzoubeidi01}@gmail.com,
{kazarokba,sbharz}@yahoo.fr

Abstract. Today the enterprise resource planning (ERP) became the cornerstone of information systems companies. This tool allows a uniform and consistent management information system of the company, especially in the management of the production chain to the sale of a product. In addition, the multi-Agents system (MAS) is a field of very active research. It has become a new modeling approach to solve complex problems. It is a branch that is interested in the collective behaviors resulting from the interactions of several autonomous and flexible entities called agents. The work we propose in this article aims to implement an agent modeling of the ERP thanks to techniques proposed by the MAS. In this article, the JADE platform based on JAVA is the most appropriate, since it is well adapted with the cognitive agents and at the same time, it provides a complete set of services and agents conform to the FIPA specifications.

Keywords: Enterprise Resource Planning (ERP), Multi-Agents system (MAS), JADE, FIPA.

1 Introduction

The companies' access to new technologies, Internet in particular, tends to alter the communication between the various actors in the business. Particularly between the company and its customers (Business to consumer, B2C), the inner workings of the business (Business To Employees, B2E) and business relationship with its various partners and suppliers (Business to Business, B2B). We also mean by "e-Business", the integration within the enterprise of tools based on information technology and communication, it's called the Enterprise Resource Planning (ERP). This software [2] helps to manage the whole process within a company by integrating all the functions such as human resources management, accounting and financial management, decision support, but also sales, distribution, procurement, electronic commerce.

Open systems such as resource planning systems are very large; whose behavior changes over time and very often undergoes significant changes due to internal and external information. The modeling of complex systems requires a thorough study

J. Sobecki, V. Boonjing, and S. Chittayasothorn (eds.), *Advanced Approaches to Intelligent Information and Database Systems*, Studies in Computational Intelligence 551,
DOI: 10.1007/978-3-319-05503-9_22, © Springer International Publishing Switzerland 2014

and analysis of the functional processes before reaching the appropriate modeling. In the case of ERP (Enterprise Resource Planning), the environment is changing continuously due to changes of on internal processes and corporate functions, and external variations of the various relations with customers, suppliers, Banks ... etc.

In addition, the multi-agent modeling has become a new modeling approach to solve problems. These models represent individual actions, interactions between agents and their consequences of these interactions on the dynamics mode of the system. In this context, we propose an agent based approach for the modeling of an ERP capable of introducing the concepts of MAS on the corporate behavior resulting from interactions of its own components between themselves and with the outside world.

2   Enterprise Resource Planning (ERP)

2.1   Concepts and Definitions of ERP

The ERP is a software package incorporating the main functions needed to manage the flows and procedures of the company (accounting and finance, logistics, payroll and human resources....etc). [3]

Moreover, the definition proposed by Willis-Brown seems to be the most complete, "ERP is an integrated system that allows the company to standardize its information system to link and automate its core processes. It provides employees with the information needed to direct and control the core activities of the company from the supply chain, to the product delivery to the customer. Employees enter the information just once, which is then made available to all systems of the company. [6]

The other principle that characterizes an ERP is the use of a so-called workflow engine (Fig. 1) and which allow, the spread of date, which has been recorded on IS, the modules they need it, according to a predefined programming. [2]

Fig. 1. Organization of an ERP

2.2   Characteristics of an ERP

Summarizes all the characteristics of an ERP as follows [13]:

1. Effective management of several domains of the company with integrated modules;
2. Existence of a single data reference;
3. Rapid adaptation to the rules (professional, legal or outcome of the internal organization of the company);
4. Uniqueness of directors of the application subsystem (applications);
5. Standardization of man-machine interfaces (same screens, same buttons, etc.);
6. Existence of development or customization tools of additional applications.

2.3   Modular Architecture of an ERP

The founding principle of an ERP is to build applications for the various functions mentioned above in a modular way knowing that these modules are independent, and at the same time, share a single and common database and common. [4]

Here is an example of a modular architecture that seeks to represent all the ERP (Fig.2):

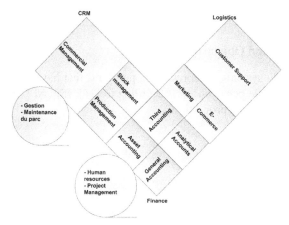

Fig. 2. Modular **architecture ERP [4]**

3   Multi-Agent Systems (MAS)

3.1   The Concept of Agent

An agent is an autonomous entity, real or abstract, which is capable of acting on itself and its environment, which, in a multi-agent universe, can communicate with other agents, and whose behavior is the result of its observations, knowledge and interactions with other agents. [9]

3.2   Multi-Agent Systems (MAS)

A Multi-Agent System (MAS) is a community of autonomous agents working together, sometimes complex ways of cooperation, conflict, competition, in order to achieve a goal: solving a problem, the establishment of a diagnosis. [9]

3.3   The Elements of a M.A.S

Ferber in [9] proposes a general description of the structure of a multi-agent system. The MAS is composed of the following:

- Environment, i.e a domain usually with a metric.
- A set of objects, meaning it is possible to associate each object, at one point a position in the environment.
- A set of agents representing entities active in the system.
- A set of relationships between objects (and therefore agents) between them.
- A set of operations enabling agents to collect, produce, consume, transform, and manipulate objects.
- Operators to represent the application of these operations and the relations of the world to this attempt of change.

3.4   Advantages and Objectives of M.A.S

The MAS are ideals systems to representing problems that have multiple methods of resolution, and multiple perspectives. According to [11], [5] and [10] MAS approach is justified by the following properties:

1. The modularity.
2. The speed with the parallelism.
3. Reliability due to redundancy.
4. Symbolic Processing in terms of knowledge.
5. Reuse and portability.
6. The involvement of sophisticated patterns of interaction (cooperation, coordination, negotiation).
    The two major objectives of research in the field of SMA are [12]:
1. Theoretical and experimental analysis of mechanisms.
2. The resolution of distributed programs.

4   Choice Justification

Today we are witnessing a rapprochement between the MAS and ERP. This reconciliation manifests itself in different directions. The main ones are:

1. Autonomy: Today, most applications require distribution of tasks between autonomous entities (or semi-autonomous) to achieve their objectives in an optimal way. Since conventional approaches are generally monolithic and their concept of

intelligence is centralized, the current applications are established based on multi-agent system.

2. Share-ability: the Multi-Agents System focuses on the model of the agent whose characteristics are: cooperation, coordination and communication. This allows you to share the database (ERP Data Base) between the different entities (Agents). Therefore we get a Sharing ability of data for the ERP.

3. Modularity: one of the most important characteristics of a multi-agent system is its ability to address issues, in a modular fashion. The basic architecture of ERP is based on this ability of managing several domains, each are with its own activities.

4. Cooperation and communication: the ERP uses a predefined system (workflow). It can be modeled by communication and cooperation models provided by the Multi Agents systems.

5. Learning: the Multi-Agents System provides agents (Ex: interface, purchase Agent ....etc) that allow recognition the of users behavior in each domains. It indicates that ERP learns to recognize users' behavior in their fields (payroll, accounting, inventory management ...).

6. Interoperability: Interoperability is well applied in the ERP systems (ERP) whose interfaces are fully identical, while sharing a single and common database.

7. The resolution of complex problems: with Multi Agents systems we can solve complex problems like the case of ERP that manages many users, entities, sometimes geographically dispersed.

8. Integration of incomplete expertise: ERP manages several domains each with its activities, which can be modeled with the agent approach that integrates incomplete expertise.

5   Approach of Multi-Agent ERP Proposed

Our approach is based on the agent model. It applies the concepts of Multi-Agents System to model an ERP. We describe more precisely the different components (agents) of our system and their interactions, showing the relationship between the various agents. In Fig. 3 the approach of multi-agent ERP is in three levels:

The first is the agent user interface that acts as an interface between the system and external agents (supplier, customer, bank ... etc). It allows you to retrieve messages from the various agents of the system and transmit them to external agents. It is also responsible for the reverse, to retrieve the messages sent by the various external agents and transmit them to agents of the system.

The second level represents the domain agents (Agent domain Purchase / supply, domain Agent Sales, domain Agent Accounting / Finance ... etc.). To ensure communication between different domains as well as the communication with the user interface agent (UIA), this allows the establishment of direct interaction with other domain agents to achieve a common task.

The third level represents service agents in each domain; it allows the achievement of well-defined functions (inventory, purchasing, accounting .... etc).

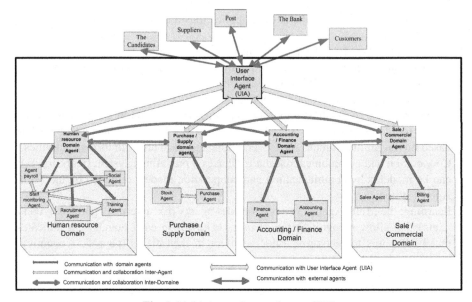

Fig. 3. Multi-Agent Approach to an ERP

For the communication model we have chosen the communication by sending messages. It allows agents to share the burden of a problem by subdividing it into sub-problems. The multi-agent systems based on communication by sending messages are characterized by the fact that each agent possesses its own and local representation of the environment surrounding it. Each agent will interrogate other agents in this environment or send information about its own perception of things. The communication by sending messages is done either in point-to-point mode, or in diffusion. [7]

In our proposal we choose the point to point mode for the transmission of information because the type of agent used is the cognitive agent. The agent issuing the message must know the precise address of the agent (s) recipients.

6   Agents and Their Roles

6.1   User Interface Agent (UIA)

This agent ensures complete interface between the system and the outside. It retrieves the query made by the external staff (customer, supplier, Bank etc ....). The agent does not represent the user to the system, but it is also the entry point for external queries to all domains agents. It can also retrieve messages sent by the various agents of the system and transmit them to external agents.

6.2   Purchase / Supply Domain Agents

6.2.1. Purchase / Supply Domain Agent

It is a principal agent that enables Purchase / Supply domain agents to communicate and collaborate with other agents of the system. It is also mandated to retrieve and respond to queries of external agents through the user agent.

6.2.2. Purchase / Supply Agent

The role of this agent is to manage purchases of the company from the purchase application to receipt and accounting.

6.2.3. Stock Agent

The agent is used to store and manage all goods or items accumulated in anticipation of future use ensure the users supply according to their needs.

6.3   Accounting / Finance Domain Agents

6.3.1. Accounting / Finance Domain Agent

It is a principal agent that enables Accounting / Finance domain agent to communicate and collaborate with other agents of the system. It is also mandated to retrieve and respond to queries of external agents through the user agent.

6.3.2. Accounting Agent

The agent can save in accounting accounts, the flows of the company (the operations of other agents in the business: Purchase Agent, Sale, Stock ... etc.) to determine the outcome of the year (profit or losses) and the present situation of the company.

6.3.3. Finance Agent

The agent determines the current financial status of the company to assist policy makers in the important decisions concerning the current and future forecasting.

6.4   Sale / Commercial Domain Agents

6.4.1. Billing Agent

The role of this agent is the preparation of bills to show customers the details of the price to pay.

6.4.2. Sales Agent

The agent presents the commercial aspects of the company with its customers and the market. It provides all major commercial operations at the company.

7   Functional Architecture of the System Agents

Among the different agents, we will only choose the user interface agent to represent her functional architecture.

The agent acts as a complete interface between the system agents and the outside. It includes a module for external communication, a user database, a processing module and a module for internal communication.

1. External communication module: used to retrieve the queries of external agents and send them to agents of the system.
2. User Database: contains the beliefs of the user interface agent on other agents and on external agents (supplier, customer, banking, etc ....).
3. Processing Module: allows the delivery of messages to the agents in question.
4. Internal communication module: provides interaction with the various domain agents.

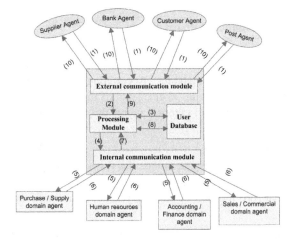

Fig. 4. Functional Architecture: User Interface Agent

The details of the operation are as follows:
1. The external agent sends the query to the system;
2. The external communication module retrieves the query and returns it to the processing module;
3. The processing module in turn analyzes and verifies the query with the user database to send it to the concerned domain agents;
4. the processing module sends the query to the module of internal communication;
5. In turn the internal communication module refers the query to the concerned domain agent (eg Purchase / Supply domain agent).
6. The domain agent retrieves the response of the query and sends it to the internal communication module;

7. The internal communication module in turn transfers the message to the processing module.
8. The processing module analyses and checks the answer with the user database to determine the concerned external agent;
9. The processing module sends the message to the external communication module;
10. The external communications module returns the message to the concerned external agent.

8 Implementation

To implement the proposed architecture, it is more convenient to use the JADE (Java Agent DEvelopement framework) platform while providing a complete set of services and agents conform to FIPA (Foundation for Intelligent Physical Agents) specifications. The JADE platform is fully implemented in JAVA, and meets the FIPA specifications.

JADE tries to optimize the performance of a distributed agent system. The JADE agent platform includes all required components that control an AMS (Agents Management System). These components are Canal of communication between agents (ACL) (Agent Communication Language), the Agents Management System (AMS) and the Directory Facilitator (DF). Any communication between agents is performed by FIPA-ACL messages. These three agents are automatically created and activated when the platform is activated. [15]

9 Conclusion and Perpectives

In this article we propose a Multi-Agent approach for the modeling of an enterprise resource planning (ERP), while justifying the choice and giving the general architecture and function of the proposed system.

Finally, to implement a multi-agent system it is necessary to use a multi-agent platform. However, in the context of this paper, the JADE platform based on Java seems to be most appropriate, because the latter is the most suitable to the cognitive agents, it also provides a complete set of services and agents conform to FIPA specifications.

Our approach is an open approach. Our work is not limited to this level and thus some perspectives come in sight. We plan to enrich and expand the approach by additional features such as the integration of other modules, such as E-Commerce Module, which offers businesses the opportunity to seek the best products and best suppliers via the Internet while reducing the time needed for research.

In addition, we want to use our approach in an environment that has heterogeneous systems, in this case it is necessary to solve the problem of interoperability between different systems.

References

1. Anne, N.: Les systèmes multi-agents, Maîtrise d'Informatique MI6, Université de Paris (2002)
2. Baudry, S.: Développement sur l'ERP OfbizNéogia, Rapport de stage, Polytechnique de l'Université de Tours (2005)
3. Benchikh, W.: L'Architecture Contractuelle des Groupements de contrats de Progiciel de Gestion Intégrée (PGI/ERP), Mémoire de D.E.E.S, Université de Paris (2004)
4. Blain, F.-A.: Présentation générale des ERP et leur architecture modulaire (2006), http://fablain.developpez.com/tutoriel/presenterp/
5. Boudina, A., Tayeb, L.: L'intelligence Artificielle Distribuée et les SystèmesMulti-Agent (2007), http://opera.inrialpes.fr/people/Tayeb.Lemlouma/Papers/IAD_Presentation.pdf
6. Chaabouniimp, A.: Lantation d'un ERP (Enterprise Resource Planning): Antécédents et Conséquences, XV ème Conférence Internationale de Management Stratégique, Annecy / Genève (2006)
7. Dailly, N.: Utilisation des systèmes multiagents dans les EIAH, Université de Technologie de Compiègne, UTC (2007), http://www.dailly.info/
8. Duro, O., Hoarau, W.: Langage de Communication pour Agents de la FIPA, Master Recherche Informatique, Université Paris Sud XI (2004)
9. Ferber, J.: Les systemes multi-agents, vers une intelligence collective, InterEditions, Paris (1995)
10. Guyet, T.: Systèmes Multi-Agents (SMA) Introduction et Applications biomédicales, Cours IMTC, TIMC/PRETA et LIG/MAGMA, grenoble (2007), http://www-sante.ujfgrenoble.fr/imtc/download/cours/SMA-T_Guyet.pdf
11. Jarras, I., Chaib-draa, B.: Aperçu sur les systèmes multi-agents, Série scientifique, Centre interuniversitaire de recherche en analyse des organisations (CIRNO), Université de Montréal (2002)
12. Kazar, O.: Conception et réalisation d'un modèle de réseau sémantique, Mémoire de magister, Université de Constantine, pp. 6–45 (1996)
13. Lequeux, J.-L.: Manager avec les ERP, Editions d'organisation (1998)
14. Paraschiv, C.: Les agents intelligents pour un nouveau commerce électronique, InterEditions, Paris (2004)
15. Saidna, S.: Plates-formes des systèmes multi-agents, Master de recherche en Informatique, Université Paris Sud XI (2007)

Design Requirements of Usability and Aesthetics for e-Learning Purposes

Aneta Bartuskova and Ondrej Krejcar

University of Hradec Kralove, Faculty of Informatics and Management,
Department of Information Technologies,
Rokitanskeho 62, Hradec Kralove, 500 03, Czech Republic
Aneta.Bartuskova@uhk.cz, ondrej.krejcar@remoteworld.net

Abstract. This paper presents principal design requirements, in terms of usability and aesthetics, for e-learning purposes. Usability and aesthetic attributes applicable for standalone materials as well as e-learning applications are identified, based on research of existing literature and experience. Design requirements are then defined from logically coherent groups of these attributes, namely: legibility, design consistency, visual presentation, content arrangement and content adjustment. Specified requirements are further discussed according to their role in e-learning, application and frequent mistakes based on experience.

Keywords: design principles, e-learning, usability, aesthetics.

1   Introduction

E-learning is still quite new but rapidly evolving area of education. Exact definition of e-learning is not established yet as various authors have different opinions on e-learning discipline, especially on its scope. Fee states that e-learning is an approach to learning and development: a collection of learning methods using digital technologies, which enable, distribute and enhance learning [4].

This research paper inquires into design requirements for e-learning purposes. Design is one of many aspects of e-learning discipline and it can be understood in more ways, differentiating in scope and complexity. One point of view is design as a complex conception of learning strategy including among many other aspects also presentation of content. Other meaning of design in e-learning, which is also a focus of this article, is concerned solely on presentation of content. Such a design in human-computer interactions consists of two major components – usability and aesthetics. These two concepts include number of design attributes concerning content organization, visual presentation, navigation, consistency etc.

General design principles and guidelines are applicable on different areas of expertise. Some of these principles are more applicable than others in various contexts. The aim of this article is to examine these principles, decide which are most applicable to e-learning in scope of presentation, categorize them into coherent groups and identify principal design requirements, in terms of usability and aesthetics.

J. Sobecki, V. Boonjing, and S. Chittayasothorn (eds.), *Advanced Approaches to Intelligent Information and Database Systems*, Studies in Computational Intelligence 551,
DOI: 10.1007/978-3-319-05503-9_23, © Springer International Publishing Switzerland 2014

2   Types of e-Learning Materials

Design requirements can be very different for various e-learning purposes. One group of these requirements is aimed at content presentation, which is the main subject of this research paper. The importance of proper content presentation is undeniable in many disciplines including e-learning. Although learners generally do not appreciate good visual design, it is noticed immediately if learners are confused or cannot find what they need. Good visual design is about solving problems, not drawing attention [1].

Brown & Voltz proposed usage of three components for effective training. These components are: content (the training material), experience activities (games, exercises) and feedback activities (comments) [17]. The content component then includes several types of materials: instructions, reminders, technical documents and manuals, schematics, reference books, programs etc. Good e-learning course usually consists of all these components; nevertheless it would be beneficial to separate the organization part from the content. Proposed components of e-learning course would be then: organization framework, content, experience and feedback.

The scope of this research paper applies to the content of e-learning, specifically embedded or standalone e-learning materials. We can roughly divide these materials to categories, as listed in [Tab 2]. All these materials have usability as well as aesthetics aspects, yet most of presented design principles does not apply to multimedia and active tasks, as this types of materials have its own set of rules for presentation. The widest range of design principles is applicable on combined media, which also benefits from complementarity of all other standalone materials. There are not included collective and cooperative types of activities in the list.

Table 1. Basic types of standalone materials

Material type	Purpose	Examples
Text documents	Thorough reading and understanding	Word or PDF documents
Presentations	Topic introduction or consolidation	Powerpoint, PDF or Flash presentations
Multimedia	Facilitation and enrichment	Pictures, flash animations, audio, video
Tasks	Repetition and self-testing	HTML form
Combined media	All mentioned above	HTML webpages

All of these types differ in technology, content and design as well. Fee also defined these aspects – technology, content and design – as the three essential component parts of e-learning [4]. These materials can be used and combined, usually in complex e-learning applications (offline) or embedded web interfaces (online). Even more design principles are applicable on such applications and interfaces, especially concerning navigation, technology or compatibility. This research paper will cover design principles for e-learning purposes which are applicable also on standalone materials, not only applications and interfaces.

3   Design Principles in Literature

Usability and aesthetics are widely researched aspects of design in human-computer interactions [18,19,20]. These concepts are considered as two principal dimensions in design [21,23]. Previous studies have shown that evaluations of usability and aesthetics are also often correlated [20,21]. Good usability is undoubtedly essential for working with any system as it includes measures of effectiveness, efficiency and satisfaction [22]. As for aesthetics, while in disciplines such as web design is aesthetics a key predictor of an overall impression or user satisfaction [20], its role in e-learning is more of a backstage unobtrusive nature. Horton argues that e-learning communicates visually, yet visual design in e-learning is often ignored or treated as a minor cosmetic detail. According to Tractinsky, visual aesthetics of an interface significantly influences users´ perceived ease of use of the entire system [21].

Many authors attempted in their publications for representative or definitive list of universal design principles for user interfaces. Among these publications belong e.g. "Universal Principles of Design" by W. Lidwell, K. Holden and J. Butler [6] and "100 Things Every Designer Needs to Know About People" by S. Weinschenk [14]. The same principles are often repeating for specific area of interest, e.g. in web design publications such as "Web Style Guide: Basic Design Principles for Creating Web Sites" by P.J. Lynch and S. Horton [13]. And finally, also several e-learning publications dedicate a space to presentation of these design principles, yet only partially without a structured organization.

S. Chapnick and J. Meloy in their "Renaissance eLearning: Creating Dramatic and Unconventional Learning Experiences" brought together several universal design principles, which has relevance to e-learning [5]. These principles were also presented by Fee [4]. Most of these principles are applicable on usability and aesthetics of standalone e-learning materials.

1) The aesthetic usability principle. This principle implies that aesthetic designs are perceived as easier to use than less-aesthetic designs [6].
2) Consistency. This means that similar parts of content should be expressed in similar ways. Two aspects of consistency are considered - aesthetic (style and appearance) and functional (meaning and action).
3) Von Restorff effect. This suggests use of different style for elements which are important and should be remembered.
4) Proximity. Gestalt principle, stating that elements which are close together are perceived to be more related than elements farther apart [6].
5) LATCH model. This principle is used for organizing and presenting information, which helps during the remembering process.

One of the standards for e-learning design quality is E-learning Courseware Certification Standards from the ASTD Certification Institute, which indicates four groups of design aspects [1]. These main groups are named interface standards, compatibility standards, production quality standards and instructional design standards [16]. Applicable for e-learning materials in relation to content presentation are especially production quality standards with these design aspects: legibility of text and graphics, formatting and consistency. Partially applicable is group of interface standards, featuring orientation, tracking features, navigational functions, navigational devices and operational support.

Horton´s "E-Learning by Design" decomposes the whole concept of design including strategic decisions and activities and tests for various types of e-learning [1]. Design for presentation of content is limited here to the issue of legibility, layout, unity, yet these issues are thoroughly explored. Especially valuable chapter about readability is based on a half-century of research. Following legibility aspects were specified: character shape (recommended: simple), type size (bigger than 10pt), color and tonal contrast (high), line spacing (1/30 of line length), line length (40-60 characters), alignment (left-aligned).

Bozarth suggests several tips for good interface, which comprise of both usability and aesthetics concepts. Some of these recommendations are similar to those already mentioned by other authors, emphasised are principles of consistency, clarity and navigation [8]. Mentioned is also role of color, especially its meaning and cultural implications. Bozarth also considers design in terms of cognitive load on learners and presents Mayer´s SOI model, SOI as "select", "organize" and "integrate" [8]. The "select" component is most interesting for the visual presentation of content and comprises of several useful design principles: remove extraneous information, chunk information into smaller pieces, be concise and use white space, font size, colors and highlighting for emphasis or importance [30]. Mayer also developed many principles of multimedia learning useful for e-learning purposes. Applicable for content presentation are following:

1) Multimedia principle. Learning is enhanced by the presentation of words and pictures rather than words alone.
2) Coherence principle. Learning is enhanced without extraneous material is omitted.
3) Contiguity principle. Learners learn better when on-screen text and visuals are integrated rather than placed apart.

Another design attributes can be deduced from related filed of knowledge such as webdesign. McCracken and Wolfe defined web design attributes as: content organization, visual organization, navigation system, colour and typography [24]. Usability of webpages influences user´s performance and can be manipulated through quality of organization, navigation or readability. Aesthetics affects first impression, visual appeal and can be manipulated through the quality of visual elements, pictures and other graphics.

4   Design Attributes and Requirements

The authors identified usability and aesthetics attributes, based on research of general design principles and relevant principles from e-learning and webdesign. These

Table 2. Identified design attributes and requirements

Requirement	Attributes	Similar concepts
Legibility	Typeface [6,13]	Font [8,14], Character shape [1]
	Type / font size [1,5,6,13,14]	
	Tonal contrast [1,6,13]	Quiet / noisy background [1,8]
	Spacing	Character / word spacing [5,6], Line spacing [1,5,6]
	Alignment [1,5,6,13]	
	Line length [1,13]	
	Media legibility	Compression rate [8,13]
Visual presentation	Aesthetic design [5,6]	Attractive interface [8]
	Color [6,8,13]	Color meaning [8,14], Color emphasis [1,13]
	Color contrast [1,13,14]	Contrast [8], Color balance
	Relevant graphics [8]	Theme graphics [1]
	Supportive graphics [1,8]	Recognition over recall [6,14]
	Visual hierarchy [13]	Visual contrast [13]
Design consistency	Functional consistency [5,6]	Consistency in iconography [8]
	Aesthetic consistency [5,6,13]	Consistent look and feel [1,8]
	Consistency in layout and structure [1,13]	Internal consistency [6], Mental models [6,14]
Content arrangement	Layout [1,13]	Z movement [8]
	Organization [6]	LATCH [5], Five Hat Racks [6]
	Navigation mechanism [1]	Navigational information [8]
	Multiple presentation media	Use of pictures [8], Depth of processing [6]
Content adjustment	Chunking [6,8,13,14]	Progressive disclosure [14], Layering [6]
	White space [8,13]	
	Gestalt Proximity [5,6,13,14]	Contiguity principle [8]
	Emphasis mechanisms [1,5,13]	Highlighting [6,8] Von Restorff effect [4,5]
	Noise reduction	Coherence principle [8] Signal to noise ration [6]

attributes are listed in [Tab 2], already logically sorted into five main groups of design requirements, defined by the authors. These design requirements are: legibility, design consistency, visual presentation, content arrangement and content adjustment. Similar concepts are listed along with assigned attributes for completeness in [Tab 2].

4.1 Legibility

Legibility is one of the key usability aspects, ensuring that a learner can read text and recognize pictures. It is even more important in e-learning courses than traditional courses, because reading on a computer screen is harder than reading paper [14]. We should generally choose typeface or font with simple character shapes, preferably san-serif. Legibility of typeface indicates the level to which characters in text are recognizable and understandable regarding their appearance. It includes factors such as x-height, character shapes, stroke contrast, the size of its counters, serifs or lack thereof, and weight [12]. Simple typeface is important not only for legibility and pattern recognition but also for e-learning process, because difficulty of reading the text transfers the diffriculty on the text itself [14]. Appropriate font size depends mostly on type of material, for text documents is minimal recommended font size 10pt [1]. Concerning legibility of pictures, the quality depends mostly on chosen format and compression rate. Appropriate line length is disputable, as people read faster with a longer line length, but they prefer a shorter line length [14].

Tonal contrast is a usability aspect with direct impact on text readability but also on recognition of pictures. Tonal contrast defines the difference between the perceived lightness of two colors [10] and should be high enough between foreground and background, and kept low in background for ensuring the quiet background [1,8].

The authors encountered several problems from experience with e-learning courses concerning text legibility. First problem arises mostly from automatic processing in PowerPoint presentations – when the text does not fit in the frame, it is made smaller. Switching of font sizes during reading is not very comfortable for learners. This is more significant issue when printing the material, especially with favourite layout of 6 frames per page. Smaller fonts are often hardly legible. Media legibility is also frequent issue in e-learning materials, graphs and schematics are often re-used from other sources and their quality is not sufficient. This is usually a consequence of repeated modifications or corrections and also automatic transformations in Word or PowerPoint. The next frequent issue is noisy background - even some of default styles of PowerPoint application have backgrounds with considerable contrast.

4.2 Visual Presentation

Attractive design is important in every human-computer interaction. Aesthetic designs look easier to use and have a higher probability of being used, whether or not they actually are easier to use [6]. Aesthetic design helps to create positive attitude towards a system or product or also e-learning material.

Color in design is very powerful tool, which influences the overall appearance and emotional impact on learner. Color should be also used for separation of different functional areas (such as header, navigation, main content, ...) to create visual hierarchy and for highlighting of featured content. Impact of color is mostly subjective, yet there are some general rules e.g. concerning warm and cold colors.

Color contrast is rather aesthetics feature, with influence on visual balance and harmony of the presentation. This corresponds closely with an overall choice on color combinations and their color or visual contrast. In general, there are combinations which are more pleasing to the eye than the others.

Supportive graphics as e.g. iconic representation makes easier to find, recognize, learn and remember objects and concepts [6]. This is in accord with a "Recognition over recall" principle, as people are better at recognizing things they have previously experienced [6,14]. Graphics have to be relevant otherwise it creates visual noise and disturbs visual hierarchy, which is important aspect of visual presentation. Visual hierarchy includes usage of colours, spacing and styles to divide blocks of content, visually separate headlines etc.

The common issue with visual presentation in e-learning materials is creation of undesired visual noise. This refers to overuse or unappropriate use of visual features such as decorative graphics or textures. Especially frequent is usage of gradient or texture on background, which also affects legibility of text and relates to issue of contrast. Visual noise can also arise as a consequence of emphasis overuse.

4.3   Design Consistency

As design consistency refers to both usability and aesthetics, two aspects of consistency are considered - aesthetic (consistency of style and appearance) and functional (consistency of meaning and action) [4-6]. Aesthetic consistency improves recognition and feeling by making the content visually coherent. Functional consistency enables people to use existing knowledge. Systems are generally more usable and learnable when similar parts are expressed in similar ways [6].

Consistency in usability also includes unity in content structure and navigation. If the unity is preserved, a reader can develop mental models according to organization of course materials. Consistent approach to layout and navigation allows users to adapt quickly to design and to predict the location of information [13]. Consistency then allows representation and recall based on mental models.E-learning materials can be taken as a user interface, where it is important to reflect the users´ expectations, because people understand and interact with systems based on their mental models. By meeting these expectations, the efficiency of interaction can be enhanced [7].

Aesthetic consistency is ensured by use of the same or similar visual appearance for the same or similar elements and thus is closely connected with functional consistency. Unity in aesthetics helps to maintain overall coherency and visually forms relevant groups of information. Aesthetic consistency contributes to visual and emotional appeal and feeling, which has great impact on reader´s motivation and progress. Another aspect of visual consistency and harmony lies in proper use of colors and their combinations. Only limited number of colors should be dominant across the learning material, usually only one or two. The dominant color should be then accompanied by complementary or neutral colors.

4.4   Content Arrangement

Structural arrangement and organization of content, usually text and pictures but also various other multimedia content, is very dependent on type and purpose of particular e-learning material. Among main criteria for our decision belong the desired goal and available resources. We also have to consider technology possibilities and limitations, which can affect content arrangement. E.g. with today´s expansion of mobile devices, we should aspire for flexible layouts [25-29].

Organization, layout and also navigation through series of materials depend greatly on particular type of e-learning materials. General design principles indicate that layout should be most of all predictable. This means that the learners can find the information and controls they seek in the display without difficulty [1]. This predictability also corresponds with other design requirement – design consistency. It is also useful to use so called Z-movement to accommodate the eye's natural movement. Important content is given center area, while supporting information like links and navigation icons are placed in positions of less emphasis [8,14].

Content should be then structured and organized logically and hierarchically, with table of contents and division e.g. by chapters. There are several strategies of organization, most frequently used in e-learning is organization by category - similarity or relatedness [6]. These categories are usually called topics or themes.

Navigation usually combines two mechanisms – scrolling and paging [1]. For e-learning applications is also essential use of a menu or other navigational panel.

Use of multiple presentation media is always encouraged, as learning is enhanced by the presentation of words and pictures rather than words alone [8,30]. Level of processing is also deeper when images accompany the text, which is essential for recall and retention of the information [6].

4.5   Content Adjustment

Chunking means division of content into logical units [13]. This is very useful principle in e-learning as the brain can only process small amount of information at a time [14]. Chunked information is then easier to process and remember [6]. Chunking works in correspondence with use of proximity principle and white space. It is also related to progressive disclosure, which is about presenting only the information the learner needs at that moment [14]. Similar to chunking and proximity principle is also layering, the process of organizing information into related groupings in order to manage complexity and reinforce relationships in the information [6]. Gestalt principle of proximity then states, that elements which are close together are perceived to be more related than elements that are farther apart [6,13].

Emphasising or highlighting should be used in moderate rate, recommended amount of highlighted content ranges from 10% [6] to 15% [1]. Among emphasis mechanisms belong usage of color, size and bold or otherwise emphasised type.

Table 3. Aesthetics and usability design requirements

Usability	Aesthetics	Requirement	Purpose / effect
x		Legibility	Readability of content
x	x	Design consistency	Representation and recall based on mental models
	x	Visual presentation	Overall appeal and feeling
x		Content arrangement	Understanding and recognition
x	x	Content adjustment	Learnability of content

5   Conclusions

This article presented principal design requirements for e-learning purposes. The motivation for this paper was the fact, that learning itself is difficult enough and design of e-learning materials should support this process and not discourage it. Aesthetics and usability have great impact on learning experience, yet often principal design principles are neglected. Findings of this research are summarized in [Tab. 4]. These design requirements were created by logically grouping individual attributes, researched and categorized from relevant sources.

Introduced design requirements refer to presentation of content in terms of usability and aesthetics. They are applicable for standalone e-learning materials as well as for e-learning applications and interfaces. The list is not complete for applications, as the authors included only concepts applicable for all types of materials. The more detailed categorization and appropriately assigned attributes can be a subject of future research. The impact of content itself and issues of motivation, retention etc. are not considered. This can be also a subject of future research, as well as revised classification based on wider range of literature.

Acknowledgment. This work was supported by project "SP/2013/03 - SmartHomePoint Solutions for Ubiquitous Computing Environments" from University of Hradec Kralove.

References

[1] Horton, W.: E-Learning by Design. Pfeiffer (2006) ISBN: 978-0787984250

[2] Brown, A.R., Voltz, B.D.: Elements of Effective eLearning Design. The International Review of Research in Open and Distance Learning 6(1), 217–226 (2005)

[3] Steen, H.L.: Effective eLearning Design. MERLOT Journal of Online Learning and Teaching 4(4) (December 2008)

[4] Fee, K.: Delivering E-Learning: A Complete Strategy for Design, Application and Assessment, 4th edn. Kogan Page (2009) ISBN:978-0749453978

[5] Chapnick, S., Meloy, J.: Renaissance eLearning: Creating Dramatic and Unconventional Learning Experiences. Pfeiffer (2005)

[6]   Lidwell, W., Holden, K., Butler, J.: Universal Principles of Design, Revised and Updated: 125 Ways to Enhance Usability, Influence Perception, Increase Appeal, Make Better Design Decisions, and Teach through Design. Rockport (2010)

[7]   Roth, S.P., Schmutz, P., Pauwels, S.L., Bargas-Avila, J.A., Opwis, K.: Mental models for web objects: Where do users expect to find the most frequent objects in online shops, news portals, and company web pages? Interacting with Computers 22(2) (2010)

[8]   Bozarth, J.: Better Than Bullet Points: Creating Engaging e-Learning with PowerPoint. Pfeiffer (2008) ISBN: 978-0787992453

[9]   Chapman, C.: Color Theory for Designers, Part 1: The Meaning of Color, http://www.smashingmagazine.com/2010/01/28/color-theory-for-designers-part-1-the-meaning-of-color (retrieved October 31, 2013)

[10] Stone, M.: Contrast Metrics Explained, http://www.stonesc.com/pubs/Contrast%20Metrics.htm (retrieved October 31, 2013)

[11] Shallbetter, J.L.: Color combinations, http://www.worqx.com/color/combinations.htm (retrieved October 31, 2013)

[12] Strizver, I.: Type Rules: The Designer's Guide to Professional Typography. John Wiley & Sons, New Jersey (2010) ISBN: 978-0470542514

[13] Lynch, P.J., Horton, S.: Web Style Guide: Basic Design Principles for Creating Web Sites. Yale University Press (2009) ISBN: 978-0300137378

[14] Weinschenk, S.: 100 Things Every Designer Needs to Know About People. New Riders (2011) ISBN: 978-0321767530

[15] Allen, M.W.: Designing Successful e-Learning, Michael Allen's Online Learning Library: Forget What You Know About Instructional Design and Do Something Interesting. Pfeiffer (2007) ISBN: 978-0787982997

[16] Sanders, E.S.: E-Learning Courseware Certification Standards. American Society for Training and Development (2003)

[17] Brown, A.R., Voltz, B.D.: Elements of Effective eLearning Design. The International Review of Research in Open and Distance Learning 6(1), 217–226 (2005)

[18] Tuch, A.N., Roth, S.P., Hornbæk, K., Opwis, K., Bargas-Avila, J.A.: Is beautiful really usable? Toward understanding the relation between usability, aesthetics, and affect in HCI. Computers in Human Behavior 28(5), 1596–1607 (2012)

[19] Lee, S., Koubek, R.J.: Understanding user preferences based on usability and aesthetics before and after actual use. Interacting with Computers 22(6), 530–543 (2010)

[20] Hassenzahl, M.: The interplay of beauty, goodness, and usability in interactive products. Human–Computer Interaction 19(4), 319–349 (2004)

[21] Tractinsky, N., Katz, A.S., Ikar, D.: What is beautiful is usable. Interacting with Computers 13(2), 127–145 (2000)

[22] Hornbæk, K.: Current practice in measuring usability: Challenges to usability studies and research. International Journal of Human-Computer Studies 64(2), 79–102 (2006)

[23] Hartmann, J., Sutcliffe, A., de Angeli, A.: Towards a theory of user judgment of aesthetics and user interface quality. Transactions on Computer-Human Interaction 15(4) (2008)

[24] McCracken, D.D., Wolfe, R.J.: User-centered Website Development: A Human-Computer Interaction Approach. Pearson Prentice Hall Inc., Upper Saddle River (2004)

[25] Wiesner, J., Kriz, Z., Kuca, K., Jun, D., Koca, J.: Influence of the acetylcholinesterase active site protonation on omega loop and active site dynamics. J. Biomol. Struct. Dyn. 28(3), 393–403 (2010)

[26] Kopacova, M., Tacheci, I., Kvetina, J., Bures, J., Kunes, M., Spelda, S., Tycova, V., Svoboda, Z., Rejchrt, S.: Wireless video capsule enteroscopy in preclinical studies: methodical design of its applicability in experimental pigs. Dig. Dis. Sci. 55(3), 626–630 (2010)

[27] Benikovsky, J., Brida, P., Machaj, J.: Proposal of User Adaptive Modular Localization System for Ubiquitous Positioning. In: Pan, J.-S., Chen, S.-M., Nguyen, N.T. (eds.) ACIIDS 2012, Part II. LNCS, vol. 7197, pp. 391–400. Springer, Heidelberg (2012)

[28] Cheng, W.C., Liou, J.W., Liou, C.Y.: Construct Adaptive Template Array for Magnetic Resonance Images. In: IEEE International Joint Conference on Neural Networks, Brisbane, Australia, June 10-15 (2012), doi:10.1109/IJCNN.2012.6252560

[29] Penhaker, M., Stankus, M., Prauzek, M., Adamec, O., Peterek, T., Cerny, M., Kasik, V.: Advanced Experimental Medical Diagnostic System Design and Realisation. Electronics and Electrical Engineering 1(117), 89–94 (2012)

[30] Mayer, R.H.: Designing instruction for constructivist learning. In: Reigeluth, C.M. (ed.) Instructional-Design Theories and Models: A New Paradigm of Instructional Theory, vol. II, pp. 141–160 (1999)

Workflow Scheduling on Virtualized Servers

Johnatan E. Pecero and Pascal Bouvry

Computer Science and Communications research unit
University of Luxembourg, Luxembourg
{firstname.lastname}@uni.lu

Abstract. Workflow applications comprise a number of structured tasks and computations featuring application services to be executed and the dependencies between these services. This paper deals with the problem of scheduling workflow applications, represented by directed acyclic graphs, on a set of virtualized servers. Each server hosts multiple virtual machines. Virtual machines sharing a host can communicate with each other, and with virtual machines hosted in different servers. The aim is to partition the application services and distribute each partition among the virtual machines in such a way that the dependencies are respected, the response time is minimized improving the quality of service and the intra- and inter-virtual machine communications are minimized. We model this problem as a workflow scheduling problem with hierarchical communications. The main contribution is to provide an evolutionary-based scheduling algorithm that considers this model when scheduling the applications. Simulation results demonstrate the effectiveness of the provided algorithm when compared with a related approach on a set of real-world applications emphasizing the interest of the approach.

Keywords: Cloud, IaaS, Workflow Scheduling, Performance of System.

1 Introduction

Virtualization is a mechanism to abstract the system resources and hardware of physical servers. It is one of the main hardware-reducing, cost-, and energy-saving technique used by cloud providers. It is done with software-based computers that share the underlying physical machine resources among different virtual machines (VMs). Using VMs that can be configured before deployment can increase resource utilization and reduce excess overhead by consolidating multiple workload streams. Other benefit of using virtualization is high availability enabling options such as fast restoration of service by providing fault isolation across components placed in separate VM containers. Virtualization also allows stricter resources partitioning for performance isolation between VMs. However, the significant increase in VM mobility have resulted in greater performance demands on the network subsystems at the server-network edge (i.e., two-tiers layer).

Workflows are parallel applications that comprise a number of structured tasks and computations that arise in business and scientific problem solving. They are

J. Sobecki, V. Boonjing, and S. Chittayasothorn (eds.), *Advanced Approaches to Intelligent Information and Database Systems,* Studies in Computational Intelligence 551,
DOI: 10.1007/978-3-319-05503-9_24, © Springer International Publishing Switzerland 2014

used to combine different computational processes into a single coherent whole. In this work, we focus on scientific workflow applications. Scientific workflows are used in different functions such as data analysis, simulations, image processing in many scientific disciplines such as bioinformatic domains (e.g., Epigenome to maps short DNA segments and SIPHT applications), seismology (e.g., Broadband application), and astronomy (e.g., the LIGO workflow and Montage). Scientific workflows are commonly represented as a directed acyclic graphs (DAG) in which nodes represent application services or tasks to be executed and edges represent dependencies and communications between these services.

Different works evaluate the benefits of executing workflow applications on a virtualized infrastructure [2,5,7,8,9,12]. Such benefits include user-customization of system software and services, performance isolation, check-pointing, migration, provenance and reproducibility, elasticity, and enhanced support for legacy applications. However, one of the main problems is the efficient resource management and workflow execution on the virtualized infrastructure. Having a collection of VMs is not sufficient to execute workflows. The performance of a workflow application executed on a virtualized infrastructure heavily depends on the scheduling of the tasks of the application onto the available virtual machines, which if not properly solved, can nullify the benefits of the workflow execution. Another problem that can lead to performance degradation is the communication [6], incurred while running workflows, not only between virtual machines hosted in a same physical resource (intra-communications), but also among VMs deployed on different hosts (inter-communications). Some solutions from the architectural and technological level at the network layer include distributing management intelligence at the physical server-network edge, flattening the two-tiers network by using technologies like intelligent resilient framework, making the network more efficient by implementing future multi-path standards [4]. However, in this work we focus at the software scheduling level. The scheduling algorithm should be aware of the intra-, inter-virtual machine communications.

In this paper, we model the problem as a workflow scheduling problem with hierarchical communications. That is, we adopt the model in which the communications between VMs on a same physical server is faster than those between VMs belonging to different physical hosts. Our main contribution is to provide a scheduling algorithm based on an evolutionary approach that considers this model when scheduling the workflow applications on the multiple virtual machines. The algorithm follows the framework of a simple genetic algorithm. Simulation results using real-world applications demonstrate the effectiveness of the proposed approach.

This paper is organized as follows. In the next section, we present the investigated scheduling problem. The proposed solution is described in Section 3. Next, in Section 4 we provide experimental results. Section 5 concludes the paper.

2   Problem Definition

The main scope of this work is to deal with the workflow task allocation and scheduling problem on a virtualized infrastructure environment. We consider a private Cloud Infrastructure-as-a-Service architecture with a set P of M servers, where each server deploys a set $VM = \{VM_1, ..., VM_m\}$ of m virtual machine images. Resources in our private Cloud IaaS model or at least those allocated to a particular application are homogeneous and reliable. Figure 1 depicts an abstraction of a private IaaS composed of two servers. Each server deploys two virtual machines that can communicate between them. Each VM implements a driver for its virtual communication that is called its net-front driver. Virtual machines from different servers communicate through Ethernet among them.

From the hypervisor, one or more VMs can be started concurrently as seen in Figure 1. Each VM runs a separate operating system within it. The virtual machines provide the compute cycles needed by the tasks of the workflows. To provide the performance required for workflows, many VMs instances must be used simultaneously. These collections of virtual machines are called virtual clusters [10].

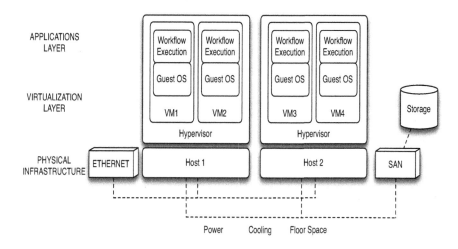

Fig. 1. Private Cloud IaaS Abstraction

We denote a workflow application by a directed acyclic graph (DAG) $G = (T, E)$, where $T = \{t_1, ..., t_n\}$ is a finite set of n tasks and E is a finite set of directed edges. An edge $(t_i, t_j) \in E$ of graph G corresponds to the data dependencies between these tasks (the data generated by t_i is consumed by t_j and must be available before t_j starts its execution). Task t_i is called the immediate predecessor of t_j which is the immediate successor of t_i. We consider that each task is an indivisible unit of work and is non-preemptive. To every task t_j, there is an associated value p_j representing the execution time estimation to complete a task on a VM. To every edge (t_i, t_j) there is a weight d_{ij} associated corresponding to the volume of data to be transmitted from task t_i to task t_j.

A given workflow application has to be executed on P, a set of mM virtual machines. The $l - th$ server is denoted P_{l*} whereas the $k - th$ VM of the $l - th$ server is denoted $P_{lk} \in P_{l*}$. The computational model we adopt corresponds to the *hierarchical communication model* [1]. The communication cost between two VMs deployed in the same server is c_{kl}, while it takes $C_{ll'}$ units of time if the two VMs belong to different servers P_l and $P_{l'}$. The aim of scheduling is to distribute the tasks among the VMs in such a way that dependencies are respected, and the response time C_{max} (i.e., makespan) is minimized. The response time corresponds to a quality of service metric and for a given workflow depends on the allocation of tasks in the VMs and scheduling policy applied in individual VMs. In terms of complexity, the problem is NP-hard as it contains NP-hard problems as particular cases [3].

3   Proposed Solution: A GA-Based Scheduling on Virtual Machines (GAS-VM)

The solution we provide is based on a Genetic Algorithm (GA). A GA is a general-purpose, stochastic search method where elements (called individuals) in a given set of solutions (called) population are randomly combined (using a recombination operator) and modified (based on mutation) until some termination condition is achieved. GAs use global search techniques to explore the search space by keeping track of a set of potential solutions of diverse characteristics.

String Representation. To encode a solution, we use two arrays of integers M' and m' of length n. The array M' encodes the server. The $i - th$ location of the array corresponds to the task t_i and its content represents the server P_{l*}, where task t_i is assigned. The array m' encodes the VM P_{lk} where task t_i is executed on server P_{l*}.

Initial Population. The initial population is randomly generated.

Fitness Function. The fitness function of each individual is a measurement of performance of the design variables as defined by the objective function and the constraints of the problem. For each offspring S_i of the population, we compute the response time $(C_{max_{S_i}})$. Hence, the fitness function is based on Eq. 1.

$$f_{s_i} = \frac{1}{C_{max_{S_i}}} \tag{1}$$

To compute the response time for a particular offspring, we implemented an algorithm based on the list scheduling principle. Informally, the principle is to schedule the workflow iteratively one time slot after the other on the assigned server and VM.

Selection. The proposed algorithm implements the proportional selection scheme based on the roulette wheel principle. We used an elitist model, where the global best individual is always kept and replace the worst individual in the new population.

Recombination and Mutation. We implemented single recombination operator; mutation randomly changes one task in the scheduling to a different server and VM.

Stopping Condition. Different conditions can be used to stop the execution of the algorithm. The proposed algorithm is executed until a fixed number of iterations has been reached.

4   Experiments

In this section, we presents some simulation results. We compare the performance of GAS-VM against a list-based scheduling algorithm, called ListS-VM in this paper. The algorithm considers hierarchical communications. The algorithm works in two phases. In the first phase, ListS-VM selects the next task to be scheduled. Thereafter, in the second phase a suitable VM that minimizes a predefined cost function is selected to schedule the task. ListS-VM maintains a list of ready tasks that are greedily scheduled on available resources. A task is ready when all its immediate predecessors have been scheduled. To make a more accurate comparison we have decided to select at random the ready task to be executed. Once the task is selected, it is assigned to the VM that provides the minimum response time.

We have selected for our simulations workflows from the *Standard Task Graphs* (STG) set [11]. The STG set provides workflows modeled from real application programs. The workflows are represented by a robot control application, a sparse matrix solver, and fpppp problem from SPEC benchmark. Table 1 summarizes the main characteristics for these applications: type (i.e., name of instance), instances size, edges amount and the ratio between tasks and edges (ETR). ETR gives information on the degree of parallelism.

Table 1. Employed benchmark from STG and their main characteristics

Type	# of Tasks	# of Edges	ETR
Robot control	88	131	1.48
Sparse matrix solver	96	67	0.69
SPEC fpppp	334	1145	3.42

For each graph, we have varied the *communication to computation cost ratio* (CCR), which is the average of communication costs (data volume) to the average estimated execution times of the tasks on the VMs. For a given workflow the CCR ratio is a measure that indicates whether an application is computation intensive (i.e., low CCR value less than 1), communication intensive (i.e., high CCR greater than 1) or moderate. We have generated three CCRs (0.5, 1, 2). The algorithms have been simulated on a private Cloud IaaS composed of two servers. For each server we have simulated 2, 3, 4, and 6 VMs to be deployed.

GAS-VM used the following parameters in the experiments: population size equal to 40 individuals, recombination probability equal to 0.85, mutation probability of 0.005, and generation limit of 50. These parameters were fixed experimentally. We use the response time generated by each algorithm as the main performance metric. We show average (avg) response time. Each algorithm has been executed 30 times, recall that the list scheduling algorithm used as a basis for comparison has been randomized.

Figure 2 depicts the results for the simulations for different CCRs. It can be observed that GAS-VM shows better performance than ListS-VM. It produces schedules with average response time 18% shorter than the related approach.

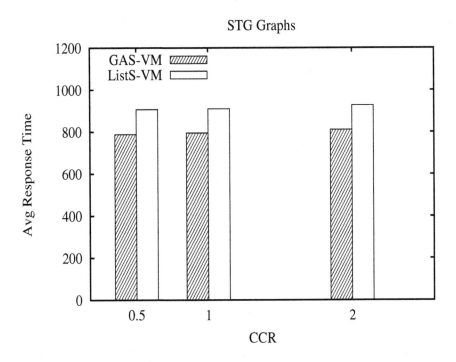

Fig. 2. Average response time for different CCRs

Figure 3 shows the results when increasing the number of VMs per sever. As we can see both algorithms show the same behavior, that is both algorithms improve the average response time when the number of VMs deployed on each server increases. The main reason is that the number of inter-virtual machines communication decreases when the number of virtual machines increases, i.e., allowing more intra-virtual machines communications. GAS-VM computes schedules with average response time 12% shorter than ListS-VM.

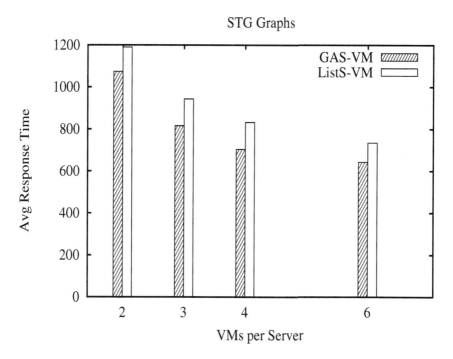

Fig. 3. Average response time when increasing the number of VMs

5   Conclusions

In this work, we have provided an evolutionary approach for scheduling work-flow applications on a private Cloud IaaS infrastructure. We have adopted the hierarchical model considering intra- and inter-virtual machine communications. We have compared the performance of the proposed algorithm with a related approach based on a randomized list scheduling algorithm using real-world work-flow applications. The preliminary results emphasize the interest of the designed algorithm.

Acknowledgment. The authors would like to acknowledge the funding from National Research Fund, Luxembourg in the framework of the FNR INTER-CNRS-11-03 Green@cloud project.

References

1. Bampis, E., Giroudeau, R., König, J.-C.: An approximation algorithm for the prece-dence constrained scheduling problem with hierarchical communications. Theoret-ical Computer Science 290(3), 1883–1895 (2003)

2. Bessai, K., Youcef, S., Oulamara, A., Godart, C., Nurcan, S.: Bi-criteria workflow tasks allocation and scheduling in cloud computing environments. In: Chang, R. (ed.) IEEE CLOUD, pp. 638–645. IEEE (2012)
3. Blachot, F., Huard, G., Pecero, J., Saule, E., Trystram, D.: Scheduling instructions on hierarchical machines. In: 2010 IEEE International Symposium on Parallel Distributed Processing, Workshops and Phd Forum (IPDPSW), pp. 1–8 (2010)
4. Burtsev, A., Srinivasan, K., Radhakrishnan, P., Bairavasundaram, L.N., Voruganti, K., Goodson, G.R.: Fido: fast inter-virtual-machine communication for enterprise appliances. In: Proceedings of the 2009 Conference on USENIX Annual Technical Conference, USENIX 2009, p. 25. USENIX Association, Berkeley (2009)
5. Figueiredo, R.J., Dinda, P.A., Fortes, J.A.B.: A case for grid computing on virtual machines. In: Proceedings of the 23rd International Conference on Distributed Computing Systems, pp. 550–559 (2003)
6. Govindan, S., Jeonghwan, C., Nath, A.R., Das, A., Urgaonkar, B., Anand, S.: Xen and co.: Communication-aware cpu management in consolidated xen-based hosting platforms. IEEE Transactions on Computers 58(8), 1111–1125 (2009)
7. Hoffa, C., Mehta, G., Freeman, T., Deelman, E., Keahey, K., Berriman, B., Good, J.: On the use of cloud computing for scientific workflows. In: IEEE Fourth International Conference on eScience, eScience 2008, pp. 640–645 (2008)
8. Huang, W., Liu, J., Abali, B., Panda, D.K.: A case for high performance computing with virtual machines. In: Proceedings of the 20th Annual International Conference on Supercomputing, ICS 2006, pp. 125–134. ACM, New York (2006)
9. Jha, S., Katz, D.S., Luckow, A., Merzky, A., Stamou, K.: Understanding Scientific Applications for Cloud Environments, pp. 345–371. John Wiley & Sons, Inc. (2011)
10. Juve, G., Deelman, E.: Scientific workflows and clouds. Crossroads 16(3), 14–18 (2010)
11. Tobita, T., Kasahara, H.: A standard task graph set for fair evaluation of multiprocessor scheduling algorithms. Journal of Scheduling 5(5), 379–394 (2002)
12. Younge, A.J., Henschel, R., Brown, J.T., von Laszewski, G., Qiu, J., Fox, G.C.: Analysis of virtualization technologies for high performance computing environments. In: Proceedings of the 2011 IEEE 4th International Conference on Cloud Computing, CLOUD 2011, pp. 9–16. IEEE Computer Society, Washington, DC (2011)

Toward a User Interest Ontology to Improve Social Network-Based Recommender System

Mohamed Frikha, Mohamed Mhiri, and Faiez Gargouri

University of Sfax - MIRACL Laboratory - Sfax, Tunisia
{med.frikha,med.mhiri}@gmail.com, faiez.gargouri@isimsf.rnu.tn

Abstract. This paper aims to improve traditional recommender systems by incorporating information in social networks, including user preferences and influence from social friends. A user interest ontology is developed to make personalized recommendations out of such information. In this paper, we present a preliminary work that sheds light on the role of social networks as sources for the development of recommendation systems. The need for user interest ontology in recommender systems and its importance as a reference to find similar items in social network is also emphasized. Finally, we describe and account for the role of user interest model based on user interest ontology to deal with the lack of semantic information in personalized recommendation system.

Keywords: Ontology, Social Semantic Web, Recommender Systems, Social Networking, User Interest Model.

1 Introduction

A social network is a set of people or groups of people whose social relationships (e.g. friendship, collaborative work, information exchange....) make them interact together. Social network analysis, therefore, could be defined as the study of social and economic entities such as people in organizations (actors) and of their interactions and relationships. Those very interactions and relationships could improve the enterprises' work processes in a collaborative environment as well as the organizations actors' competencies. Generally speaking, those networks are represented by the graphs' theory and the matrix technology. In fact, the emergence of new communication tools and networks' platforms, especially the social network services of web 2.0, has presented new opportunities to explore the power of social networks both inside and outside the organizations [4].

The study of social networks, as a result, has been used as a powerful tool in the organizations in order to better understand the connections and their influence on the performance of the principal processes. It has also been used as a method of obtaining analytical results on whatever group interactions in which the social entities are represented. Furthermore, the social network can be a source for the development of recommendations, finding an expert in a given field, suggesting products to sell, and offering a friend, etc. This development may be based on paths exploration algorithm and degree analysis [10].

J. Sobecki, V. Boonjing, and S. Chittayasothorn (eds.), *Advanced Approaches to Intelligent Information and Database Systems*, Studies in Computational Intelligence 551,
DOI: 10.1007/978-3-319-05503-9_25, © Springer International Publishing Switzerland 2014

However, since social networks are developed by several different kinds of relationships, it could be impossible for the graphs' edges and the numerical values to explain all the semantic relationships. Having mentioned this problem, we propose, as a solution, a method that makes it possible to represent a social network based on ontology. Using an ontology-based method could allow us to describe all the semantic relationships and the interactions in a social network. The importance of the use of ontologies to represent the types of actors and the relationships in a social network for the purpose of semantically visualizing the databases has been demonstrated by [2].

The objective of this paper is determined based on the fact that ontologies can be used to semantically describe social relationships and interactions and that social networks can generate and suggest several varied recommendations with reference to the user needs. This paper aims at generating indications and recommendations for social network users through ontologies. This could be made possible thanks to a recommender system. We seek to improve the recommender systems that determine decision making in a semantic social network.

Based on mainstream literature on recommender systems in social networks [3], social networks already propose to their users recommendations based on their profiles. In a similar fashion, they help users to find people for sharing common social activities and preferences. Being aware of the importance of the user and of his/her preferences in recommendations, we study, in this paper, the effect of incorporating semantic user profile derived from past user behavior and preferences on the accuracy of a recommender system.

Different devices and techniques have been suggested by researchers in the field of personalization to recommend services that could catch the users' interest [18], [7]. For example, collaborative and content-based filtering are two of the most developed and used tools in recommendation systems. Collaborative filtering is based on the assumption that similar users can recommend and infer similar items. Therefore, the major mechanism underlying this approach is to find out similar users by comparing every user's interests with the interest histories of other users (neighbors) [13]. Accordingly, these historical interests are then gathered and used to give items to users. This approach, however, has been subject of criticism because of its limitation in handling cold-start and first-start problems.

Unlike the Collaborative-filtering, Content-based filtering's major goal is to recommend items based on users' previous interests. It seeks to implicitly or explicitly capture user's interests and then suggest items that could be similar to the captured ones. The Spreading Activation Mechanism is a technique used in content-based filtering. This mechanism has been employed, for example, to explore the user's ontological profile and to infer items and services that could be interesting to users [7].

These studies on recommender systems suffer from many issues. For example, in order to measure item similarity, content-based methods rely on explicit item descriptions. However, such descriptions may be difficult to obtain for items like ideas or opinions. Collaborative filtering has the data sparsity problem and the cold-start problem [1]. In contrast to the huge number of items in recommender systems, each user normally only rates a few. Therefore, the user/item rating matrix is typically very sparse. It is difficult for recommender systems to accurately measure user similarities from those limited number of reviews. A related problem is the cold-start problem. Even for a system that is not particularly sparse, when a user initially joins, the system has none or perhaps only a few reviews from this user. Therefore, the system cannot accurately interpret this user's preference.

As we can see, traditional collaborative filtering and its variations do not utilize the semantic friend relations among users in recommender systems. However, this is essential to the buying decisions of users. In the following sections, we are going to present a new paradigm of recommender systems that improves the performance of traditional recommender systems by using the information in social networks.

The rest of the paper is organized as follows. Section 2 describes and accounts for the use of a new paradigm of recommender systems by utilizing information in social networks, while Section 3 presents our method about personalized social network-based recommender system with user interest ontology. Section 4 explains a work preview in ontology-based recommendations. Finally, section 5 concludes the paper and describes the orientations of our future work.

2   Social Network-Based Recommender Systems

The impact that social networks has on product marketing has been recognized by many researchers [20], [24]. When we think of buying a particular and unfamiliar product, we most often tend to seek immediate advice from some of our friends who have come across this product or experienced it. We, similarly, tend to accept and welcome a friend's recommendation because we trust them [9].

Integrating social networks in recommender systems can result in more accurate recommendations. That is to say, the information, interests, and recommendations retrieved from social networks can improve the prediction accuracy. Additionally, the information obtained about the users and their friends makes it unnecessary to look for similar users and measure their rating similarity as the fact that two people are already friends can imply that they have things in common [9]. The data sparsity problem can be solved in this case. The cold start issue can also be overcome because even if a user has no interests' history, recommendations can still be made based on a friend's preferences.

If we recall the decisions that we make in our daily life, such as finding restaurants, buying a house, and looking for jobs, many of them are actually influenced by some factors. Intuitively, a customer's buying decision or rating is decided by both his/her own preference for similar items and his/her knowledge about the characteristics of the target item [9]. Fig. 1 further illustrates how these three factors impact customers' final buying decisions.

All of these intuitions and observations motivate us to adopt this approach of social network-based recommender system that can take advantage of information in social networks. Our goal is to represent all these information extracted from social network in a user interest ontology that can help us in the recommendation process.

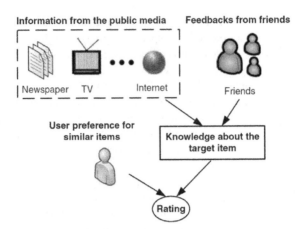

Fig. 1. The three factors that influence a customer's buying decision: user preference for similar items, information regarding the target item from the public media, and feedback from friends [9].

3   User Interest Ontology to Personalize Social Recommender System

Recommender systems have emerged as one successful approach that can help tackle the problem of information overload [12]. They exploit patterns in item metadata and reviews posted by groups of people to find new items that might be interesting to a user. Ontologies are increasingly used within the field of recommender systems, allowing knowledge-based techniques to supplement classical machine learning and statistical approaches. Added to that, ontologies help extend recommender systems to a multi-class environment, allowing knowledge-based approaches to be used alongside classical machine learning algorithms [19]. In fact, different vocabulary items, edited based on ontologies, offered a richer representation of social networks if compared to other classical models [11].

Ontology is a formal specification of shared conceptualizations of a domain, and it can analyze the domain knowledge and make domain assumptions explicit [19]. Ontology facilitates shared common understanding among people or software agents, and it enables the reuse of domain knowledge. [23] shows that the modelling of social networks can be aided by ontologies for several reasons. First, ontologies are commonly deployed for the specification and explication of concepts and relationships related to a given domain. In other words, social networks have the same purpose but with the focus on social relations and entities, hence domain ontologies related to social entities and relations can be designed and deployed.

Second, through reasoning and inference, ontologies do not allow the modelling of contradictory or inconsistent information. Modelling social networks via ontologies ensures the validity of the information encoded.

Third, ontologies, together with the inference mechanism, enable information gain through deploying rules to infer new information. Inference mechanism can be facilitated thanks to   ontology based social networks by coming up with new relations and concepts out of the already existing ones between the social entities i.e. people, locations, organizations and events.

Contrary to research information systems where a request has to be produced in every search, recommender systems help provide users with several recommended resources without explicitly expressing what they are looking for. Additionally, recommender systems should be able to know the most important interests of users and to follow their evolutions over time. This would necessitate the creation of a user interest ontology that helps recommender systems find new items that might be interesting to a user.

User Interest Ontology is a subset of domain ontology of users' goal domain, and is built around the classification of hierarchical ontology tree of users' interests [21]. In our work, user interest ontology is created based on a real social network. Added to that, user interest ontology can be used to classify recommendation items, and is composed of the basic elements of user interest model. It contains all data about preferences as well as old and current user's interest. This ontology is composed of all user's concepts, including the user's preferences and the relationships between concepts. A key factor in this ontology is the semantic relationship used to determine sense relations' types. In other words, the user interest ontology is composed of a set of concepts, a set of relationships and some axioms using logical language. Every concept in our ontology has an interesting degree representing real numbers between 0 and 1 to express the degree of interest of this concept for a user.

User Interest Model is the key and foundation to provide personalized recommendation, and is the formal expression of a user's interest preference in the particular domain. User interest model is the basis and core of a personalized search engine and personalized recommendation system. A good user interest model can express interest of individual users, improve search quality, and help the personalized recommendation system to provide the more accurate referral service [25]. In our work, we establish initial user interest model using user interest ontology.

In order to integrate ontology knowledge into social network-based recommender system and to improve the personalized recommendations performance, we propose the personalized recommendation framework based on user interest model, which is shown in Fig. 2:

When a user logs on the system, the user's personal information is got. Then, based on the personal information, the user interest model is picked. According to the user interest ontology and user interest model, we select the k-nearest similar friends of the user. Based on the opinions of their friends on the social network, the best recommendations are produced. Even if a user has no past reviews, the recommender system still can make recommendations to the user based on the preferences of his/her friends if it integrates with social networks. In the other cases, if we can't find a similar item with his/her friends, the system will try to find the most similar items with the other users in the user interest ontology. Finally, the k-nearest similar items will be recommended to the user. The k-nearest denotes k users that are most similar to the target user. Social network-based recommender system generates the recommendation items according to the opinion of the target user's k-nearest friend.

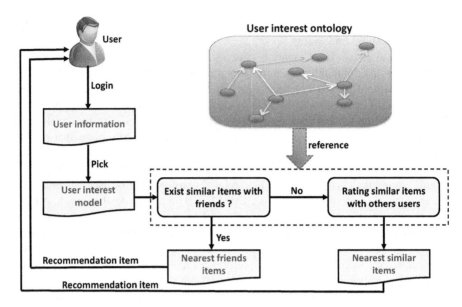

Fig. 2. Personalized social network-based recommender system with user interest ontology

The most important characteristic of our framework is to design a new paradigm of recommender systems that can take advantage of information in social networks and exploit the semantic relations of the user interest ontology and also take the advantage of collaborative filtering when we can't find a friend that is similar to the user. In many cases the type of approach adopted will depend heavily on how much metadata is available about the items and how much user feedback is available, both implicit and explicit. Social network-based recommender system can't work well if we don't have similar friends. However, if the user does not have many friends and there are no similarities to the target item with friends, we try to find similar items with other users in our user interest ontology.

Traditional algorithms for recommender systems suffer from many issues. For example, in order to measure item similarity, content-based methods rely on explicit item descriptions. However, such descriptions may be difficult to obtain for items like ideas or opinions. Collaborative filtering has the data sparsity problem and the cold-start problem. In this paper, we try to solve these problems from a different perspective. In particular, we propose a new paradigm of recommender systems by utilizing information in social networks, especially that of social influence.

In our framework, we have used some semantic similarity measurements [6] in the recommender system algorithm. The module "similarity measurement" aims at calculating similarities between user interest model and the user interest ontology for a target item. In our user interest model, each user has different interest degrees of the same concept. We turn the ontological user interest model into flat vectors of interest degree over the space of concepts. Thus, the similarity between two users is calculated by the Euclidean distance of the interest degrees, not involving the rating of the items. The formula is defined as follows:

$$UserSimilarity(u,v) = \frac{1}{\sqrt{1 + \sum_{i \in C}\left(D^u(C_i) - D^v(C_i)\right)^2}} \qquad (1)$$

Where, u is the target user, v is the friend. C is the set of all concepts in the user interest ontology. $D^u(C_i)$ and $D^v(C_i)$ are the interest degrees of the concept C_i for user u and user v respectively.

According to the calculation result of (1), the most similar k friends are selected to generate the k-nearest friends of the target user u. in the other case, where we have not found a similar friend; our system calculates the similarity between the target user and the other users in the ontology with the same formula (1). Our system gives more importance to social network relations like friendliness. The feedback from friends is another source of knowledge regarding the item, and it is often more trustworthy than neighbors in collaborative filtering. When a user starts considering the feedback from his/her friends, he/she is then influenced by his/her friends.

After the k-nearest friends of the target user have been selected, we have to accurately predict the target user's rating for an item he has not seen before according to the opinion of their friends. The prediction formula is shown as follows [16]:

$$p_{u,i} = \bar{r}_u + \frac{\sum_{v \in friends} UserSimlarity(u,v) * (r_{v,i} - \bar{r}_v)}{\sum_{v \in friends} UserSimlarity(u,v)} \qquad (2)$$

Where, $\bar{r}_u$ is the mean rating of the target user for the target item, friends v is the set of k-nearest friends, $\bar{r}_v$ is the mean rating of a friend, $UserSimilarity(u, v)$ is the user similarity between user u and user v described above, $r_{v,i}$ is the rating of user v for the target item i. Based on the result of (2), we can get the top-N items according to the prediction scores. These items construct the top-N recommendation set.

This paper aims at improving recommender systems regulating decision making in a semantic social network in which semantic indexing of resources will be performed automatically based on the principal behavioral traits of the members of this social network. We try to use the users' preferences published on the social web to enrich the recommender system.

4   Related Work

Traditional methods of recommendation rely on keywords to model user's interests and information requirements [22]. Due to the vagueness of keywords, the user interest model cannot represent user's interests and the relationships between different items of the target domain accurately. Besides there are cold-start and data sparsity problems [17], resulting in poor performance. Many studies have used ontology to design user interest model. Their work has, however, been restricted to using it in traditional recommendation methods.

Pretschner and Gauch [15] are among the first to make use of ontology to build user model and provide personalized document access. Domain ontologies have been employed to organize documents. Based on a user's surfing history, the user model is built and the personalized document access is performed by referring to the interest degree of each document's concept. The semantics of concept relations and ontology structure, however, is not taken into consideration during the calculation of the user's concept interest degree.

The creation of a user model based on ontology to make improvement for information recommendation has also been suggested by Middleton et al. [14]. They advocate that when a low-level concept in the concept hierarchy is selected by the user, this implies that the user may also be interested in a high-level concept. The major limitation of this method is that the inference algorithm is rather simple, which makes it unable to handle complex ontologies.

Jiang and Tan [8] suggest the concept of user ontology, and develop a set of statistical methods to relate individual user ontology to domain ontology. When building user interest model, a value is assigned to every concept and relation. It contains more abundant semantics and more precise interest description. Nevertheless, dealing with complex ontologies requires a longer learning cycle to gain the user ontology.

El-Korany and Khatab [5] propose a social recommender system that follows user's preferences to provide recommendation based on the similarity among users participating in the social network. In this work, ontology is used to define and estimate similarity between users and accordingly being able to connect different stakeholders working in the community field such as social associations and volunteers. This approach is based on integration of content based and collaborative filtering techniques.

Kadima and Malek [10] study the effect of incorporating user semantic profile derived from past user's behaviors and preferences on the accuracy of a recommender system. They present a preliminary work which aims at tackling technical issues due to the integration of an ontology-based semantic user profile within a hybrid recommender system.

Su and al. [21] propose a user interest model based on user interest ontology. In this work, recommendation process is presented by using the ontological user interest model. The biggest shortcoming of this method is that it does not use the advantage of social network in the recommendation process. [21] use only collaborative recommendation in the process of recommendation, but collaborative recommender systems fail to help in cold-start situations, as they cannot discover similar user behavior because there is not enough previously logged behavior data upon which to base any correlations.

5   Conclusion and Future Work

In this work, we have presented an overview of recommendation approaches and their utility to define how to extract information from social networks. We present the role of social networks as sources for the development of recommendation systems. In addition, we proposed a method that personalized social network-based recommender system with user interest ontology. Finally, we described and accounted for the role of user interest model based on user interest ontology to deal with the lack of semantic information in personalized recommendation system nowadays.

This paper applies ontology knowledge to accurately describe users and items to generate personalized recommendation. Based on user interest ontology, we established user interest model and then described the process of social network-based recommendation. The updating of our user interest ontology could be developed in future research work to encompass user interest changes. Furthermore, we are going to improve semantic similarity measurements in order to calculate the similarity between the user interest ontology and the user interest model with more accuracy.

References

1. Adomavicius, G., Tuzhilin, A.: Toward the next generation of recommender systems: A survey of the state-of-the-art and possible extensions. IEEE Transactions on Knowledge and Data Engineering 17(6), 734–749 (2005)
2. Correa, C.D., Ma, K.L.: Visualizing social networks. In: Aggarwal, C.C. (ed.) Social Network Data Analytics, 1st edn., pp. 307–326. Springer (2011)
3. Zhou, X., Xu, Y., Li, Y., Josang, A., Cox, L.: The state-of-the-art in personalized recommender systems for social networking. Artificial Intelligence Review 37(2), 119–132 (2012)
4. Hernâni, B.F.: Social Networks in Information Systems: Tools and Services. In: Manuela, C.M. (ed.) Handbook of Research on Social Dimensions of Semantic Technologies and Web Services, 1st edn. IGI Global (2009)
5. El-Korany, A., Khatab, S.M.: Ontology-based Social Recommender System. IAES International Journal of Artificial Intelligence (IJ-AI) 1(3), 127–138 (2012) ISSN: 2252-8938
6. Frikha, M., Mhiri, M., Gargouri, F.: Extraction of semantic relationships starting from similarity measurements. In: 9th International Conference on Enterprise Information Systems (ICEIS), Funchal-Madeira, Portugal (2007)
7. Gao, Q., Yan, J., Liu, M.: A Semantic Approach to Recommendation System Based on User Ontology and Spreading Activation Model. In: IFIP, pp. 488–492 (2008)
8. Jiang, X., Tan, A.-H.: Ontosearch: A full-text search engine for the semantic web. In: Proceedings of the 21 National Conference on Artificial Intelligence, pp. 1325–1330 (2006)
9. He, J., Chu, W.W.: A Social Network-Based Recommender System (SNRS). In: Memon, N., Xu, J.J., Hicks, D.L., Chen, H. (eds.) Data Mining for Social Network Data. Annals of Information Systems, vol. 12, pp. 47–74 (2010)
10. Kadima, H., Malek, M.: Toward ontology-based personalization of a Recommender System in social network. International Journal of Computer Information Systems and Industrial Management Applications 5, 499–508 (2013) ISSN 2150-7988
11. Lacassaigne, P.: Analyse des Réseaux sociaux: Tendances, Méthodes et Outils libres et open source. Master 2 Prisme (2010)
12. Mabroukeh, N.R.: SemAware: An Ontology-Based Web Recommendation System. A Dissertation in the University of Windsor, Ontario, Canada (2011)
13. Mobasher, B., Jin, X., Zhou, Y.: Semantically Enhanced Collaborative Filtering on the Web. In: Berendt, B., Hotho, A., Mladenič, D., van Someren, M., Spiliopoulou, M., Stumme, G. (eds.) EWMF 2003. LNCS (LNAI), vol. 3209, pp. 57–76. Springer, Heidelberg (2004)

14. Middleton, S.E., Shadbolt, N.R., Roure, D.C.D.: Ontological user profiling in recommender systems. ACM Transactions on Information Systems 22(1), 54–88 (2004)
15. Pretschner, A., Gauch, S.: Ontology based personalized search. In: Proceedings of the 11th IEEE International Conference on Tools with Artificial Intelligence, pp. 391–398 (1999)
16. Resnick, P., Iacovou, N., Suchak, M., Bergstrom, P., Riedl, J.: Grouplens: An open architecture for collaborative filtering of netnews. In: Proceedings of the ACM Conference on Computer Supported Cooperative Work, CSCW 1994, New York, NY, pp. 175–186 (1994)
17. Sarwar, B., Karypis, G., Konstan, J., Riedl, J.: Item-based collaborative filtering recommendation algorithms. In: Proceedings of the 10th International Conference on World Wide Web, New York, NY, pp. 285–295 (2001)
18. Sieg, A., Mobasher, B., Burke, R.: Improving the effectiveness of collaborative recommendation with ontology-based user profiles. In: Proc. of Intl. WIHFR, pp. 39–46 (2010)
19. Middleton, S.E., Roure, D.C.D., Shadbolt, N.R.: Ontology-based Recommender Systems. In: Staab, S., Studer, R. (eds.) Handbook on Ontologies, 2nd edn. Springer (2009)
20. Subramani, M.R., Rajagopalan, B.: Knowledge-sharing and influence in online social networks via viral marketing. Communications of the ACM 46(12), 300–307 (2003)
21. Su, Z., Yan, J., Chen, H., Zhang, J.: Improving the preformance of personalized recommendation with ontological user interest model. In: Seventh International Conference on Computational Intelligence and Security (2011)
22. Tan, A.-H., Teo, C.: Learning user profiles for personalized information dissemination. In: Proceedings of International Joint Conference on Neural Networks, pp. 183–188 (1998)
23. Wennerberg, P.O.: Ontology Based Knowledge Discovery in Social Networks. Final Report (2005)
24. Yang, S., Allenby, G.M.: Modeling interdependent consumer preferences. Journal of Marketing Research 40, 282–294 (2003)
25. Zhu, L., Yan, J., Ling, H., Qian, H.: An Improved Ontology-Based User Interest Model. Modern Applied Science 6(6) (2012) ISSN 1913-1844 E-ISSN 1913-1852, Published by Canadian Center of Science and Education

Measurement and Simulation
of Pacemaker Supply Management

Jan Kubicek[*] and Marek Penhaker

VSB–Technical University of Ostrava, FEECS, K450
17. listopadu 15, 708 33, Ostrava–Poruba, Czech Republic
{jan.kubicek,marek.penhaker}@vsb.cz

Abstract. This paper deals with the design of mathematical algorithms for measuring and predicting the status of batteries used in pacemakers. In clinical practice, the status of pacemaker batteries can be measured using a programmer. A disadvantage of this measurement is the fact that it is available only in large hospitals because the respective financial costs are quite high. The aim of this work is the development of numerical algorithms that can measure battery status after applying a magnet with an induction greater than 1mT. This methodology would be significantly cheaper and would allow determining the battery status in clinical practice. The proposed methodology is tested on historical patient data provided by the University Hospital in Ostrava.

Keywords: Pacemaker, MATLAB, battery, peak detector, ECG.

1   Introduction

Currently, the number of people who suffer from heart diseases leading to permanent cardiac pacing is increasing. Therefore, implantation of pacemakers controlling the activities of the heart in the event of its failure is common practice. A pacemaker's service life, however, depends on the durability of its battery, which is an integral part. Today, pacemaker power supplies are provided by lithium-iodine batteries whose lifespan ranges from 5 to 10 years (or more). Pacing frequency declines and pacing pulse extends with drops in battery capacity. This results in more frequent medical controls and the subsequent exchange of the pacemaker itself. Battery status is determined either through telemetric connections via a programmer or by the speed of the output pacing when applying a magnet. Nevertheless, programmers are available only in large hospitals due to their high cost.

2   Pacemaker Power Supplies

In principle, implantable pacemakers can be supplied from a high-frequency external source or battery placed in a pacemaker's case. A primary requirement is independence of any external sources for a theoretically unlimited period of time. Primary or

---

[*] Corresponding author.

J. Sobecki, V. Boonjing, and S. Chittayasothorn (eds.), *Advanced Approaches to Intelligent Information and Database Systems*, Studies in Computational Intelligence 551,
DOI: 10.1007/978-3-319-05503-9_26, © Springer International Publishing Switzerland 2014

secondary cells have therefore been used over many years. With regard to the desired service life (7-10 years), emphasis is placed on a low level of self-discharge current, the greatest possible capacity per source volume, optimal discharge characteristics and the possibility of sealing. For many years, pacemakers used mercury cells with a terminal voltage of 1.35 V, a capacity of 1-1.8 ampere-hours with a self-discharge current of 10 to 23 μA and a durability of about three years. Their main disadvantage consisted in releasing gases (due to chemical reactions) that had to be collected by an absorber. [1], [2], [4], [6]

2.1   Pacing Pulse

Stimulation requires a negative voltage pulse. Its amplitude and width are optional parameters. The pulse amplitude is measured as a leading edge voltage of the device output pulse. It is an independently programmable parameter whose values can range from zero (stimulation off) up to 7.5 V or 8.4 V or more, depending on the device type.

The pulse width is an independently programmable parameter and determines how long the output stimulation amplitude is applied between the stimulation poles of electrodes. The pulse width is programmable anywhere from 0.05 to 2 ms (typically from 0.2 to 0.5 ms). According to ČSN EN 45502-2-1, the pulse width is measured as one-third of the pulse leading edge. [10]

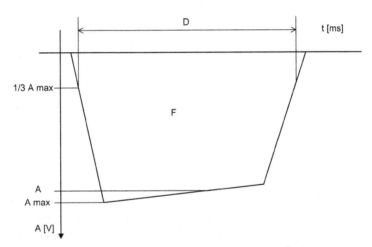

Fig. 1. Measurement of the pulse amplitude and width according to ČSN EN 45502-2-1 (D – pulse width; pacing pulse amplitude A is determined as a ratio of area F to pulse width D) [5], [7], [11]

2.2   A Drop in Power Source Capacity

Upon depletion of the power source (battery), the output of pacing pulses may fail. This is manifested by a decrease in the frequency of pacing pulses and by their extension. [3]

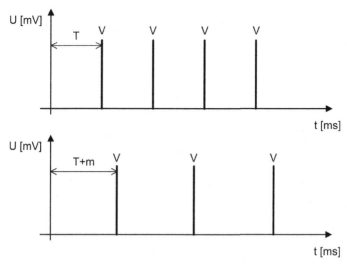

Fig. 2. Pacing frequency induced by a magnet when a pacemaker is functioning properly and when a pacemaker's battery capacity drops (single-chamber pacemaker - VVI) [6]

In single-chamber pacemakers, we monitor the distance (time response) between two pacing pulses; then, the battery capacity decrease is signalled by the increased distance of these pulses. [12]

Regarding double-chamber pacemakers, in addition to distances AA and VV (for which AA=VV), we also have to consider the distance AV. [6], [8], [9]

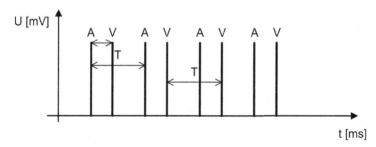

Fig. 3. Pacing frequency induced by a magnet when a double-chamber pacemaker (DDD type) is functioning properly [6]

3   A Proposal of Algorithms for Measuring Pacemaker Battery Status

As stated above in the previous chapter, a drop in pacemaker battery status may occur for two reasons:

- A decrease in pacing frequency (the distance between pacing pulses increases).
- A change in pacing pulse duration.

These two factors have been used as a basis for the proposed algorithms for measuring battery status. The algorithms were processed and tested in MATLAB simulation software.

3.1 Algorithms for Single-Chamber Pacemakers

This pacemaker type emits a pacing pulse in either atria (A) or ventricles (V). The function for detecting the pulse width and determining the distance between pulses is based on a selected layer and its detection when passing through the signal. The mentioned level is the pacing pulse height from which the pacing frequency after applying a magnet and pacing pulse width could be deducted. This value can be selected at will based on the range of y (amplitude) axis. Detected pulse widths and distances between pacing pulses are then tabulated.

Given the possibility of defining the mentioned level, we designed two functions: one for the selected level for positive values (pulse width 1) and the other for the selected level for negative values (pulse width 2).

3.2 Functions for Double-Chamber Pacemakers

This type of stimulation is applied to atria as well as ventricles (these pacemakers emit 2 pacing pulses); functions for detecting the distance between pacing pulses are therefore solved in another way. One table column measures the distances between all recorded pacing pulses (column 4); distances AA and VV (given by summing the respective distances), and AV (these correspond to the shortest distances in the above-mentioned table column), are then solved based on these distances. This function, however, is based on the assumption that the first pacing pulse is applied to atria (A). The function for detecting the pacing pulse width is solved in the same manner as with single-chamber pacemakers.

3.3 Functions for Detecting Pacing Frequency

Part of the algorithm is a function that searches for local maxima and minima. The calculated coordinates are then stored in two vectors. Through this process, it is possible to estimate pacing frequency. Based on these data, pacing frequency is calculated according to the following relationship:

$$STF = \frac{60}{t} \tag{1}$$

The output parameter STIM_FREK is the pacing frequency and t indicates the duration of the interval between individual pacing pulses (in seconds). Part of this algorithm is a function that – based on the identified local minima – determines the optimal value of the level that can be entered into the pulse width function. This level is defined based on the fact that the pacing pulse width is measured at approximately one-third of this pulse height.

4   Battery Durability Test in Pulse Mode

In order to understand and demonstrate how the batteries of implanted pacemakers behave under stimulation, we took measurements of the durability of batteries that were first discharged through drawing a constant current and then through the pulse mode. We did not take measurements of pacemaker batteries (lithium-iodine), but on other batteries due to the high cost of the mentioned battery type.

4.1   The Implementation of Measurements

The measurements were taken using a functional device that cyclically charges and discharges batteries with a balancing system to determine the number of cycles in batteries until they have the appropriate characteristics. These were adapted for our needs.

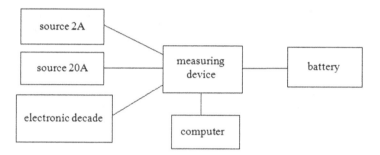

Fig. 4. Block diagram of the measuring device

We used a laboratory power source to charge the batteries and an electronic decade to discharge them. The entire process was controlled by a PC equipped with Labview software and a measuring card that operated a measuring jig and relay for switching between charging and discharging. To ensure proper battery charging, each battery was fitted with a balancing system. The measuring jig contained a relay for switching the charging sources, FET transistors for connecting the load to discharge batteries and a measuring card that handled communication with the PC. It switched the relay and FET transistors from commands by the program running on the computer. We used an NI USB 6008 measuring card.

4.2   Discharging Batteries by Drawing a Constant Current

Within this type of measurement, the batteries were gradually charged and discharged in several consecutive cycles. Discharging was performed using a constant current of 200 mA.

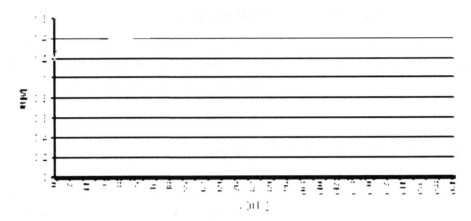

Fig. 5. Example of battery charging

The battery was discharged to 1.171 V; its charge at 1.496 V lasted approximately 2.75 hours (2 hours and 45 minutes).

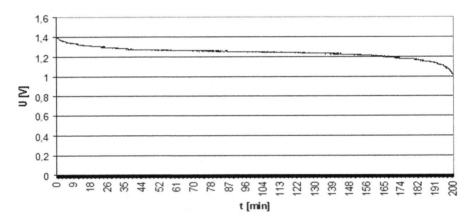

Fig. 6. Battery discharging

The graph shows that a battery discharged to 1.008 V lasted approximately 3.33 hours (3 hours and 19 minutes and 45 seconds).

5   Testing Algorithms with Real Patient Data

Proposed algorithms for measuring pacemaker battery status were tested using ECG records. The tested record was synthetically combined with pacing pulses. The record below was made with a sampling frequency of 10 kHz. Pacing pulses were always placed at the beginning of the P-wave.

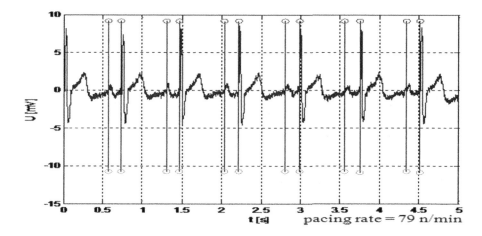

Fig. 7. Testing an ECG signal with detection of local extremes for calculating the pacing frequency – single-chamber stimulation

In the following table, the first and second columns show the indices (rows) from the source file recording the level value when passing through the signal (-4.66 in this case). The third column shows the width of each pacing pulse, 0.1 ms in our case. The last column presents the relative distances between pacing pulses in seconds (the first value is 0 because it indicates the first pacing pulse).

Table 1. Results for single-chamber cardiac pacing when using the pulse width 2 function

Detected A	Detected V	Difference
5694	5699	0
13036	13041	0.7333
20381	20386	0.7336
28046	28051	0.7656
35528	35533	0.7473
43467	43472	0.7930

6   Conclusion

This work results in the identification of a possible method that can be used for assessing the measurement of pacemaker battery status while applying a magnet. In practice, this is done by a programmer, but the respective measurement is very expensive.

The proposed algorithms stem from the fact that a drop in pacemaker battery capacity leads to extensions of the width of used pacing pulses as well as distances between them.

The objective is to apply algorithms to a prototype clinical ECG device in the future. A prototype ECG monitor was used to sense an ECG signal. On this signal, we created pacing pulses in order to test and demonstrate the functionality of the proposed functions. The only function whose results could not be used was that for finding the optimal level (which corresponds to 1/3 of the pacing pulse height). This value was lower than the S-wave value; when using this optimum value, we could also detect these S-waves in addition to pacing pulses. The functions were also tested on data from the City Hospital in Ostrava, sensed by a DASH4000 device using 4 electrodes (I., II., III. standard lead + V_5 lead). After applying the functions to these data, we found only a part of them to be usable. This was the part that calculated the distance between pacing pulses. These data could not be used for assessing the pacing pulse width due to the low sampling frequency (240 samples / second).

To get an idea how the batteries of implanted pacemakers behave during stimulation, we measured the battery service life; the batteries were first discharged through drawing a constant current and then through the pulse mode. Lithium-iodine batteries were not used for taking measurements because of their high cost. In the pulse mode, the total discharge time was approximately 3.09 times greater than when discharging using a constant current. Battery discharge characteristics were similar because the measurement method, which we developed, was not entirely correct (due to the measurement system used).

Acknowledgment. The work and the contributions were supported by the project SP2013/35 "Biomedical engineering systems IX" The paper has been written within the framework of the IT4 Innovations Centre of Excellence project, reg. no. CZ.1.05/1.1.00/02.0070 supported by the Operational Programme 'Research and Development for Innovations' funded by Structural Funds of the European Union and the state budget of the Czech Republic. The paper has been elaborated in the framework of BIOM (reg. č. CZ.1.07/2.3.00/20.0073). This paper has been elaborated in the framework of the project „Support research and development in the Moravian-Silesian Region 2013 DT 1 - International research teams" (RRC/05/2013). Financed from the budget of the Moravian-Silesian Region.

References

1. Penhaker, M.: Lékařské terapeutické přístroje, 1.vydání, VŠB – Technická univerzita Ostrava, Ostrava, 2007 – 216s (2007) ISBN 978-80-248-1558-9
2. Penhaker, M., a kol: Lékařské diagnostické přístroje, 1.vydání. VŠB – Technická univerzita Ostrava, Ostrava, 2004 – 320s (2004) ISBN 80-248-0751-3
3. Khan, M.G.: EKG a jeho hodnocení, 1.vydání, Grada Publishing, Praha, 2005 – 348s (2005) ISBN 80-247-0910-4
4. Hampton, J.R.: EKG stručně, jasně, přehledně, 2.vydání, Grada Publishing, Praha, 2005 – 152s(2005) ISBN 80-247-0960-0
5. Aschermann, M.: Kardiologie. Díl 1, 1.vydání. Galén, Praha, str. 212 (2004) ISBN 80-7262-290-0
6. Korpas, D.: *Kardiostimulační technika.* 1.vydání; Praha: Mladá fronta, 2011 – 208s (2011) ISBN 978-80-204-2492-1

7. Extramiana, F., Maison-Blanche, P., Badilini, F., Pinoteau, J., Deseo, T., Coumel, P.: Circadian modulation of QT rate dependence in healthy volunteers. Gender and age differences. Journal Electrocardiol. 32, 33–43 (1999)
8. Halamek, J., Jurak, P., Leinverber, P., Vondra, V., Lipoldová, J.: Gender Dependency od QT Parameters. In: Wessel, N. (ed.) Proceedings of the 6th ESGCO, pp. 1–14. Humboldt – Universität zu Berlin, Berlín (2010a)
9. Jeong, G.-Y., Yu, K.-H.: Design of Ambulatory ECG Monitoring System to detect pattern change. In: SICE-ICASE International Joint Conference, Bexco, Busan, Korea, pp. 5873–5877 (2006)
10. Kicmerová, D., Jurak, P., Halamek, J., Provaznik, I.: Verification of QT Interval Detectors. In: Proceedings, J. (ed.) Proceedings of the 18th Biennial International Eurasip Conference Biosignal 2006, pp. 163–165. Brno University of Technology VUTIUM Press, Brno (2006)
11. Lund, K., Perkiomaki, J.S., Brohet, C., Elming, H., Zaidi, M., Trop-Pedersen, C., Huikuri, H.V., Nygaard, H., Kristein Padersen, H.: The prognostic accuracy of different QT interval measures. Ann. Noninvasive Electrocardiol. 7, 10–16 (2002)
12. Vanus, J., Novak, T., Koziorek, J., Konecny, J., Hrbac, R.: The proposal model of energy savings of lighting systems in the smart home care. In: IFAC Proceedings Volumes (IFAC-Papers Online), vol. 12 (PART 1), pp. 411–415 (2013) ISSN: 14746670, ISBN: 9783902823533

Part IV

Digital Multimedia Systems Methods and Applications

Mixed Performance Controller Design of Dynamic Positioning Systems for Ships

Cheung-Chieh Ku, Rui-Wen Chen, and Min-Ta Li

Department of Marine Engineering, National Taiwan Ocean University
Keelung 202, Taiwan, R. O. C.
ccku@mail.ntou.edu.tw

Abstract. The mixed controller design problem of Dynamic Positioning Systems for Ships (DPSS) is discussed in this paper. Based on the H_2 control scheme, passivity theory and Lyapunov function, the sufficient conditions are derived into Linear Matrix Inequality (LMI) problems which can be solved by convex optimization algorithm. Therefore, the mixed performance controller can be designed via solving proposed sufficient conditions for guaranteeing the mixed performance of DPSS with minimized output energy and initial conditions. Finally, the simulation results are presented to demonstrated the effectiveness and useful of the proposed design method.

Keywords: DPSS, Mixed Performance, Passivity Theory, H_2 method, LMI.

1   Introduction

It is varying important as handling the ship to achieve destination since the environment on the ocean is changing and strange. For the reason, there are many articles [1-3] in discussing and researching the stability and stabilization problems of DPSS. The DPSS stabilization problems are usually solved with the assumption that the kinematic equations can be linearized about a constant yaw angle such that linear theory and gain scheduling techniques can be applying. Most of them [1-3], the two issues are investigated separately for proposing the controller design method. Although the presented design methods deal with the stabilization problems of DPSS, only one of concerned performances can be guaranteed specially with initial conditions. In issue of discussing disturbance attenuation performance [4-5], the consideration as minimizing output energy of DPSS is not investigated at the same time. For stability problems [6-7], the attenuation performance is not considered in those researches. Thus, single performance issue is usually discussed in present literatures.

For powerful design method, the mixed performance controller design technique [8-10] is proposed for taking account of both stability performance and disturbance attenuation performance. For achieving both of performances, the stability analysis and controller synthesis are more difficult than stabilization problems dealing with single performance. Referring to [8-10], the mixed performance design method is proposed with H_∞ and H_2 control schemes. In the present literatures, the H_∞

J. Sobecki, V. Boonjing, and S. Chittayasothorn (eds.), *Advanced Approaches to Intelligent
Information and Database Systems*, Studies in Computational Intelligence 551,
DOI: 10.1007/978-3-319-05503-9_27, © Springer International Publishing Switzerland 2014

control scheme is applied to discuss the disturbance attenuation performance of DPSS. And, the H_2 control scheme is employed to guarantee the stability of system and minimize the output energy of system. Thus, the attenuation performance and stability of considered system can be guaranteed via designed mixed performance controller with initial conditions and minimized output energy.

For attenuating the disturbance effect, the passivity theory [11] provides a general and flexible tool to analyze stability of system. Based on the definition of power supply function [11], the passivity theory includes several categories of performance constraints. According to settings of power supply function, one can find that the passivity theory is more generalization than the H_∞ control scheme to discuss the energy change between the system and external disturbance.

In order to discuss the mixed performance issue of DPSS, the sufficient conditions are derived via employing Lyapunov function, passivity theory and H_2 control scheme. Based on the Lyapunov stability theory, the common definite matrix is needed to be found for satisfying the conditions to guarantee the attenuation performance and stability of DPSS. Furthermore, the conditions are converting into LMI problem [12]. For the LMI conditions, the feasible solutions can be directly obtained via optimal convex programming algorithm [12] and the controller can also be designed with state feedback scheme. Through the design controller, the stability and disturbance attenuation performance of DPSS can be achieved with minimum output energy and initial conditions.

2   Linear Model of DPSS

Referring to [1], the DPSS is controlled by means of thrusters, exclusively. And, the ship can also be supplied by anchors with considering thruster-assisted mooring. Furthermore, the damping force is assumed to be linear because of the speed of the ship is quite small during ship steering systems. Thus, the dynamic equations of DPSS can be described as follows:

$$\dot{\eta}(t) = \mathbf{J}(\mathbf{\eta})v(t) \tag{1a}$$

$$\dot{v}(t) = \mathbf{R}_1\eta(t) + \mathbf{R}_2v(t) + \mathbf{R}_3\tau(t) \tag{1b}$$

where $\eta(t) = \begin{bmatrix} m(t) & n(t) & \psi(t) \end{bmatrix}^{\mathrm{T}}$ and $v(t) = \begin{bmatrix} s(t) & v(t) & r(t) \end{bmatrix}^{\mathrm{T}}$. The elements of the vector can be express as the earth-fixed positions $(m(t), n(t))$ and yaw angle $\psi(t)$ of the ships. And, the surge, sway and yaw modes are represented as $s(t)$, $v(t)$ and $r(t)$, respectively. The control and moment are proposed with thruster system as $\tau(t) = \begin{bmatrix} \tau_1(t) & \tau_2(t) & \tau_3(t) \end{bmatrix}^{\mathrm{T}}$. Moreover, the $\mathbf{R}_1$ is the state matrix of the earth-fixed positions and yaw angle $\psi(t)$ of vessel, the $\mathbf{R}_2$ is the state matrix of the body-fixed velocities, and the $\mathbf{R}_3$ is the state matrix of force and moment which are provided by the thruster system. The $J(\eta)$ is the rotation matrix in yaw angle of ship. And the above matrices can be presented as follows.

$$J(\eta) = \begin{bmatrix} \cos(\psi(t)) & -\sin(\psi(t)) & 0 \\ \sin(\psi(t)) & \cos(\psi(t)) & 0 \\ 0 & 0 & 1 \end{bmatrix}, \quad R_1 \equiv -N^{-1}T = \begin{bmatrix} r_{1_{11}} & r_{1_{12}} & r_{1_{13}} \\ r_{1_{21}} & r_{1_{22}} & r_{1_{23}} \\ r_{1_{31}} & r_{1_{32}} & r_{1_{33}} \end{bmatrix},$$

$$R_2 \equiv -N^{-1}Q = \begin{bmatrix} r_{2_{11}} & r_{2_{12}} & r_{2_{13}} \\ r_{2_{21}} & r_{2_{22}} & r_{2_{23}} \\ r_{2_{31}} & r_{2_{32}} & r_{2_{33}} \end{bmatrix} \quad \text{and} \quad R_3 \equiv -N^{-1} \begin{bmatrix} r_{3_{11}} & r_{3_{12}} & r_{3_{13}} \\ r_{3_{21}} & r_{3_{22}} & r_{3_{23}} \\ r_{3_{31}} & r_{3_{32}} & r_{3_{33}} \end{bmatrix} \tag{2}$$

Starboard-port symmetry of ships implies that N and Q that are constructed such as

$$N = \begin{bmatrix} n_{11} & 0 & 0 \\ 0 & n_{22} & n_{23} \\ 0 & n_{32} & n_{33} \end{bmatrix} \quad \text{and} \quad Q = \begin{bmatrix} q_{11} & 0 & 0 \\ 0 & q_{22} & q_{23} \\ 0 & q_{32} & q_{33} \end{bmatrix} \tag{3}$$

The N is the inertia matrix including hydrodynamic added inertia. Form the (3), one can find that N is not symmetrical matrix since $n_{23} \neq n_{32}$ and the properties of hydrodynamic added inertia [3]. The Q is damping matrix which also belongs to non-symmetrical matrix in most cases.

Since the coupling is not caused between surge and sway-yaw subsystem, the anchor forces and moment are usually represented by a diagonal matrix, such as

$$T = \begin{bmatrix} t_{11} & 0 & 0 \\ 0 & t_{22} & 0 \\ 0 & 0 & t_{33} \end{bmatrix} \tag{4}$$

For considering the stern environment, the external disturbance is added in $m(t)$. With setting the new variables, the dynamic equation (1) can be converted as follows.

$$\dot{x}_1(t) = \cos(x_3(t)) x_4(t) - \sin(x_3(t)) x_5(t) + 0.1w(t)$$

$$\dot{x}_2(t) = \sin(x_3(t)) x_4(t) + \cos(x_3(t)) x_5(t)$$

$$\dot{x}_3(t) = x_6(t)$$

$$\dot{x}_4(t) = r_{1_{11}} x_1(t) + r_{1_{12}} x_2(t) + r_{1_{13}} x_3(t) + r_{2_{11}} x_4(t) + r_{2_{12}} x_5(t) + r_{2_{13}} x_6(t) + r_{3_{11}} u_1(t) + r_{3_{12}} u_2(t) + r_{3_{13}} u_3(t)$$

$$\dot{x}_5(t) = r_{1_{21}} x_1(t) + r_{1_{22}} x_2(t) + r_{1_{23}} x_3(t) + r_{2_{21}} x_4(t) + r_{2_{22}} x_5(t) + r_{2_{23}} x_6(t) + r_{3_{21}} u_1(t) + r_{3_{22}} u_2(t) + r_{3_{23}} u_3(t)$$

$$\dot{x}_6(t) = r_{1_{31}} x_1(t) + r_{1_{32}} x_2(t) + r_{1_{33}} x_3(t) + r_{2_{31}} x_4(t) + r_{2_{32}} x_5(t) + r_{2_{33}} x_6(t) + r_{3_{31}} u_1(t) + r_{3_{32}} u_2(t) + r_{3_{33}} u_3(t)$$

$$(5)$$

where $w(t)$ is an external disturbance chosen as zero mean white with variance one. And the state vectors are shown as follows.

$$x(t) = \begin{bmatrix} x_1(t) & x_2(t) & x_3(t) & x_4(t) & x_5(t) & x_6(t) \end{bmatrix} = \begin{bmatrix} m(t) & n(t) & \psi(t) & s(t) & v(t) & r(t) \end{bmatrix}$$
$$u(t) = \begin{bmatrix} u_1(t) & u_2(t) & u_3(t) \end{bmatrix} = \begin{bmatrix} \tau_1(t) & \tau_2(t) & \tau_3(t) \end{bmatrix}^T.$$

For obtaining the linear model of DPSS (5), the Teixeira-Zak's linearization formula is applied with equilibrium point $x_{EP}(t) = \begin{bmatrix} 0 & 0 & 0 & 0 & 0 & 0 \end{bmatrix}$. Then, one has

$$\dot{x}(t) = A_2 x(t) + B_2 u(t) + E_2 w(t)$$

$$(6)$$

where
$$A = \begin{bmatrix} 0 & 0 & 0 & 1 & 0 & 0 \\ 0 & 0 & 0 & 0 & 1 & 0 \\ 0 & 0 & 0 & 0 & 0 & 1 \\ r_{1_{11}} & r_{1_{12}} & r_{1_{13}} & r_{2_{11}} & r_{2_{12}} & r_{2_{13}} \\ r_{1_{21}} & r_{1_{22}} & r_{1_{23}} & r_{2_{21}} & r_{2_{22}} & r_{2_{32}} \\ r_{1_{31}} & r_{1_{32}} & r_{1_{33}} & r_{2_{31}} & r_{2_{32}} & r_{2_{33}} \end{bmatrix}, B = \begin{bmatrix} 0 & 0 & 0 & r_{3_{11}} & r_{3_{21}} & r_{3_{31}} \\ 0 & 0 & 0 & r_{3_{12}} & r_{3_{22}} & r_{3_{32}} \\ 0 & 0 & 0 & r_{3_{13}} & r_{3_{23}} & r_{3_{33}} \end{bmatrix}^T \text{ and}$$

$E = \begin{bmatrix} 0.1 & 0 & 0 & 0 & 0 & 0 \end{bmatrix}^T$. Otherwise, the measured output vector $y(t)$ and controlled output vector $z(t)$ are chosen as following forms.

$$y(t) = C_1 x(t) + D_1 w(t) \quad \text{and} \quad z(t) = C_2 x(t) + D_2 u(t)$$

$$(7)$$

where $D_1 = 1$, $C_1 = \begin{bmatrix} 1 & 0 & 0 & 0 & 0 & 0 \end{bmatrix}$, $C_2 = \begin{bmatrix} 0 & 1 & 1 & 0 & 0 & 0 \end{bmatrix}$ and $D_2 = \begin{bmatrix} 0 & 1 & 1 \end{bmatrix}$. In this paper, the state feedback controller is considered to deal with the mixed performances of DPSS. Thus, the controller can be structured as follows.

$$u(t) = Fx(t)$$

$$(8)$$

where F is the feedback gain matrix. Substituting (8) into (6) and (7), the closed-loop system can be obtained such as

$$\dot{x}(t) = (A + BF)x(t) + Ew(t) = Gx(t) + Ew(t)$$

$$(9a)$$

$$y(t) = C_1 x(t) + D_1 w(t)$$

$$(9b)$$

$$z(t) = (C_2 + D_2 F)x(t) = Hx(t)$$

$$(9c)$$

where $G = A + BF$ and $H = C_2 + D_2 F$.

The linear system (9) is concerned in this paper for analyzing the stability and designing the controller. And, with designed controller, the stability and attenuation performance of DPSS can be achieved with initial conditions and minimum output energy. Thus, the proposed theorem is stated as next section.

3   Mixed Performance Controller Design for DPSS

In this section, the mixed performance controller design method is proposed for DPSS. Before discussing the mixed performance, some definitions are firstly presented for deriving the proposed theorem of this paper. For discussing the attenuation performance, the definition of passivity theory is introduced as follows.

3.1   Definition 1 [11]

The closed–loop system (9) with disturbance input $w(t)$ and measured output $y(k)$ is said to be strictly input passivity if there exists a scalar $\gamma \geq 0$ such that

$$2 \int_0^{t_p} w^{\mathrm{T}}(s) \, y(s) \, ds \geq \gamma \int_0^{t_p} w^{\mathrm{T}}(s) \, w(s) \, ds \tag{10}$$

for all $t_p \geq 0$ and for all solution of (9).   #

On the other hand, the H_2 performance inequality is defined as follows.

3.2   Definition 2   [8]

The controller (8) is an H_2 performance measure which can be defined as follow.

$$\int_0^{\infty} z^{\mathrm{T}}(t) z(t) < \alpha \tag{11}$$

where the positive value α is the upper bounded of H_2 inequality.   #

In (11), one can minimize the upper bound of α for minimizing the value of inequality in Definition 2.

With the above definitions, the sufficient conditions can be derived via Lyapunov function. Through the conditions, the proposed theorem can be obtained to guarantee the stability and performance of DPSS with initial conditions. And the theorem is proposed as follows.

3.3   Theorem 1

With a positive scalar γ, the mixed performance of closed-loop system (9) is achieved if positive definite matrices **P** and feedback gain **F** can be found with minimization α such that

$$\begin{bmatrix} \mathbf{G}^T\mathbf{P}+\mathbf{PG} & * \\ -\mathbf{C}_1+\mathbf{E}^T\mathbf{P} & \gamma\mathbf{I}-\mathbf{D}_1^T-\mathbf{D}_1 \end{bmatrix} < 0, \quad \mathbf{G}^T\mathbf{P}+\mathbf{PG}+\mathbf{H}^T\mathbf{H} < 0 \quad \text{and} \quad x(0)^T \mathbf{P}x(0)-\alpha < 0$$

(12)

where * denotes the transposed elements or matrices for symmetric position.

- *Proof:* Let us choose the Lyapunov function as $V(x(t)) = x^T(t)\mathbf{P}x(t)$ to represent the energy of system. And, the sufficient conditions can be derived via applying the Lyapunov function, passivity theory and H_2 control scheme. Then, one can find the feasible solutions for satisfying the conditions (12) such that the DPSS achieves the definitions of this paper. Since the limitation of space, the proof of this theorem is omitted.   #

Through applying the Lyapunov function and passivity theory, the BMI condition (12) is derived to guarantee the mixed performance. Unfortunately, the BMI conditions cannot be calculated by optimal convex programming algorithm [12]. For this reason, the BMI conditions of this paper can be directly converted into LMI problems with using similar process of [3]. And then, the common positive definite matrix **P** and feedback controllers **F** can be obtained for satisfying the proposed conditions.

4   Simulation Results for DPSS

Referring to [1, 3], the tanker system matrices can be obtained as follows.

$$\mathbf{N} = \begin{bmatrix} 1.0852 & 0 & 0 \\ 0 & 2.0575 & -0.4087 \\ 0 & -0.4087 & 0.2153 \end{bmatrix}, \quad \mathbf{Q} = \begin{bmatrix} 0.0865 & 0 & 0 \\ 0 & 0.0762 & 0.151 \\ 0 & 0.0151 & 0.0031 \end{bmatrix} \text{ and}$$

$$\mathbf{T} = \begin{bmatrix} 0.0389 & 0 & 0 \\ 0 & 0.0266 & 0 \\ 0 & 0 & 0 \end{bmatrix}$$

(13)

With above matrices, the matrices of DPSS (1) can be calculated as follows.

$$\mathbf{A} = \begin{bmatrix} 0 & 0 & 0 & 1 & 0 & 0 \\ 0 & 0 & 0 & 0 & 1 & 0 \\ 0 & 0 & 0 & 0 & 0 & 1 \\ -0.0358 & 0 & 0 & -0.0797 & 0 & 0 \\ 0 & -0.0208 & 0 & 0 & -0.0818 & -0.1224 \\ 0 & -0.0394 & 0 & 0 & -0.2254 & -0.2648 \end{bmatrix}$$

$$\text{and } \mathbf{B} = \begin{bmatrix} 0 & 0 & 0 & 0.9215 & 0 & 0 \\ 0 & 0 & 0 & 0 & 0.7802 & 1.4811 \\ 0 & 0 & 0 & 0 & 1.4811 & 7.4562 \end{bmatrix}^{\mathrm{T}} \tag{14}$$

Thus, one can employ the optimal convex programming algorithm [12] in the LMI-toolbox of MATLAB software to calculate the sufficient conditions for finding feasible solutions with given $\gamma = 1$ and initial condition as $x(0) = \begin{bmatrix} 1 & 1 & 10^{\circ} & 0 & 0 & 0 \end{bmatrix}$. Thus, the common positive definite matrix $\mathbf{P} > 0$, feedback gain $\mathbf{F}$ and upper of minimum output energy can be obtained as follows.

$$\mathbf{P} = \begin{bmatrix} 0.079 & -0.0015 & -0.0024 & 0.0015 & 0.0012 & -0.0017 \\ -0.0015 & 0.0018 & 0.0013 & 0 & -0.0002 & 0.0009 \\ -0.0024 & 0.0013 & 0.0072 & -0.0001 & -0.0051 & 0.0041 \\ 0.0015 & 0 & -0.0001 & 0.0002 & 0.0001 & -0.0001 \\ 0.0012 & -0.0002 & -0.0051 & 0.0001 & 0.0066 & -0.0035 \\ -0.0017 & 0.0009 & 0.0041 & -0.0001 & -0.0035 & 0.0043 \end{bmatrix}$$

$$\mathbf{F} = \begin{bmatrix} -365.0953 & 9.4116 & 14.7045 & -8.1764 & -7.1939 & 10.2641 \\ 1.9568 & -0.8636 & 0.2945 & 0.0188 & -1.1246 & 1.3585 \\ -1.84794 & -0.1453 & -1.3248 & -0.0177 & 1.1449 & -1.3947 \end{bmatrix}$$

$$\text{and } \alpha = 0.0776 \tag{15}$$

Based on the obtained feedback gain matrix in (15), the mixed performance controller can be structured as (8). And then, the simulation results of DPSS (1) are stated in Fig. 1 to Fig. 6. Form the simulation results, the following equations are calculated for proving that the mixed performances of DPSS with designed controller are achieved.

$$\frac{E\left\{ 2 \int_0^{t_p} y^{\mathrm{T}}(t) \mathbf{S} \, v(t) \, dt \right\}}{E\left\{ \int_0^{t_p} v^{\mathrm{T}}(t) \, v(t) \, dt \right\}} = 2.023 \quad \text{and} \quad \int_0^{T_f} z^{\mathrm{T}}(t) z(t) = 0.01154 \tag{16}$$

The ratio value in (16) is bigger than determined dissipation rate $\gamma = 1$, one can find that the inequality (10) of Definition 1 is satisfied. On the other hand, the output energy of DPSS in simulation can be obtained and is smaller than the $\alpha = 0.0776$. Therefore, the considered DPSS (1) with external disturbance can achieve desired mixed performances by the designed fuzzy controller.

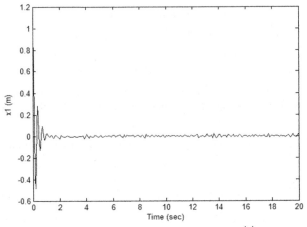

Fig. 1. Responses of earth-fixed position $m(t)$

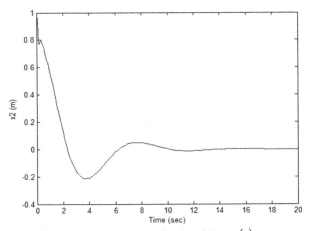

Fig. 2. Responses of earth-fixed position $n(t)$

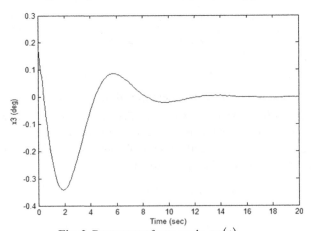

Fig. 3. Responses of yaw angle $\psi(t)$

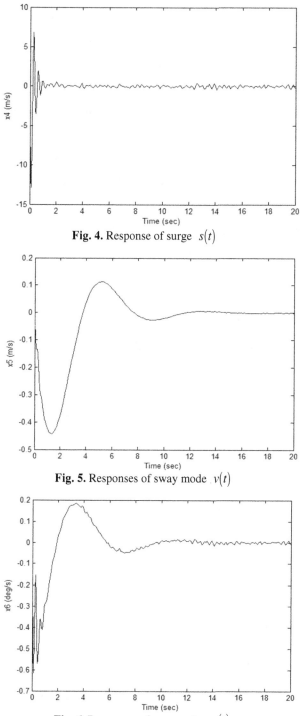

Fig. 4. Response of surge $s(t)$

Fig. 5. Responses of sway mode $v(t)$

Fig. 6. Responses of yaw mode $r(t)$

5   Conclusions

The problems of stability and synthesis for DPSS are investigated and studied focused on mixed performance. For solving the problems of this paper considering, the stable condition is developed by passivity theory and Itô's formula. Based on the LMI technique, the BMI condition is converted into LMI problem that can be calculated by LMI toolbox in MATLAB. Particularly, the PDC concept is employed to design the fuzzy controller for the multiplicative noise T-S fuzzy model to achieve mean square stable and passivity performance.

Acknowledgements. This work was supported by the National Science Council of the Republic of China under Contract NSC 102-2218-E-019-003.

References

1. Fossen, T.I.: Guidance and Control of Ocean Vehicles. Wiley, New York (1994)
2. Fossen, T.I., Grovlen, A.: Nonlinear output Feedback Control of Dynamically Positioned Ship Using Vectorial Observer Backstepping. IEEE Trans. on Control Systems Technology 6, 121–128 (1998)
3. Chang, W.J., Chang, K.Y.: Variable Structure Control with Performance Constraints for Perturbed Ship Steering Systems. J. of the Society of Naval Architects and Marine Engineers 17, 45–55 (1998)
4. Kim, J.H., Park, H.B.: H_∞ State Feedback Control for Generalized Continuous/Discrete Time-Delay Systems. Automatica 35, 1443–1451 (1999)
5. Iglesias, P.A., Glover, K.: State-Space Approach to Discrete-Time, H_∞ Control. International Journal of Control 54, 1031–1073 (1991)
6. Yu, L., Chu, J.: An LMI Approach to Guaranteed Cost Control of Linear Uncertain Time-Delay Systems. Automatica 35, 1155–1159 (1999)
7. Guan, X.P., Chen, C.L.: Delay-Dependent Guaranteed Cost Control for T-S Fuzzy Systems with Time Delays. IEEE Trans. on Fuzzy Systems 12, 236–249 (2004)
8. Kim, J.H.: Robust Mixed H_2/H_∞ Control of Time-Varying Delay Systems. International Journal of Systems Science 32, 1345–1351 (2001)
9. Khargonekar, P.P., Rotea, M.A.: Mixed H_2/H_∞ Control: A Convex Optimization Approach. IEEE Trans. on Automatic Control. 36, 824–837 (1991)
10. Karimi, H.R., Gao, H.: Mixed H_2/H_∞ Output-Feedback of Second-Order Neutral Systems with Time-Varying State and Input Delays. ISA Trans. 47, 311–324 (2008)
11. Lozano, R., Brogliato, B., Egeland, O., Maschke, B.: Dissipative Systems Analysis and Control Theory and Applications. Springer, London (2000)
12. Boyd, S., Ghaoui, L.E., Feron, E., Balakrishnan, V.: Linear Matrix Inequalities in System and Control Theory. SIAM, Philadelphia (1994)

The Approach to Image Classification by Hybrid Linguistic Method

An-Zen Shih

Dept. of Information Engineering, Jin-Wen Technology University, Taiwan
anzen325@just.edu.tw

Abstract. Image retrieval has become a popular field in computer science. Nevertheless, there are many efforts needed to be done to solve the problems that to retrieval images by using logical inference or abstract attributes. In this paper a group of linguistic descriptors and textures set is formed. This word-texture set are then used to select images from database. The experiment results were successful and suggested that our hypothesis is very promising and practical.

Keywords: content-based image retrieval.

1 Introduction

Image retrieval has become a popular field in computer science. To date, there are hundreds of millions on-line images with diverse contents. To find something exactly meet people's need becomes very difficult. Thus, image retrieval draws a lot of attentions.

Image retrieval can be either text-based or content-based.[1] The text-based image retrieval, however, has two main disadvantages. One is the vast labor required in manual image annotation, the other is the perception subjectivity and annotation could be impreciseness. Because of these disadvantages, content-based image retrieval, or called CBIR, has been introduced.

Content-based image retrieval (CBIR) is any technology which can help us to organize digital picture archives by their visual content. It can be classified into three levels: retrieval by primitive features; retrieval by objects with some degree of logical inference; and retrieval by abstract attributes.[2] Many technologies have been introduced to solve the first level problems. Some methods are proposed for the problems or retrieval images with logical inference or attributes. Nevertheless, there are many effort needed to be done to solve the problems in second and third levels.

In this paper we propose a method to solve the problem. A group of linguistic descriptors and textures set is formed by using a priori knowledge. This knowledge was based on pervious study of the Brodatz image database[8]. This word-texture set are then used to select images, which contained nature scene, from database. Our results suggested that our hypothesis is very promising and practical. With the intermediate link of texture images, we can search images by using key word which described certain texture existed in the image.

J. Sobecki, V. Boonjing, and S. Chittayasothorn (eds.), *Advanced Approaches to Intelligent Information and Database Systems*, Studies in Computational Intelligence 551,
DOI: 10.1007/978-3-319-05503-9_28, © Springer International Publishing Switzerland 2014

The paper is arranged as following: In the next section we will present some basic state-of-art knowledge, which include content-based image retrieval theory experimental image datasets. Secondly we will describe our experiment and show and results. The third part covers the discussion of our results derived from our experiment and a short conclusion is followed.

2   Basic State-of-Art Knowledge

Here we will address some basic knowledge in this section which will help people understanding our research. In the first place, we will give the reader a clear overview of content-based image retrieval(CBIR), and then, in the second place, we will introduce the Brodatz image database. With these knowledge, people can understand the detail of the experiment.

2.1   Content-Based Image Retrieval

Content-based image retrieval (CBIR) helps people to organize digital picture archives by their visual content. The term "content-based" implies that computers have to analyze the content of an image during the search and the term "content" refers to colors, shapes, textures, or any other information which can be extracted from the image. In addition, anything ranging from an image similarity function to a robust image annotation engine belongs to the area of CBIR. People from different fields, such as, computer vision, machine learning, information retrieval, human-computer interaction, database systems, Web and data mining, information theory, statistics, and psychology contributing and becoming part of the CBIR community.[3] Two problems exist in CBIR. The first is how to mathematically describe an image. The second is how to access the similarity between images based on description. The first problem lies in the difficulty of .computers have in understanding the image data. When an image is presented, people can usually see beyond the shapes and colors on the image to the real content of that image. However, computers can't understand the content of the image if we don't program the computers. This is because an image data is just an array of numbers for the computers. So we hope to find a mathematic description of the image, which is sometimes called signature, in order that computers can understand the semantic meaning of an image. After we find the mathematic description of an image, computers can possibly use the signature to compare different images and select interesting ones.

Three levels of queries are used in CBIR to retrieval images.[2] The levels are firstly retrieval by primitive feature, secondly retrieval of objects of given type identified by derived features, with some degree of logical reference, and finally retrieval by abstract attributes. Level 2 and 3 are called semantic image retrieval. The gap between level 1 and level 2 is referred to as semantic gap. People need to give an image as a query example in level 1 retrieval while in level 2 and 3 people must offer a semantic keyword or a sentence as the query.

A lot of efforts have been made to solve out the problems of level one. There are, however, many researchers focus on level two and level three and generated many research works. In level one, people tend to extract color, texture, shape and spatial location to be the features of an image and use it for image retrieval. In level 2 and level 3, people will give a more meaningful query such as ?find pictures of a monkey? or "find pictures of a young student.?

There are many studies trying to bridge the semantic gap. They may be divided into five types: (1) object ontology, which define high-level concepts, (2) machine learning methods, which associate low level features with query concepts, (3) incorporate relevant feedback with retrieval loop for continuous learning the users' intensions, (4) generating semantic template (ST) to support high level image retrieval (5) use visual content and texture information to obtain images from the Web.

The techniques listed above mostly perform retrieval at level two. They comprise three elemental parts in their systems: (1) image feature extraction, (2) similarity measure, (3) semantic gap reduction.

Although a lot of research works have been generated to solve problems of CBIR, it is still a very difficult area to be working on. In this paper, we try to combine the technique of level 1 and level 2 to abridge the semantic gap. We hope that our method will find the similar texture between the predefined image database and the real world images. If this can be proved, we can, in the future, use this property in the field of CBIR.

2.2   Brodatz Image Database

Brodatz image database is chosen for our experiment because the Brodatz textures are the most commonly used texture data set. Many people in the computer vision and signal processing prefer to use the Brodatz textures in their works. So there have been a lot of studies using the Brodatz textures and the results showed they are a very good and stable testing data. In addition, there are studies applying using semantic terms to describe the Brodatz textures [5][6][7][8].

The reason why the Brodatz textures are so popular is that, firstly, it contains 111 different texture classes and can be download from WWW. Secondly, for each class, it is represented by only one sample, which is then divided into 9 sub-images non-overlappingly to form the database. Thirdly, most of the Brodatz textures are photographed under controlled lighting conditions, so the images are of very high quality. In addition, they expose the most amount of textures so that irrelevant information such as noise and non-texture stuff are not there.

Based on the thoughts above, we are satisfied with using the Brodatz textures for our experiment.

3   Experiment and Results

Here we will address our experiment below. In the beginning we will describe our hypothesis and give it a clear explanation. Next, we will stated that how do we prepare the linguistic term sets for the experiment. The samples will be described and the experimental system is discussed in the next part. Finally the searching method will be discussed.

3.1   Experiment

Hypothesis. Our experiment involves using linguistic terms to describe the textures of Brodatz images and applying these word-texture terms sets to select images from real world image database. Since real world are full of textures,[4] it is possible to use textures patterns to choose interested pictures among the pool of images. In this experiment, we argue that, since Brodatz image database has collected a great amount of texture images, it is reasonable that these textures might appear in a real picture. We hope that we can prove that the textures appear in the Brodatz image database can also be found in real world pictures, or nature scenes. If we can prove this feature, this can be in terms to help us, in the future, to find images from database. This might be a good way in CBIR.

Several documents [5][6][7] have applied different linguistic terms to describe the textures of Brodatz images. Nevertheless, they didn't use the results for advance application in CBIR. In other words, to our best knowledge, few people use these results as a good tool for further application in CBIR. They simple contend themselves in use word terms to find similar texture images in Brodatz image database. Thus we want to go one step further to use these linguistic terms to search wanted pictures in image database. For example, we can use the sentence "Find the picture which has cloud in it." Then the system will pick up the key word "cloud" from the sentence and choose the possible images.

Linguistic-texture sets preparation. The preparation of the linguistic-texture sets is as below. In the first place, three linguistic descriptions were chosen from [8] for our system. They are cloud, landscape and waves. Each word describes a kind of texture images in Brodatz image database. The reason we choose these terms is that the scenes depicted by these terms can be seen everywhere around us in real world. So it is reasonable that the textures in these Brodatz images, i.e. cloud, landscape and waves, will be selected as well in the test images. If this does, it will suggest that our hypothesis is correct.

In the second step, totally six texture images are chosen from Brodatz image database. These images are pre-evaluated and selected for these linguistic words mentioned above. These six images are worked as representative images for a specific texture such as cloud, landscape or waves. These images will be used later to be target images for the examination for pictures with similar textures in the testing sample database. In other words, we use these images as an intermediate media between a king of texture description word and the nature scene images which contained the kind of texture in it.

The samples and the system. A total 100 samples are deliberately chosen from our personal collected image database for the experiment. The samples include those images with interested natural objects, such as clouds, landscape or waves, and those images which don't have interested textures. In the 100 Samples, we have deliberately put 15 images contained cloud texture, 12 images contained landscape texture, and 10 images contained wave texture. With this priory knowledge, we want to test that whether or not these images, which we put them into the test image database, can be picked up by our algorithm. These samples are put into our system for the examination.

Our algorithm was written in C and executed on a PC with 2.33GHz, 1.95GB RAM, and Microsoft XP. In the beginning, we will ask the user to give the system a sentence, which will contain, or not contain, important key word, such as cloud, landscape or wave in the sentence. This sentence is treated as the input data. The database includes two parts, one is the pre-defined word texture database, and the other is the 100 sample images. The output data will be 20 images.

Searching method. At the beginning of searching, the system will ask the user to input a short sentence such as "Find the picture with cloud." After the user had input such kind of sentence, the system will read through the whole sentence to see whether there is a word (words) that existed in our linguistic- texture term sets. In other word, our program will read through the whole sentence to look for three words, cloud, landscape and wave. If the system cannot find such words, a warning message will be announced on the screen to the user to mentioned that we can't find the kind of texture images he/she wants. If it does exist, the system will search for the word-texture database to find the correlated texture image. This texture image will be used to compare with the images to find the similar targets.

The comparison step between the texture images and the sample images are using traditional image comparison method. In the beginning, the image is divided into 16 non-overlapped squares. And, in the second step, each part is taken to be compared with the texture images. If more than 4 squares are found to be similar, we regard that the image contains the texture we are interested. After that, the image will be output.

The output image data will contain 20 images. The output list is arrange by the percentage that the system judge how similar the image to the texture. The order is from highest similarity to lowest similarity.

Here we put the diagram to describe our system. (see figure 1)

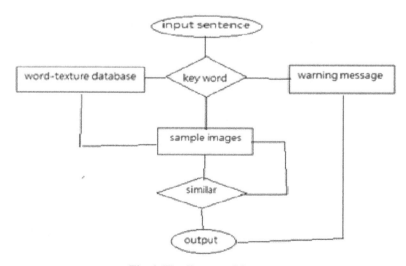

Fig. 1. The diagram of the system

3.2   The Results

The data obtained in previous studies[5][6][7] focused on applying different linguistic terms to select a king of texture from Brodatz database. However, they didn't go further to use the results to apply on the real images. That is to say, they confined themselves with applying linguistic method on Brodatz texture image set which is a well-defined texture set. These results didn't show us the possibility whether or not they can work on real image as well. Our experiment will cover this field. In this section we will present and describe the results of our system.

Our experiment results have suggested that our hypothesis is working. The data presented a great amount of recognition rate, which included recall rate and precision rate. In other words, the system can use the combination of Brodatz texture and linguistic terms to select real world images with the same texture patterns.

After the user input his query, the system will check the images and select top 20 possible images from the 100 images dataset. The system will then output the 20 images which seems to match our need, that is, how many images contained the interested texture. Here we proposed that the recognition rate can be evaluated by using the following formulas:

$$recall = \frac{retrieval \ and \ relevant}{relevant \ obiects} \qquad (1)$$

The above formula computes the recall rate which addresses how many of the chosen meets our need when we choose image among the relevant images in the database.

$$precision = \frac{retrieval \ and \ relevant}{retrieved \ obiects} \qquad (2)$$

The above formula computes the precision rate which states that how many images we chosen from database contained the correct object we are interested.

The above two formula can gives us a very clear view of the recognition rate of our algorithm, not only about the ability of finding the images that we know, with priory knowledge, contained the interested texture, but also the correctness rate of our algorithm.

The Table 1 lists our results of using different linguistic terms to select images from database.

Table 1. Recall and precision rate

	Recall rate	Precision rate
cloud	40%	30%
landscape	41%	25%
wave	30%	15%

As can been seen in table 1, the racall rate and precision rate of the experiment results gave us a considerable satisfaction rates. Among the 100 experimental images data, the mean recall rates of our system are: cloud texture-40%, landscape texture-41%, and wave texture-30%. The mean precision rates of our system are: cloud texture-30%, landscape texture-25%, and wave texture-15%. In the 100 Samples, we have deliberately put 15 images contained cloud texture, 12 images contained landscape texture, and 10 images contained wave texture. The meaning of these images is a test to see whether or not they can be pick out from testing images by our method. So we can understand the recall rate of our experiment. And we can compute how many images chosen have the interested texture in the image. This can give use the precision rate.

Generally speaking, we are satisfied with these results. Perhaps some people might questioned that the system could not find out most texture images, we argued that we could pick up nearly half of the interested texture images in the first 20 possible images. This is a considerable good result under the thought that we just use the raw Brodatz texture without any moderation. That is why we argue that the results are satisfied and acceptable.

Nevertheless, we must point out a minor drawback of the system. We have noticed that, if the image's texture was very distinct and clear, the system could easily select it as the wanted image. If, however, the image's texture was not so clear, the system will neglect it as the unwanted image. In addition, we found that, our system often misjudge some natural scene, such as forest or bushes and treated them as the interested feature. Below we put the data of waves as an example. (see figure 2) But we want to stress and argue that this is simply a drawback about the codes rather than the hypothesis and, in the future work, this drawback can be corrected with a modification of the system.

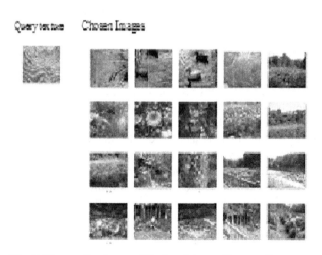

Fig. 2. The results of query "Find a picture with waves"

In this sample, the query sentence is "Find a picture with waves." We can see in the left in the example that the typical texture image found in the word-texture image database is a texture image in Brodatz textures image. The system used this image as a sample to search the test image file and selected 20 "similar" images. The output images are then presented to the users.

After the result images are presented, the system will output the recall rate and the precision rate. The user will know the total recognition rate. Nevertheless, in this experiment, we didn't let the user to put his/her personal feedback because that is not our experiment purpose. However, it will be implemented in the future works.

4   Discussion

To date, CBIR becomes a popular issue in computer science. Many techniques have been proposed to improve its efficiency and ability. It helps people to find images from enormous data from network.

However, there are still a, lot of problems need to be solved in this field. One of them existed in how to bridge the "semantic gap". That is, from retrieval images by primitive feature to retrieval images by using derived features, with some degree of logical reference. Compare to primitive feature, such as color, texture, shape and spatial location of images, people tends to prefer select images with a more meaningful query such as "find pictures of a monkey? or "find pictures of a young student.?

There are a lot of effort have been made to do the task, such as [1]-[8]. Nevertheless, to use semantic sentence to withdraw pictures still a difficult task.

In this paper we propose a way to solve the difficult problem. Our hypothesis is that textures can be seen everywhere in the real world around us, it is possible to use textures patterns to choose interested pictures among the pool of images. In this paper, textures of Broadaz image database are chosen for the experiment. The experiment involves using linguistic terms to describe the textures of Brodatz images and applying these word-texture terms sets to select images from real world image database. Before the experiment, we must build a link between a specific word   and a kind of texture. Such groups of linguistic descriptors and textures set is formed by using a priori knowledge in [8].   So when we mentioned about a descriptive word, for example, cloud, a certain kind of texture forms are there for our usage. We can use them for further application of CBIR. In this experiment, these word-texture sets are then used to select images from database.

The results, seen in precious section, suggest themselves that our hypothesis is reasonable and practical. It can be seen in table 1 that all three key words can find a considerable amount of images contained the interesting textures are selected from the pool of images. Although there are still images with interested texture are left and unfound, we considered it is a program's problem rather than the problem of our hypothesis.

Here we want to summarize several important points about our experiment in the following.

Firstly, we have showed that the Brodatz textures can be found in real images. The fact is not only confirm the idea of [4], but also that we can use these texture for further application of content-based image retrieval.

Secondly, we argued that, compared with previous studies[5][6][7][8], The data obtained in previous studies[5][6][7] focused on applying different linguistic terms to select a king of texture from Brodatz database. Nevertheless, they didn't go further to use the results to apply on the real images. In other words, they confined themselves with applying linguistic method on Brodatz texture image set which is a well-defined texture set. These results didn't show us the possibility whether or not they can work on real image as well. Here we step further from using linguistic terms to select Brodatz textures to applying linguistic terms to the real world images.

Thirdly, we suggested that, that, with improvement, these results could be used in CBIR in the future. The query used in CBIR are either by primitive feature, by derived features, with some degree of logical reference, or by abstract attributes. In this experiment we use a way to combine primitive features and abstract attributes to solve the semantic problems. Useing texture in Brodatz image database as a link, people can find interested nature features, i.e. waves, landscapes and cloud, in real image datasets.

Our result provided a possibility that we could combine the primitive features we know in the present, and some logical or abstract attributes to give the computer system a certain degree of selective judgments in content-based image retrieval.

5 Conclusion

In this paper we propose a method to solve the problem. A group of linguistic descriptors and textures set is formed by using a priori knowledge. This knowledge was based on pervious study of the Brodatz image database[8] . This word-texture set are then used to select images, which contained nature scene, from database. Our results suggested that our hypothesis is very promising and practical. With the intermediate link of texture images, we can search images by using key word which described certain texture existed in the image.

References

1. Rui, Y., Huang, T.S., Chang, S.-F.: Image Retrieval: Current Techniques, Promising Directions, and Open Issues. Journal of Visual Communication and Image Representation 10, 39–62 (1999)
2. Liu, Y., Zhang, D.S., Lu, G., Ma, W.Y.: A Survey of Content-based Image Retrieval with high-level Semantics. Pattern Recognition, 262–282 (2006)
3. Datta, R., Joshi, D., Li, J., Wang, J.Z.: Image Retrieval:Ideas, Influence, and Trends of New Age. ACM Computing Survey 40(2), Article 5 (April 2008)
4. Mandelbrot, B.: Fractals-forms, chance and dimension. W. H. Freeman and Company (1997)
5. Kulkarni, S., Verma, B.: Fussy Logic based Texture Queries in CBIR. In: Proceedings of the Fifth International Conference on Computational Intelligence and Multimedia Applications, ICCIMA 2003 (2003)
6. Chu, C.-Y., Lin, H.-C., Yang, S.-N.: Texture Retrieval Based on Linguistic Descriptor. In: International Conference on Image Processing, ICIP (October 1998)
7. Chang, S.-F., Chen, W., Sundaram, H.: Semantic Visual Template: Linking Visual Features to Semantics. In: International Conference on Image Processing, ICIP (October 1998)
8. Shih, A.-Z.: A New Approach of texture examination by 1-D similarity. In: International Conference on Machine Learning and Cybernatics (2013)

Measuring the Distance from the Rear Surface of the Cornea Leas

M. Augustynek[1], T. Holinka[1], and L. Kolarcik[2]

[1] VSB – Technical University of Ostrava/Department of
Cybernetics and Biomedical Engineering,
Faculty of Electrical Engineering and Computer Science, Ostrava, Czech Republic
[2] Faculty Hospital of Ostrava, Ophthalmic Clinic, Ostrava, Czech Republic
martin.augustynek@vsb.cz

Abstract. The aim of this paper was to base cooperation with the Eye Clinic University Hospital Ostrava to compare the precision of optical biometry by paramedical personnel (General Nurse) with the results of ultrasound biometry performed by an experienced physician (ophthalmologist) using ultrasound Biometry and evaluate the need for routine measurement of duplicity. This work is focused on the comparison of biometric methods for measuring lengths of the eye in patients with cataract.

Keywords: eye, optical biometry, ultrasound biometry.

1 Introduction

Biometrics is the method used for measuring different parameter lengths of eye. Most biometrics are associated with cataract surgery. Cataracts causes failure of transparency and scattering of light passing through the lens. Measured values are required to calculate values of the diopter intraocular lens (IOL), which are replaced instead clouded lens. Calculating the value of the IOL lenses are made by using various formulas.

The work deals with the two most common ways of measuring the axial length of the eye. The first method is the acoustic biometrics, which uses the reflection of sound waves from the echogenic interface. Ultrasound biometry is divided into the contact and immersion method. The second method of measurement is using optical biometry, which works on the principle of partial coherence interferometry.

To achieve good post-operative result is needed to measure the biometric value with high precision. However, it also depends on the quality of operations performed by the operator, and the quality of intraocular lenses.

Good results of calculating the IOL power lens is achieved in patients with the average axial length of the eye and the average corneal power. More difficult is the situation in patients with very short eye, with a short loop and a flat cornea, patients with excessively long eye, with a long lug and a steep cornea. A more complex situation may also be patients who underwent refractive surgery on the cornea prior to cataract

J. Sobecki, V. Boonjing, and S. Chittayasothorn (eds.), *Advanced Approaches to Intelligent
Information and Database Systems*, Studies in Computational Intelligence 551,
DOI: 10.1007/978-3-319-05503-9_29, © Springer International Publishing Switzerland 2014

surgery, patients requiring a corneal transplant while cataract surgery or patients while cataract surgery and silicone oil removal.

2   Discussed Problems

Biometrics eye helps us to determine the optical power of the implanted lens. In order to properly determine the cardinality of artificial lenses is used precise formula for calculating the lens. Into formulas are entered values, which measure a biometric device. Mostly these devices contain software that automatically calculates the IOL. Basic parameters that measure these devices are the axial length of the eye and the optical power of the cornea. There are two methods of measurement - acoustic or optical biometry. [1]

2.1   Ultrasound Biometry of the Eye

In ophthalmology majority of A-scan and B-scan ultrasound probes use frequency of approximately 10 MHz. The speed of sound is entirely dependent on the density of the media through which passes. Sound travels faster in solid environments than in a liquid media. The eye is comprised of both worlds. The A-scan biometry sound passes through the fixed cornea, aqueous water, fixed lens, vitreous humor, choroid, sclera and orbital tissue, so the speed is constantly changing. The speed of sound through the cornea and lens is 1641 m / s, the velocity in the aqueous humor and the vitreous body is 1532 m / s. Average speed before phakic eyes is 1550 m / s, aphakic 1532 m / s average velocity in pseudophakic eyes is 1532 m / s plus a correction factor for an intraocular lens (IOL) material. [9]

2.2   Optical Biometry of the Eye

Optical coherence biometry, which is also sometimes called partial coherence interferometry (PCI), laser interference biometry (LIB) or laser Doppler interferometry (LDI) is an innovative optical method for measuring the axial length of the eye, keratometry, pachymetry, anterior chamber depth, but also lens thickness. It uses the infrared light of a higher wavelength (820 nm) and newer algorithms averaging, thereby increasing their ability to measure eyes with posterior subcapsular cataract. OCB is used as an alternative to ultrasound biometry. Measurement is non-contact, which is more convenient for patients than the contact method by ultrasound biometry. Instruments include software that allows to calculate the power of the intraocular lens according to various formulas.

In this measurement method is used the interference beam reflected by the various optical interfaces. These devices consist of a Michelson interferometer having a light source of a small coherence length, confocal optical detector of light in the focus of a computer, which controls the movement of free mirror interferometer while calculating the visibility of interference depending on the feed. After this, the analysis yields the thickness of the individual layers. The assumption is that this is a layer of at least diffusing light beam and the layer of known refractive index. [7]

The rays are reflected at each interface and include different path differences. All beams can interfere with each other when feed Michelson interferometer path difference offset. The graph visibility for 6 rays appears until 15 highs. The relationship with the thickness of the cornea, lens or retina analysis must be carried out based on the knowledge of the refractive index and coherence length. Maximum visibility value of the peak depends on the intensity of the interfering beams, which in turn depends on the reflectivity of the optical interface. [7]

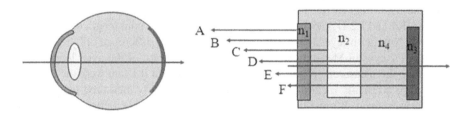

Fig. 1. Diagram of the eye and each interface, where n1 denotes the cornea, lens n2, n3, n4 sclera and vitreous

2.3   Comparison of Ultrasound and Optical Biometry of the Eye

Among the advantages of optical biometry is mainly that the examined person examined do not touch because it is a non-contact method. This substantially reduces the risk of infection, while the optical biometry is more comfortable for the patient than ultrasound biometry. Optical biometry is easy to use and properly trained personnel, would be able to perform measurements even people with less medical expertise. The measuring time is shorter than the ultrasound method, at around 0.3 to 0.4 seconds and is suitable for children and patients who are difficult to work together. The main advantage is to enable the optical biometry measurements eyes with silicone filler after vitreoretinal surgery and myopic eyes back staphyloma where the optical axis is different from the axis of vision.

The disadvantage of optical biometry is primarily the inability to measure the axial length of the eye during occlusion of the media, resulting in a dense cataract, vitreous haemorrhage, corneal scarring or retinal thickening of the macula. In such cases it is necessary to measure the parameters of ultrasound biometry. [11]

3   Analysis of Measured Data

The patient population consisted of 52 women and 33 men. In some patients, performed cataract surgery was on both eyes, some only one eye. Of the 85 patients, was measured the parameters of the eye 122. Patients underwent a standard examination at the University Hospital Ostrava, before cataract surgery. Measurements were carried out by IOL Master optical biometry, keratometry measurement was carried out on an automatic refractometer, or the unit Pentacam and ultrasound biometry on the unit OcuScanTM.

The initial measurement results were taken of general nurses and doctors. The data set does not include the measured corrected parameters.. In case of significant differences, after consultation of the individual patient records, whether there was a remeasurement.

The average difference in corneal curvature in the meridian K1 is 0.22459 D and K2 meridian with a value of 0.26443 D. The average value of the difference between the axial lengths of the eye measured the optical and ultrasound biometry is 0.03277 mm. The maximum difference was measured by the curvature of the cornea in the meridian K1 and R2 D 3.05 1.69 D. The biggest difference axial length of the eye was 0.17 mm. Number of eyes, which was measuring a larger difference of 0.5 D in Meridian K1 is 7 eyes in Meridian K2 were 12 eyes in the case of the axial length of the eye, where the measurement of the difference greater than 0.1 mm were the two eyes. There were originally measured in 122 eyes, 21 eyes poorly measured, which had to be re-measurement. It must therefore be re-measurement of the optical biometry and keratometry, or the Pentacam. In the case of permanent bad results, the patient may suffer from any eye diseases or have higher astigmatism, for which the value cannot be measured accurately on keratometer. Always is consulted with the doctor to choose the right method.

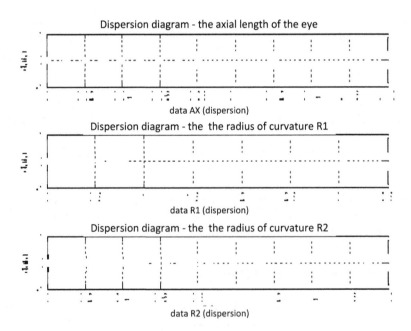

Fig. 2. Diagrams dispersion eye axial length, radius of curvature R1 and the radius of curvature R2

In Fig. 2 are diagrams of dispersion parameters. On the first chart, you can see two differences measured results axial length of the eye, beyond the 0.1 mm. Other values are below this value. The next chart distraction in Fig. 2, can be seen in one eye, there was a measurement of the difference in Meridian K1 by more than 3 D. Also, you can see some excess measured meridian K1, despite the difference of 0.5 D, but most of the measured correctly and the values are below this difference. The third diagram showing the measured differences meridian K2 can see the largest number of bad readings, exceeding the difference 0.5 D.

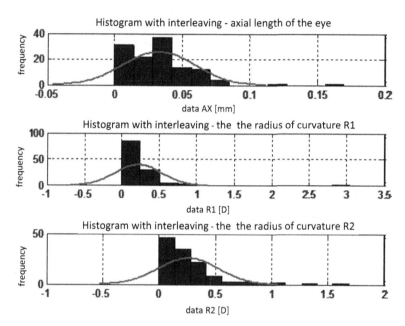

Fig. 3. Histogram with interleaving eye axial length, radius of curvature R1 and the radius of curvature R2

In Fig. 3, are histograms of curve probability function of the normal distribution. The first column shows the histogram of the same size and the occurrence of the difference between the axial lengths of the eye at each interval. Most differential values ranging from 0.028 mm to 0.042 mm, which is the third column and is present in it unlike the 37 eyes. Also, you can see two outlier values difference. The second diagram shows the frequency meridian K1 intervals. In the interval from 0 to 0,254 D is much difference values with a total of 85 eyes. The third diagram shows the frequency meridian K2. In the interval from 0 to 0.14 D is the most measured difference with the number of 46 eyes. To assess normality is presented Gaussian curve.

Bar graphs of the parameters are shown in Fig. 4. The first bar graph shows the individual differences in axial lengths eyes. On the x-axis shows the sequence in the order in which the patient is registered in the database (excel). The y-axis shows the magnitude of the difference. It can be seen that there are two eyes that the difference value exceeds 0.1 mm. The second bar graph for the radius of curvature R1 can be seen in the eye that exceeds the difference of the 3-D in the third graph radius of curvature R2 is seen several values exceeding 0.5 D.

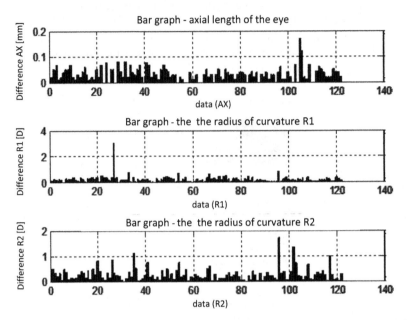

Fig. 4. Bar graph eye axial length, radius of curvature R1 and the radius of curvature R2.

4   Conclusion

From a set of 122 eyes was found 21 primary poorly measured parameters, which had to remeasure again for accuracy targeted results.

We evaluated a difference of initial measurement of the parameters measured by a general nurse and doctor. There were found minimal differences in the measured axial length of the eye. Only in two cases the difference exceeded 0.1 mm. It is obvious that the measurement of optical and ultrasound biometry is a measurement of very similar results. The average value of the difference of axial length of the eye is 0.03277 mm. Greater differences were observed in different meridians, when measuring the corneal curvature to keratometry. In total, 122 eyes of 12 found greater differences than 0.5 D in R2 and 7 meridian at meridian R1. In all of these very different eyes was measured additional correcting measurement, either general nurse or doctor. Some patients had to be tested for the Pentacam.

In this work was not rated preoperative and postoperative visual acuity. It was due to the fact that patients do not go to an eye clinic, where there is a preoperative examination to check the operation usually go to their doctor or the district to another ophthalmologist. There was no detectable extent to which the measurement was correct, because that measurement may perform poorly multiple users on multiple devices.

A proposal for solutions is the measurement of optical biometry, keratometry and comparing the measured values after displaying the measurement results. When the difference would be a measurement in case of unsatisfactory results, would be passed on to another unit of measurement such as Pentacam. Measurement could, in addition to general nurses also perform medically trained person such as biomedical engineer. It depends on the correct train and rehearses the measurement. Measurement of the ultrasound biometry would be undertaken only by uncertain results obtained with the optical biometry and also in case if it did not go because of a disease to measure the axial length of the eye to optical biometry. Because of the limitations of measurement on ultrasound biometry, which is measured by your doctor, it may save time for other doctor's activities and saving finance Eye Clinic. For patients would limit ultrasound biometry measurements to reduce preoperative stress, reducing stress in the examination and could also occur in some patients to improve cooperation by examination.

Acknowledgment. The work and the contributions were supported by the project SP2013/35 "Bio-medical engineering systems IX" and TACR TA01010632 "SCADA system for control and measurement of process in real time". The paper has been elaborated in the framework of the IT4Innovations Centre of Excellence project, reg. no. CZ.1.05/1.1.00/02.0070 supported by Operational Programme 'Research and Development for Innovations' funded by Structural Funds of the European Union and state budget of the Czech Republic. The work is also supported by project ESF OP VK, CZ.1.07/2.2.00/15.0113, 2010-2013 "Innovation of study branch Measurement and Control Systems on FEI, VSB-TU Ostrava" and by project OPVK CZ.1.07/2.2.00/15.0112, "Increasing competitiveness in the field of biomedical engineering at the Technical University of Ostrava", project ESF OP RLZ, CZ.1.07/2.2.00/28.0322, 20011-2014 "Informatics and Telemedicine".

References

1. Kuchynka, P.: Oční lékařství. 1.vyd. Grada, Praha, 768 s (2007) ISBN 978-802-4711-638
2. Kraus, H.: Kompendium očního lékařství. 1. vyd. Grada Publishing, Praha, 341 s (1997), ISBN 80-716-9079-1
3. Anton, M.: Refrakční vady a jejich vyšetřovací metody. 3. přeprac. vyd. Národní centrum ošetřovatelství a nelékařských zdravotnických oborů, Brno, 96 s (2004), ISBN 80-701-3402-X
4. Ganong, W.F.: Přehled lékařské fyziologie. 20. vyd. Galén, Praha, xx, 890 s (2005), ISBN 80-726-2311-7
5. Synek: Svatopluk a Šárka SKORKOVSKÁ. Fyziologie oka a vidění. 1. vyd. Grada, Praha, 93 s, [8] s. obr. příl (2004), ISBN 80-247-0786-1
6. Kohnen, T.: Modern cataract surgery, viii, vol. 34, p. 244. Karger, New York (2002), Developments in ophthalmology, ISBN 38-055-7364-2
7. Dostupné z: Optické principy počítačových skenovacích a jiných metod užívaných v oftalmologii. Lékařská fakulta Masarykovy univerzity (2002), http://www.med.muni.cz/ocnipek/aplikoptII.ppt (cit. November 16, 2012)

304   M. Augustynek, T. Holinka, and L. Kolarcik

8. Dostupné z: B-Scan Ocular Ultrasound. Diseases & Conditions - Medscape Reference (2012), http://emedicine.medscape.com/article/1228865-overview#showall (cit. November 16, 2012)
9. Dostupné z: A-Scan Biometry. Diseases & Conditions - Medscape Reference (2012), http://emedicine.medscape.com/article/1228447-overview#showall (cit. November 16, 2012)
10. Hrazdira, I.: Úvod do ultrasongrafie. Lékařská fakulta Masarykovy univerzity (2008), Dostupné z: http://www.med.muni.cz/dokumenty/pdf/uvod_do_ultrasonografie1.pdf (cit. November 16, 2012)
11. Srovnání metod optické a ultrazvukové biometrie, Brno, Diplomová. Masaryková univerzita v Brně (2011), Dostupné z: https://is.muni.cz/th/214899/1f_m/srovnani_metod_opticke_a_ultrazvukove_biometrie.pdf
12. Vošícký, Z.: Vladimír LANK a Miroslav VONDRA. Matematika a fyzika. 1. vyd. Havlíčkův Brod: Fragment s.r.o., 328 s (2007) ISBN 978-80-253-2
13. KolarčíK, L., Krestová, V.: Optika a refrakce [prezentace] (2010)
14. CeláRková, D., Němčanský, J.: Zákaly cocky [prezentace] (2007)
15. Byrne, S.F.: A-Scan Axial Eye Length Measurements: A Handbook for IOL Calculations. Mars Hill, Grove Park Publishers, NC (1995)
16. Dr.Agarwal. Dr.Agarwal Eye Hospitals: Eye hospitals in india|best Eye Hospitals in Bangalore| Chennai|Andhra|Jaipur |Secundrabad|Madurai|Dilsukhnagar|Kukatpally|Cuttack| Major Eye Centre of the World (2012), Dostupné z: http://www.dragarwal.com/bio_medical/IOL_master.php (cit. April 22, 2013)
17. Alcon® OcuScan® RxP Ophthalmic Ultrasound System | Ophthalmic ultra-sound. Machinery, equipment and tools directory(2011), Dostupné z: http://www.toreuse.com/alcon%C2%AE-ocuscan%C2%AE-rxp-ophthalmic-ultrasound-system/ (cit. April 4, 2013)
18. Augustynek, M., Penhaker, M., Vybiral, D.: Devices for position detection. Journal of vibroengeneering 13(3), 531–523
19. Havlík, J., Uhlíř, J., Horcík, Z.: Thumb motion classification using discrimination functions. In: International Conference on Applied Electronics, AE 2006 , art. no. 4382963, pp. 55-57 (2006), DOI: 10.1109/AE.2006.4382963, ISBN: 8070434422;978-807043442-0
20. Hozman, J., Zanchi, V., Cerny, R., Marsalek, P., Szabo, Z.: Precise Advanced Head Posture Measurement. In: Book Challenges in Remote Sensing - Proceedings of the 3rd Wseas International Conference on Remote Sensing (REMOTE 2007), pp. 18–26 (2007) ISSN: 1790-5117, ISBN: 978-960-6766-17-6
21. Cerny, M.: Movement Activity Monitoringof Elederly People – Application in Remote Home Care Systems. In: Proceedings of 2010 Second International Conference on Computer Engineering and Applications, ICCEA 2010, Bali Island, Indonesia, March 19- 21, vol. 2NJ. IEEE Conference Publishing Services (2010) ISBN 978-0-7695-3982-9
22. Machacek, Z., Hajovsky, R., Ozana, S., Krnavek, J.: Experiments of Thermal Fields of Sensors Supported by Digital Image Processing. In: 2008 Mediterranean Conference on Control Automation, Ajjacio (2008)

Agent-Oriented Meta-model for Modeling and Specifying Transportation Systems: Platoon of Vehicles

Mohamed Garoui[1], Belhassen Mazigh[2],
Béchir El Ayeb[3], and Abderrafiaa Koukam[4]

[1] Prince Research Unit, ENSI, Mannouba, Tunisia
garouimohamed2010@gmail.com
[2] Department of Computer Sciences, FSM, Monastir, Tunisia
belhassen.mazigh@gmail.com
[3] PRINCE Research Unit, FSM, Monastir, Tunisia
ayeb_b@yahoo.com
[4] IRTES-SET, EA 7274, UTBM, F-90010 Belfort cedex, France
abder.koukam@utbm.fr

Abstract. In order to assist the development of multi-agent systems, agent-oriented methodologies (AOM) have been created in the last years to support modeling more and more complex applications in many different domains. By defining in a non-ambiguous way concepts used in a specific domain, Meta modeling may represent a step towards such interoperability. In the transport domain, this paper propose an agent-oriented meta-model that provides rigorous concepts for conducting transportation system problem modeling. The aim is to allow analysts to produce a transportation system model that precisely captures the knowledge of an organization so that an agent-oriented requirements specification of the system-to-be and its operational corporate environment can be derived from it. To this end, we extend and adapt an existing meta-model, Extended Gaia, to build a meta-model and an adequate model for transportation problems. Our new agent-oriented meta-model aims to allow the analyst to model and specify any transportation system as a multi-agent system.

Keywords: Agent technology, Transport domain, Meta-model, Multi-Agent System.

1 Introduction

The purpose of the Agent-Oriented Software Engineering is the creation of a path towards integration and interoperability of methodological approaches for multi-agent systems (MAS) development. This involves the definition of a common framework for MAS specification, which includes the identification of a minimum set of concepts and methods that can be agreed in the different approaches. The tool for defining this framework is meta- modeling. The principle of meta-modeling has been already used in other fields of software engineering, for instance, in the specification of UML [1] by OMG, to describe the elements of

J. Sobecki, V. Boonjing, and S. Chittayasothorn (eds.), *Advanced Approaches to Intelligent Information and Database Systems,* Studies in Computational Intelligence 551,
DOI: 10.1007/978-3-319-05503-9_30, © Springer International Publishing Switzerland 2014

the language, their constraints and relationships. In platooning systems research such as [2] and [3], each vehicle determines its own position and orientation only from its perceptions of the surrounded environment. In this context, the reactive multi-agent paradigm is well adapted to specify and analyze this system. The interest of those approaches results from their adaptability, simplicity and robustness. In this case, platoon configuration can be considered as the result of the self-organization of a reactive multi-agent system (RMAS). A platoon multi-agent system can then be defined as a set of agents, each one corresponding to a vehicle. The agents interacts with its environment (road, obstacles, ...).

Our problem here is when we model a vehicular system, we need an agents-oriented meta-model that gives us a set of basic concepts. These concepts are necessary to model the entire of transport system problem in different **environment** (Urban, Agricultural, and Military) and with various **navigation policies** and its **behavior**. In addition, as soon as we obtain the system model, it will be easy to implement our multi-agent system by using agent oriented programming.

In this paper, our contribution is to provide an agent-oriented meta-model adequate to transportation system problem which allowed us to model the vehicular system in their navigation environment. Our proposed meta-model has been built by adopting and extending the existing meta-model *Extended Gaia meta-model* [4] and thus we define two levels of models inspiring from PASSI meta-model [5]. Our meta-model is interesting for modeling any transportation system with their different navigation scenarios in their dynamic navigation environment by against the others meta-model does not satisfy such what we want to model. This seems to us coherent with the most accepted definition of meta-model: a meta-model is a model of a model, and it provides an explicit representation of the constructs and relationships needed to build specific models within a domain of interest. This proposition arises by remarking that in the field of transport doesn't occur any Agent oriented meta-model to clearly specify and analyze any transport system in the form of multi-agent systems.

We choose to use the Extended Gaia meta-model as it is well adapted to organizational structures such as teams, congregations and coalitions which are used in clustering and collaborative missions of the platoon entities. Furthermore, the proposed approach must take into account, in their meta-model, the concept of environment and different social structures associated with different application areas (Urban, Agricultural, Military) as indicated in Table 1. Extended Gaia specifies the notion of the environment by Environment concept. The abstraction of the environment specifies the set of entities and resources of a multi-agent system can interact with, limiting interactions using the authorized shares.

Table 1. Social structure according to the application areas

Application Area	Suitable Social Structure
Urbain	Congregations, Coalition
Agricole	Congregations, Teams, Coalition
Military	Teams, Congregations, Coalition

The Extended Gaia meta-model adds some organizational based concepts. The organization itself is represented with an entity, which models a specific structure (or topology). The organizational rules are considered responsibilities of the organization. They include safety rules (time-independent global invariants that the organization must respect) and a liveliness rules (that define how the dynamics of the organization should evolve over time).

This paper is structured as follow: in Section 2, we present a state of the art about the existing agent-oriented meta-model used for modeling and specify multi-agent systems. Section 3 presents our Proposed Agent-oriented Meta-model for transportation systems. Then, Sections 4 illustrate our Agent-oriented Meta-model with an application of urban public transportation systems. Finally, Section 5 concludes by giving a list of possible future works.

2 Related Works

Many agent-oriented meta-model have been proposed for modeling of multi-agent system. The first version of the *Gaia methodology*, which modeled agents from the object-oriented point of view, was revisited 3 years later by the same authors in order to represent a MAS as an organized society of individuals [6,7]. Agents play social roles (or responsibilities) and interact with others according to protocols determined by their roles. With that approach, the overall system behavior is understood in terms of both micro- and macro-levels. The former explains how agents act according to their roles, and the latter explains the pattern of behavior of those agents. These constraints are labeled organization rules and organization structures respectively.

The Extended Gaia meta-model adds some organizational based concepts. The organization itself is represented with an entity, which models a specific structure (or topology). The organizational rules are considered responsibilities of the organization. They include safety rules(time-independent global invariants that the organization must respect) and liveliness rules (that define how the dynamics of the organization should evolve over time).

ASPECS (Agent-oriented Software Process for Engineering Complex Systems) provides a holonic perspective to design MAS [9]. Considering that complex systems typically exhibit a hierarchical configuration, on the contrary to other methodologies, it uses holons instead of atomic entities. Holons, which are agents recursively composed by other agents, permit to design systems with different granularities until the requested tasks are manageable by individual entities. Being one of the most recent methodologies, it takes the experience gained from previous approaches (such as PASSI [8] and RIO [10] as the base to define the meta-model and the methodology.

The goal of the proposed meta-model is to gather the advantages of organizational approaches as well as those of the holonic vision in the modeling of complex systems. A three layer meta-model, with each level referring to a different aspect of the agent model, is proposed: The Problem domain covers the organizational description of the problem. An organization is composed by roles which interact within scenarios while executing role plans. Roles achieve organizational goals by means of their capacities (i.e., what a behavior is able to do). The organizational context is defined by means of ontology. This meta-model layer is used mainly during the analysis and design phases. The Agency domain defines agent-related concepts and details the holonic structure as a result of the refinement of the elements defined in the Problem domain. Each holon is an autonomous entity with collective goals and may be composed by other holons. Holonic groups define how members of the holon are organized and how they interact in order to achieve collective goals. At the finest granularity level, holons are composed by groups and their roles are played by agents, which achieve individual goals. A rich communication between agent roles (which are instances of organizational roles) is also supported, specifying communicative acts, knowledge exchange formalized by means of the organizational ontology, and protocols specifying sequences of messages.

All of these meta-models (Extended Gaia and ASPECS) model the system as an organization and set of agents and their interactions. The downfall is that these meta-models do not take into account the environment changing, different system scenarios and the system behavior into this environment. For this reason, comes our suggestion to propose an agents-oriented meta-model for the transport system that satisfies our needs.

3   Our Proposed Meta-model: Platooning Meta-model

The UML is based on the four-level meta-modeling architecture. Each successive level is labeled from M3 to M0 and are usually named meta-meta-model, meta-model, class diagram, and object diagram respectively. A diagram at the Mi-level is an instance of a diagram at the Mi+1-level. Therefore, an object diagram (an M0-level diagram) is an instance of some class diagram (an M1-level diagram), and this class diagram is an instance of a meta-model (an M2-level diagram). The M3-level diagram is used to define the structure of a meta-model, and the Meta Object Facility (MOF) belongs to this level. The UML meta-model belongs to the M2-level.

After studying the Extended Gaia meta-model, we observe how this explicit and useful models of the social aspect of agents. Although it was not designed for open systems, and provides little support for scalability, simplicity allows improvements to facilitate with a relative. it models both the macro and micro aspects of the multi-agent system. Gaia believes that a system can be regarded as a company or an organization of agents.

In this section, based on the instantiation checking Model (Fig. 1), we try to solve our contributions mentioned from the start. It manifests itself to extend and adapt an existing meta-model to build a meta-model and an adequate model for transport problems.

In Figure 2, the classes present in black color are the base classes of Extended Gaia meta-model. For against, the blue classes are the classes added to the existing meta-model to be adapted to platooning applications and then help us to implement our own methodology for modeling and dependability analysis. Table 2 presents the definition of the added new concepts.

Table 2. Definition of the added new concepts

Concept	Description
Functional Requirement	A function that the software has to exhibit or the behavior of the system in terms of interactions perceived by the use.
Non-Functional Requirement	A constraint on the solution. Non-functional requirements are sometimes known as constraints or quality requirements.
AgentModel	Abstract description of a formal model which gives an abstract view about *the agent behavior*.
OrganizationModel	Abstract description of a formal model which gives an abstract view about *the organization behavior*.

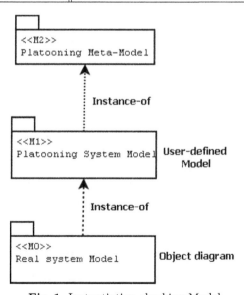

Fig. 1. Instantiation checking Model

The concept *Functional Requirement* is a function that the software has to exhibit or the behavior of the system in terms of interactions perceived by the use. This concept allowed us to identify our system requirements. The *Non-Functional*

Requirement concept provides a constraint on the solution. Non-functional requirements are sometimes known as constraints or quality requirements. *AgentModel*
concept gives an abstract view about the Agent behavior. *OrganizationModel* gives
an abstract view about the organization behavior. Behavior is described by formal
state-based models [11]. By inspiring from PASSI [8] and ASPECS meta-models
[9], we tried to organize our meta-model in two areas: Problem Domain and Agent
Domain. Problem Domain involves elements (Figure 2) are used to capture the requirements problem and perform initial analysis. Agent Domain includes elements
(Figure 2) are used to define an agent-oriented solution to the problem described
in the previous step.

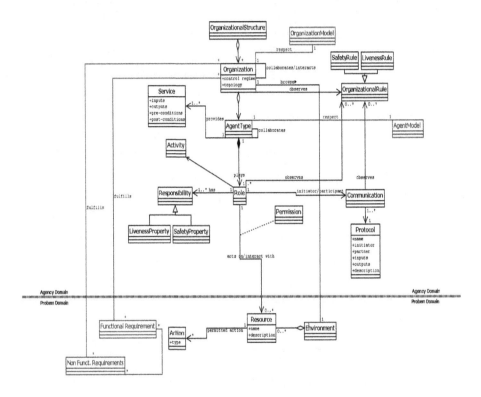

Fig. 2. Our Meta-model: Platooning Meta-model

After this, we pass to M1-level describes the *Platooning System Model* (see
Figure 3)which constitutes of instance of the concepts of M2-level model. This
model includes all the basic concepts and necessary for us to model any type
of application to platooning with their bodies, interaction, environment, their
geometric configuration and formal models associated with each component
platoon.

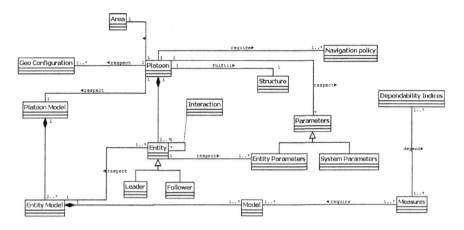

Fig. 3. Platooning System model

The figure 3 gives an idea about the basic concepts of Platooning System Model (Figure 3) which is instances of our meta-model that shown in the Figure 2. The *Platoon* concept represents the main element in our model which is an instance of meta-concept *Organization*. Any Platoon is modeled as a set the Entity. There are two kind of entity: *Leader* and *Follower* which are modeled by the two concepts *Leader* and *Follower*. The tow concepts *Entity_Model* and *Platoon_Model* are used to describe the behavior of entity and platoon in the environment. The concept Area model the environment notion. In our transportation problem, there are three types: Urban, Agricultural, and Military. The concepts *Parameters*, *Entity_Parameters* and *System_Parameters* provided a general idea about the parameters of the entities and of the system. These parameters are necessary and useful for dependability Evaluation in our future work.

4   A Case Study: Urbain Public Transportation System

According to the agent-oriented meta-model, we try to specify a transport applications in urban environment. The convoy adopts a line configuration with inter-distance between vehicle 0 meters and 2 meters in longitudinal gap. For these scenarios, the convoy will have a fixed number of three vehicles and will move on a track with a radius of curvature ranging from 15 m to infinity.

The platoon moves at a maximum speed of 15 km/h with an acceleration of $1m/s^2$ and a deceleration of $-3m/s^2$ on a maximum distance of 1000 meters. From these settings, two scenarios are proposed (see Fig. 4). The S4.1 is to evolve a convoy of vehicles with fixed-line configuration. While the scenario is S4.2, from a convoy vehicle configuration level, to evolve into a line configuration.

Through the case study, we seek to validate our meta-model on a real example: Urbain Public Transportation System.

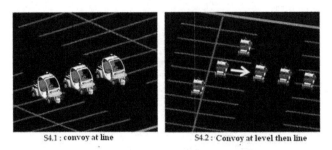

S4.1 : convoy at line   S4.2 : Convoy at level then line

Fig. 4. Urban Public Transportation System

In Figure 5, we present the object diagram which is an instantiation of the *Platooning System Model* (Figure 3). The object diagram in the data modeling language UML used to represent instances of classes, that is to say objects. As the class diagram, it expresses the relationship between objects, but also the state of objects, thereby expressing execution contexts. In this sense, this pattern is less general than the class diagram. Object diagrams are used to show the state of object instances before and after the interaction, i.e. it is a photograph at a specific time and attributes existing object. It is used in the exploratory phase.

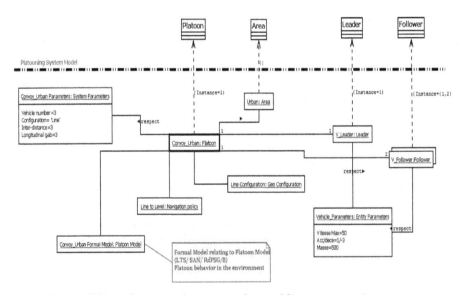

Fig. 5. Object diagram relating to urban public transportation systems

The object diagram of our study is a set of objects that have the attributes that characterize the system. The object *Convoy_Urban* is an instance of the concept Platoon which is its an instance of Organization concept of the meta-model. The convey adapt a line configuration and Line to Level navigation policy therefore we find an instance of *Geo_Configuration* named *Line_Configuration* and an instance of Navigation policy named *Line_to_Level*.

Our transportation system is constitutes of three intelligent vehicles: one Leader and two follower, thus the object diagram contains two items: *V_Leader* instance of the concept Leader with cardinality equal to 1 and *V_Follower* instance of Follower concept with cardinality equal to 2.Our system has parameters regrouped in the two tables 3 and 4. These parameters are divided into two kinds: *Vehicle_Parameters*and *Convoy_Urban_Parameters* which respectively represent convoy entities parameters and the parameters of the overall system. They are used for dependability evaluation in our future works. Transport system behavior is modeled by *Convoy_Urban_Formal_Model* object. The behavior is described by state-based models which are used in system dependability evaluation. This model and parameters are used in our future work to the dependability evaluation.

Table 3. Vehicles Parameters

Parameters	Values
Max speed	50 km/h
Acceleration/deceleration	1 m/s^2 -3 m/s^2
Weight	500 kg

Table 4. Convoy Parameters

Parameters	Values
Vehicle Number	3
Configuration	Line
Inter-distance	3 m
longitudinal gap	0 m

5 Conclusion

In this paper, we have proposed an Agent-Oriented meta-model adequate to any Transportation Systems with multi-configuration ability problem. The aim is to allow analysts to produce a transportation system model that precisely captures the knowledge and the behavior of an organization so that an agent-oriented requirements specification of the system-to-be. We illustrated our meta-model on urban public transportation system. We have tried to model our system as multi-agent system based on our proposed meta-model.

Future works will be devoted to several key points aimed at improving the proposed solution. On the one hand, we will work to provide a generic model for a methodology and will be suitable for all platonning application with different scenario in different transportation field. This model is used for the dependability evaluation. On the second hand, efforts have to be made in order to implement a tool for having an interface to aid the modeling and automate the analysis.

References

1. Gaevic, D., Djuric, D., Devedic, V.: Model Driven Engineering and Ontology Development. Springer, Heidelberg (2009)
2. Contet, J.-M., Gechter, F., Gruer, P., Koukam, A.: Application of reactive multi-agent system to linear vehicle platoon. In: Proceedings of the 19th IEEE International Conference on Tools with Artificial Intelligence, ICTAI 2007, vol. 02, pp. 67–70. IEEE Computer Society, Washington, DC (2007)
3. El-Zaher, M., Gechter, F., Gruer, P., Hajjar, M.: A new linear platoon model based on reactive multi-agent systems. In: Proceedings of the 2011 IEEE 23rd International Conference on Tools with Artificial Intelligence, ICTAI 2011, pp. 898–899. IEEE Computer Society, Washington, DC (2011)
4. Cernuzzi, L., Juan, T., Sterling, L., Zambonelli, F.: The gaia methodology: Basic concepts and extensions. In: Methodologies and Software Engineering for Agent Systems. Kluwer Academic Publishers (2004)
5. Cossentino, M., Gaglio, S., Sabatucci, L., Seidita, V.: The PASSI and agile PASSI MAS meta-models compared with a unifying proposal. In: Pěchouček, M., Petta, P., Varga, L.Z. (eds.) CEEMAS 2005. LNCS (LNAI), vol. 3690, pp. 183–192. Springer, Heidelberg (2005)
6. Cernuzzi, L., Zambonelli, F.: Experiencing auml in the gaia methodology. In: ICEIS (3), pp. 283–288 (2004)
7. Zambonelli, F., Jennings, N.R., Wooldridge, M.: Developing multiagent systems: The gaia methodology (2003)
8. Cabrera-Paniagua, D., Cubillos, C.: PASSI Methodology in the Design of Software Framework: A Study Case of the Passenger Transportation Enterprise. In: Luck, M., Gomez-Sanz, J.J. (eds.) AOSE 2008. LNCS, vol. 5386, pp. 213–227. Springer, Heidelberg (2009)
9. Cossentino, M., Gaud, N., Hilaire, V., Galland, S., Koukam, A.: ASPECS: an agent-oriented software process for engineering complex systems. Autonomous Agents and Multi-Agent Systems 20(2), 260–304 (2010), doi:10.1007/s10458-009-9099-4.
10. Gaud, N., Hilaire, V., Galland, S., Koukam, A., Cossentin, M.: A Verification by Abstraction Framework for Organizational Multiagent Systems. In: Jung, B., Michel, F., Ricci, A., Petta, P. (eds.) Proc. of the Sixth International Workshop AT2AI-6: "From Agent Theory to Agent Implementation", of the Seventh International Conference on Autonomous agents and Multiagent Systems (AAMAS), Estoril, Portugal, pp. 67–73 (2008)
11. Atlee, J.M., Gannon, J.D.: State-Based Model Checking of Event-Driven System Requirements. IEEE Trans. Software Eng, 24–40 (1993)

A New Technique for Edge Detection of Chromosome G-BAND Images for Segmentation

Saiyan Saiyod* and Pichet Wayalun

Department of Computer Science, Faculty of Science, Khon Kaen University,
Khon Kaen 40002, Thailand
saiyan@kku.ac.th, pichet.w@kkumail.com

Abstract. Chromosome edge detection of G-band type images are an important for segmentation. Generally, the chromosome G-band images consisted of the noise, lack of contrast and hole in the images. The chromosome edge can mislead of the edge detection method, particularly the chromosome overlaps and the chromosome touches. It's difficult a clear edge of the chromosome edge images. The edge detection method is difficult when the noise appear. This paper proposed approach the chromosome edge detection in the chromosome segmentation system. The chromosome edge detection method applied the FloodFill, Erosion and Canny based on the chromosome G-band images. The experimental results give the best performance for chromosome segmentation system. A success rate of proposed method achieved 98.43%.

Keywords: Edge Detection, Chromosome, Chromosome Analysis, Karyotype, Chromosome G-band.

1   Introduction

Human chromosomes are contained important information for cytogenetic analysis in which compared their patient's chromosome images against the prototype human to chromosome band patterns. Chromosome images were acquired by microscopic imaging of metaphase or prophase cells on specimen slides. Karyotype analysis is a widespread procedure in cytogenetic to assess the possible presence of genetic defects that is useful tool for detecting deviation from normal cell structure. The abnormal cells consist of an excess or deficiency of a chromosome and structural defects [1].

Nowadays, the karyotyping analysis is manually performed in most cytogenetic laboratories. However, the processes spend the time consuming and expensive procedure. Fig. 1(a) show a typical G-band normal chromosome spread and a karyotype of all the chromosomes from that cell. Fig. 1(b) shows the chromosomes classification which is used to diagnosing genetic disorders.

Currently, the karyotyping analysis utilizes the chromosome image edge for fundamental feature. The chromosome edge is caused by changes in some physical properties

---

* Corresponding author.

J. Sobecki, V. Boonjing, and S. Chittayasothorn (eds.), *Advanced Approaches to Intelligent Information and Database Systems*, Studies in Computational Intelligence 551,
DOI: 10.1007/978-3-319-05503-9_31, © Springer International Publishing Switzerland 2014

of surfaces of the image. Most of research on image edge feature are always focusing on the detection and extraction method. The goal of using edge detection is to recover the information about shape and reflectance, or transmittance in images[2].

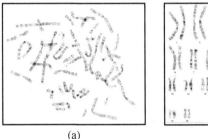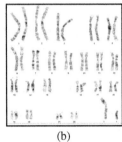

(a)   (b)

Fig. 1. (a) The metaphase chromosome images. (b) The chromosomes classification which used to diagnosing genetic disorders.

Heydarian et al. [3] proposed a novel method to detect the object edge in the MR and the CT images that applied an edge detection method by using the canny algorithm. Mondal et al. [4] applied the canny algorithm to detect the image edge of the cephalograms images. Yu and Acton [5] proposed an automatic edge detection in ultrasound imagery by using the normalized gradient and Laplacian operators.

AI-Kofahi et al. [6] proposed an automatic detection and segmentation of cell nuclei in histopathology images that used the multiscale Laplacian of Gaussian filtering constrained by distance map based adaptive scale selection for the nuclear seed points detection. Brummer [7] presented the automatic detection of the longitudinal fissure in tomographic scans of the brain that used the sobel magniture edge detection. Chen et al. [8] presented a novel multispectral image edge detection of tuberculosis CT image that used the Clifford algebra. Shichun et al.[9] proposed a fast EMD method for edge detection of medical images that used the image pyramid technique and Intrinsic Model Function (IMFs). Han et al. [10] applied the robust edge detection method by using the independent component analysis (ICA) for medical image. Gudmundsson et al. [11] proposed the edge detection technique by using the Genetic Algorithm (GAs) to accurately locate thin and continuous in medical image. Their approach also improves the performance of edge detection techniques by using the explicitly considering the local edge structure in the neighbourhood of the hypothesized edge pixels.

Accurate representation of edge transitions, or image gradient level, is also important for many applications. A binary classification designating edge / no-edge does not accurately reflect the rich set of transitions between intensity levels that is presented in most natural images. A more general approach to representing edges information is to allow the edge map to contain symbols that reflected the intensity changed, or image gradient, at each edge point.

In this paper presented the chromosome images edge detection algorithm (CEDA), which consist of the flood-fill, Erosion, and Canny method. The CEDA method realized an accuracy chromosome edge detection, and effectively suppressed the effect of noise, and ensure the continuity, integrity, and location accuracy. The experimental results show that our method gives satisfactory results.

2   Edge Detection Algorithm

The proposed method applliled the Flood-fill[12, 13], Morphology Erosion [14, 15] and Canny[11] for the chromosome edge detection algorithm (CEDA) that followed the same core steps as shown in Fig. 2.

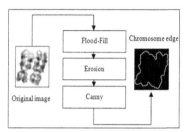

Fig. 2. The chromosome edge detection algorithm (CEDA)

2.1   Flood-Fill

Flood-Fill [12, 13] is the classical method of the enhancement technique. It can determine the connection an areas that given node in a multi-dimensional array. Then it searches for all nodes in the connected array to the start node (called seed point) by a path of the target color, and changes them to the replacement color. Figure 3 shown the sample areas surrounding by the black pixel, (a) shown the sample hole, (b) represents the seed point that is already changed to the replacement color by gray, (c) shown the final result after Flood-Fill.

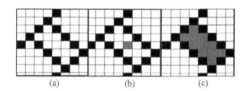

Fig. 3. The process of Flood-Fill method

2.2   Morphology Erosion

Morphology Erosion (ME) [14, 15] is one of image processing operation that performed   based on the shapes of images. It's the fundamental morphological operations, inwhich removes the pixels of boundaries objects, and can be painted the images if the damaged region is regarded as a region to be eroded.

The ME can be defined as follows, if A is an input binary images and B is a structuring element, the erosion of A by B denoted by $A \odot B$, is gives by

$$A \odot B = \{x \in A \,|\, (B)_x \subseteq A\}, \tag{1}$$

Where $(B)_x$ designates the structuring element B centered at pixel x.

The structuring elements apply to process the input images that is one of an essential part of the erosion operations that is a matrix consisting of only 0's and 1's, and can has any arbitrary structure and size , and the structuring elements are typically much smaller than the image being processed. The structuring element pixels called the original that identified. The center pixel of the structuring element, called the origin, identifies the pixel of interest of the pixel being processed that contained 1's define the neighborhood of the structuring element.

The ME steps are illustrated in Fig. 4. This figure, only one layer of pixel is eroded at every loops of erosion operation.

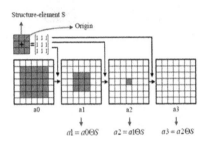

Fig. 4. The ME steps of the binary images

2.3   Canny Operator

The Canny Edge Detection[16, 17] is a classical method of edge detection and most commonly used of image processing tools that has four steps as follows. (I) smoothed images by using gauss filter. (II) calculated gradient amplitude and direction by using the first order finite difference. (III) Non-Maximum suppression is scanned along the image gradient direction. (V) edge detects and connects in double thresholds algorithm. The mathematical specific described as follows.

1. 2-D Gaussian function is

$$G(x, y) = \frac{1}{2\pi\delta^2} \exp\left(-\frac{(x^2+y^2)}{2\delta^2}\right).$$ (2)

In some direction n, the first-order directional derivative $G(x,y)$ is

$$G_n = \frac{\partial G}{\partial n} = n\nabla G$$ (3)

$$n = \begin{bmatrix} \cos\theta & \sin\theta \end{bmatrix}^T$$ (4)

$$\nabla G = \begin{bmatrix} \frac{\partial G}{\partial x} & \frac{\partial G}{\partial y} \end{bmatrix}^T$$ (5)

Where n is directional vector, ∇G is gradient vector. Let image $f(x,y)$ convolute with G_n, and simultaneously changes the direction of n, then n is orthogonal to the direction of testing edge when $f(x,y) * G_n$ obtains the maximum.

2. Strength and the direction of edge

$$E_x = \frac{\partial G}{\partial x} * f(x, y) \tag{6}$$

$$E_y = \frac{\partial G}{\partial y} * f(x, y) \tag{7}$$

$$A(x, y) = \sqrt{E_X^2 + E_Y^2} \tag{8}$$

$$\theta = \arctan(E_X / E_Y) \tag{9}$$

$A(x,y)$ reflects the edge strength of points (x,y) in the images. θ is an normal vector of point (x,y) in images.

3. Just get global gradient is not enough to determine the edge, so for sure, must keep local maximal gradient points, and suppress the Non-Maximum.

4. Typical method of reduction false edge number is used by threshold. All values will be zero lower than the threshold. Double threshold values method will connect the edge into contour in $G_2(x,y)$, and when get to the end of the contour, this algorithm searches the edge in eight adjacent point of $G_1(x,y)$ which can be connected to the contour. So, this algorithm collects the edge unceasingly in $G_1(x,y)$, until connected $G_1(x,y)$ so far.

3   Experimental Results

In this paper used the total of 1,380 chromosomes from Pki described in Ref. [18] The performance of all the edge detection methods are evaluated the chromosome images segmentation accuracy [21-23], define as

$$Accuracy = \frac{Chromosome\_segmentation}{Overall\_segmentation} x100 \tag{10}$$

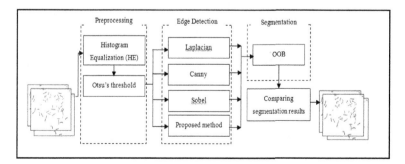

Fig. 5. Flowchart of the evaluate system for edge detection of chromosome G-band

320   S. Saiyod and P. Wayalun

Fig. 5 illustrates the performance evaluation system of all the chromosome edge detection methods that consist of three steps: (I) preprocessing by Histogram equalization [19] and the Otsu's [15] (II)edge detection of all methods, and (III) segmentation by the oriented bounding boxes (OOBs)[20], and the result can be counted and confirmed by an expert.

The proposed method will be compared with the Canny [11], Laplacian [21], and Sobel methods. Fig. 6 shown the results compared between the four methods for edge detection on chromosome cells.

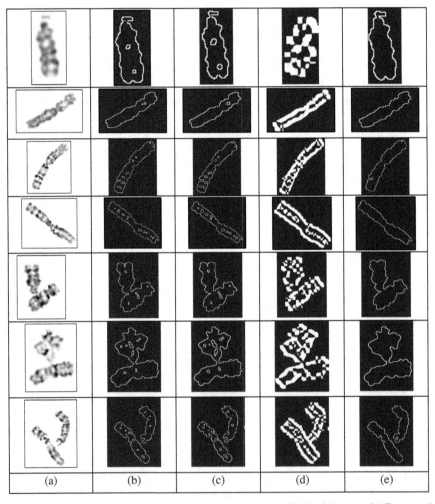

Fig. 6. Result of the edge detection on single chromosome (a) Original images. (b) Extract edge of the Canny operators. (c) Extract edge of the Laplacian operators. (d) Extract edge of the Sobel operators. (e) Extract edge of proposed method.

The edge detection performance of Canny operator, Laplacian operator, Sobel operator, and proposed methods are show in Table 1.

Table 1. Accuracy of edge detection for four classical algorithms compare with proposed method

Method	Accuracy (%)
Canny	96.54
Laplacian	51.2
Sobel	14.53
Proposed	**98.43**

Table 1 shown the performance comparison of different edge detection methods. The average accuracy rate of proposed method, Canny, Laplacian, sobel, and proposed method are 96.57%, 51.2%, 14.53%, and 98.43%, respectively.

(a)   (b)   (c)   (d)

Fig. 7. The chromosomes that failed to be detected by both modified and unmodified CEDA. (a) The original image. (b) The chromosome that unmodified by CEDA. (c) The chromosome that modified by CEDA. (d) The result that failed to be edge detection.

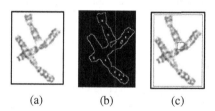

(a)   (b)   (c)

Fig. 8. The chromosome that failed to be detected by unmodified CEDA. (a) The original image. (b) The chromosome that unmodified by CEDA. (c) The result that failed to be edge detection.

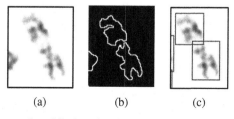

(a)   (b)   (c)

Fig. 9. The chromosomes that failed to be detected by modified CEDA. (a) The original image. (b) The chromosome that modified by CEDA. (c) The result that failed to be edge detection.

The Fig. 7-9 showed examples of the chromosomes that failed to be detected by both modified and unmodified CEDA (see Fig. 7), unmodified CEDA (see Fig. 8), and modified CEDA (see Fig. 9). It's chromosomal morphologies that difficulty to detect algorithmically and possibly lead to modifications of CEDA that can detect them.

4   Discussion

In this paper proposed approach the edge detection algorithms on chromosome images, in which consist of Floof-Fill, Erosion, and Canny that investigated   by using chromosome segmentation system. The performance improvement compared the classical edge detection algorithms. The experimental results demonstrated the performance of the proposed method in which detected edges of chromosome images, filled hole in the chromosome images and removed noises around the chromosome region. It also observed that on incorporating using the flood-fill and erosion further results in improved performance.

Acknowledgments. The chromosome images were supported by the Center of Medical Genetics Research Rajanukul Institutes of Health.

References

1. Wayalun, P., Laopracha, N., Songrum, P., Wanchanthuek, P.: Quality Evaluation for Edge Detection of Chromosome Gband Images for Segmentation. Applied Medical Informatics 32(1), 25–32 (2013)
2. Grisan, E., Poletti, E., Ruggeri, A.: Automatic Segmentation and Disentangling of Chromosomes in Q-Band Prometaphase Images. IEEE Trans. Inform. Tech. Biomed. 13(4), 575–581 (2009)
3. Heydarian, M., Noseworthy, M.D., Kamath, M.V., Boylan, C., Poehlman, W.F.S.: Optimizing the Level Set Algorithm for Detecting Object Edges in MR and CT Images. IEEE Trans. Nucl. Sci. 56, 156–166 (2009)
4. Mondal, T., Jain, A., Sardana, H.K.: Automatic Craniofacial Structure Detection on Cephalometric Images. IEEE Trans. Image Process. 20(9), 2606–2614 (2011)
5. Yu, Y., Acton, S.T.: Edge detection in ultrasound imagery using the instantaneous coefficient of variation. IEEE Trans. Image Process. 13(2), 1640–1655 (2004)
6. Al-Kofahi, Y., Lassoued, W., Lee, W., Roysam, B.: Improved Automatic Detection and Segmentation of Cell Nuclei in Histopathology Images. IEEE Trans. Biomed. Eng. 57(4), 841–852 (2010)
7. Brummer, M.E.: Hough transform detection of the longitudinal fissure in tomographic head images. IEEE Trans. Med. Imag. 10(1), 74–81 (1991)
8. Chen, X., Hui, L., WenMing, C., JiQiang, F.: Multispectral image edge detection via Clifford gradient. Science China Information Sciences 55, 260–269 (2012)
9. Shichun, P., Jian, L., Guoping, Y.: Medical Image Edge Detection Based on EMD Method. Wuhan University Journal of Natural Sciences 11, 1287–1291 (2006)
10. Han, X.-H., Chen, Y.-W.: A robust method based on ICA and mixture sparsity for edge detection in medical images. Signal, Image and Video Processing 5, 39–47 (2011)

11. Gudmundsson, M., El-Kwae, E.A., Kabuka, M.R.: Edge Detection in Medical Images Using a Genetic Algorithm. IEEE Transactions on Medical Image 13, 469–474 (1998)
12. Lee, J., Kang, H.: Flood fill mean shift: A robust segmentation algorithm. International Journal of Control, Automation and Systems 8(6), 1313–1319 (2010)
13. Hsieh, S.-L., Hsiao, Y.-J., Huang, Y.-R.: Using margin information to detect regions of interest in images. In: IEEE Int. Conf. Systems, Man, & Cybernetics, pp. 3392–3396 (2011)
14. Kimori, Y.: Mathematical morphology-based approach to the enhancement of morphological features in medical images. J. Clin. Bioinformatics 1, 1–10 (2011)
15. Guo, H., Ono, N., Sagayama, S.: A structure-synthesis image inpainting algorithm based on morphological erosion operation. In: Congress Image & Signal Processing, pp. 530–535 (2008)
16. Canny, J.: A Computational Approach to Edge Detection. IEEE Trans. Pattern Anal. Mach. Interll. PAMI-8, 679–698 (1986)
17. Zhang, X., Guo, Y., Du, P.: The Contour Detection and Extraction for Medical Images of Knee Joint. In: Int. Conf. Bioinformatics & Biomedical Eng., pp. 1–4 (2011)
18. Ritter, G., Schreib, G.: Profile and feature extraction from chromosomes. In: Proceedings of the 15th International Conference on Pattern Recognition, pp. 287–290 (2000)
19. Yang, J., Yang, J., Zhang, D.: Median Fisher Discriminator: a robust feature extraction method with applications to biometrics. Frontiers of Computer Science 2, 295–305 (2008)
20. Ding, S., Mannan, M.A., Poo, A.N.: Oriented bounding box and octree based global interference detection in 5-axis machining of free-form surfaces. Computer-Aided Design 27, 1281–1294 (2004)
21. Karvelis, P.S., Tzallas, A., Fotiadis, D.I., Georgiou, I.: A Multichannel Watershed-Based Segmentation Method for Multispectral Chromosome Classification. IEEE Trans. Med. Imag. 27(5), 697–708 (2008)
22. Karvelis, P.S., Fotiadis, D.I., Syrrou, M., Georgiou, I.: Segmentation of Chromosome Images Based on A Recursive Watershed Transform. In: The 3rd European Medical and Biological Engineering Conference, pp. 20–25 (November 2005)
23. Karvelis, P.S., Fotiadis, D.I., Tsalikakis, D.G., Georgiou, I.A.: Enhancement of Multichannel Chromosome Classification Using a Region-Based Classifier and Vector Median Filtering. IEEE Transactions on Information Technology in Biomedicine 13(4), 561–570 (2009)

System-Level Implementation of the Algorithms Used in Content-Based Multimedia Retrieval

Paweł Forczmański, Piotr Dziurzański, and Tomasz Mąka

West Pomeranian University of Technology, Szczecin,
Faculty of Computer Science and Information Technology,
Żołnierska Str. 52, 71–210 Szczecin, Poland
{pforczmanski,pdziurzanski,tmaka}@wi.zut.edu.pl

Abstract. In the paper we address the problem related to hardware implementation of Content-Based Multimedia Retrieval, which is understood as a recognition of fused audio/visual descriptors. The proposed approach consists of several adopted low-level media descriptors (both visual and audio) and classification methods that are implemented in a system-level C-based hardware description language ImpulseC. The specific tasks of such implementation include conversion from floating-point into fixed-point representation and some improvements aimed at parallel execution of loop instances. The modified algorithms has been implemented as cores in a NoC-based MPSoC. We have provided an FPGA implementation characteristic together with some results of preliminary experiments.

1 Introduction

Continuously growing volume of multimedia data causes that processing them with the help of traditional programming techniques encounters a barrier that limits the performance of general-purpose systems. Currently, one of the most interesting yet difficult problems related to the processing of data, such as images, sounds and video streams, is indexing and recognition (in particular classification), which is used in various modern applications. These tasks often constitute Content-based Multimedia Retrieval. From the practical point of view, the increase of the efficiency of related computation is a must.

A significant progress towards fully automated systems that perform the above mentioned tasks can be associated with the development of MPEG-7 standard, which defines various descriptors for multimedia data. Software implementations of MPEG-7 mechanisms are common and applied in practice. At the same time, hardware-based implementations are rather rare. It should be also noticed, that some of the most promising approaches aimed at hardware-based multimedia processing, are multi-core system-on-chip (MPSoC), and network-on-chip (NoC).

In the paper, we present some results of an implementation of MPSoC that makes it possible to improve the effectiveness of multimedia data classification by increasing the number of parameters for the classification with potential diminishing the computation time (due to the parallelization process).

J. Sobecki, V. Boonjing, and S. Chittayasothorn (eds.), *Advanced Approaches to Intelligent Information and Database Systems*, Studies in Computational Intelligence 551, DOI: 10.1007/978-3-319-05503-9_32, © Springer International Publishing Switzerland 2014

Usually, the high accuracy of recognition (retrieval) is associated with a significant increase of system response time. The specificity of large collections of multimedia data makes it highly probable that the use of just a single descriptor increases the probability of unexpected results (e.g. [1,2]). Hence, it is profitable to combine different descriptors and classifiers to enhance the effectiveness of the whole process. In the case of software implementation of such algorithms, there are problems arising from the sequential nature of processing, in particular the time-consuming process of creating the feature space used in the classification process. At the same time, in order to implement these algorithms in hardware structures, it is necessary to extract data dependencies between various stages of processing and to determine memory access schedule.

The aim of the work described in this paper is to develop a hardware implementation of selected algorithms used in the content-based multimedia retrieval. Here we focus on selected descriptors related to color and edge representations of still images and audio characteristics that can be extracted from video sequences. Further we provide some results of implementation of three classifiers, namely Naive Bayes (NBC), k-Nearest Neighbours (kNN) and Gaussian Mixture Models (GMM) used at the data retrieval stage. The general scheme of processing in this scenario is presented in Fig.1.

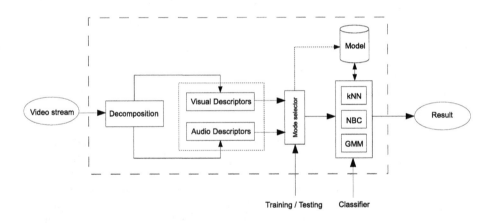

Fig. 1. Scheme of processing using audio-visual descriptors

2   Related Works

The majority of the works related to image analysis are focused on the visual descriptors connected with the shape and color due to relatively high performance. Also, the specificity of human visual system (HVS) construction increases the importance of texture-based descriptors.

In the case of audio parametrization, the descriptors should capture the time-frequency structure of acoustical scenes to describe sound sources and their properties. Due to high variability of audio signals, a feature set is dependent on the particular acoustical class. Therefore, in practice, the features based on filter banks [11] are often used in typical tasks like speech recognition, speaker identification and verification, etc.

In the scientific literature, the procedure of hardware implementation of multimedia data recognition algorithms is divided into two sub-problems: feature extraction and classification. There are some works related to the hardware implementation of selected MPEG-7 descriptors, e.g. [3] and [4]. Another trend concerns the implementation of classifiers. According to [5] and [6], their hardware realizations require relatively more computational resources in comparison to the previously mentioned sub-tasks.

To the best of the authors' knowledge, publications describing the implementation of both stages of processing (extraction and classification) are difficult to be found, what is caused by a relatively high computational complexity of both these stages. Moreover, most of the presented approaches deal with a traditional FPGA programming at a behavioral level, which is time consuming and makes Content-Based Multimedia Retrieval (CBMR) difficult to implement in hardware efficiently (see [6]). Therefore, we use C-based hardware description languages to describe computational cores dedicated to FPGA structures, connected with a mesh Network on Chip.

Despite the increasing popularity of GPU (Graphics Processing Unit) usage in various multimedia applications, FPGA is still worth considering in this domain. The authors of [7] have shown that FPGA has similar or better performance in comparison with GPU, but it is characterized with higher energy efficiency. According to the results in [8], FPGA provided 11 times faster performance than GPU in majority of testing scenarios, offering the best energy efficiency (20 watt FPGA versus 144.5 watt GPU).

In order to compute various descriptors concurrently, it is possible to implement them in separate hardware cores of a multi-core System on Chip (MPSoC). Since a number of these cores can be significant, they can be connected with network, so called Network on Chip (NoC). This connection approach is known for offering high throughput and favorable communication, as well as effective flow control mechanisms in order to avoid problems in access to resources [9]. In a typical NoC each core is equipped with a router, which is typically connected to the adjacent nodes.

In our work, we use a typical mesh-based architecture with wormhole switching, where each packet is divided into smaller portions of data called flits (flow control units). These flits are transmitted from a source to a destination node through a number of routers implementing a simple request-acknowledgement-based asynchronous communication. In the described system we also follow this technique, as described, e.g., in [10].

3   Adopted Algorithms

For the implementation we selected several algorithms of computing visual and audio descriptors. Visual descriptors include modified Dominant Color Descriptor (DCD) and Edge Histogram Descriptor (EHD). Both of them are parts of MPEG-7 Standard. They give a possibility to compare images regarding their geometrical transformations (scaling and rotation), as well as they are invariant to some extent to cropping and noising. The set of audio descriptors contains features computed in time and frequency domains. The first type include chroma, autocorrelation (AC), linear prediction coefficients (LPC) and linear prediction cepstral coefficients (LPCC) [11]. Descriptors calculated in frequency domain include mel-frequency cepstral coefficients (MFCC) and a group of typical features (FD) used in audio information retrieval systems (spectral slope, centroid, harmonicity, flatness, spectral envelope properties, etc.) [12].

The total length of feature vector exploited in our experiments is equal to 246 elements (where 152 elements are dedicated to still images classification and 94 for audio classification).

As the classification engine we use one of three typical classifiers: Naive Bayes, k-Nearest Neighbor and Gaussian Mixture Models. They were adopted to use fixed-point arithmetic.

Since all above algorithms use several stages that are independent (i.e., color quantization, convolution, color/edge histogram calculation, MFCC descriptor and linear prediction), they can be parallelized in order to create highly effective software or hardware implementation [13]. The following section describes some details of the modifications applied.

The general structure of the stages employed in the presented design is presented in Fig. 2. Input image dimensions are $M \times N \times 3$ elements (in RGB colorspace), while after grayscale conversion - $M \times N$. The length of audio frame is denoted with H. The numbers in braces represent the dimensionality of feature vectors.

4   Hardware Implementation

The original algorithms described in the previous section has been realized in Matlab (at the stage of modelling) and C language (at the stage of prototyping). Because of the limitations described above, several important programming rules has to be followed. For example, we have not used any kind of dynamic memory allocation and two-dimensional tables. Moreover, all procedures are inserted in the main code. We have chosen one of the existing C-based hardware description languages to implement the hardware-targeted counterpart of the source code, namely ImpulseC [14]. It is an extension of ANSI C with new data types, aimed at hardware synthesis, and new functions and directives for steering the hardware implementation.

The computations are divided into so-called processes, which communicate with each other using conventional means, namely streams, signals, shared memory and semaphores. The process communication using streams is considered as a

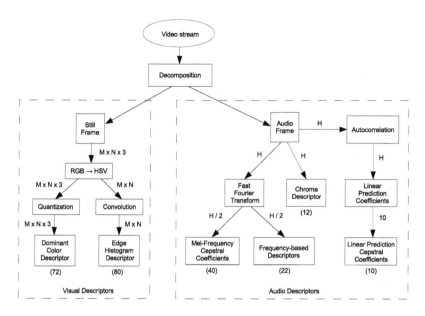

Fig. 2. Scheme of multimedia data parametrization

primary technique, and it is particularly suitable for multi-core SoCs, where each SoC core can be viewed as a single process. As the processes are to be realized in hardware, they may benefit from various ImpulseC optimization techniques, such as loop unrolling or pipelining. The first of these techniques are quite important in our system, as numerous computations in data-dominated algorithms are independent of each other and can be computed in parallel.

The only task of the designer is then to find loops whose iterations are data-independent, and put an appropriate ImpulseC directive to compute them concurrently. We should also remember, that there is usually not enough resources for generating a hardware for each iteration, thus we are forced to perform a certain clustering. This operation is a trade-of between the computation time and a target chip. In our case, this procedure has been realized during a series of experiments.

One of the most important problems with hardware-based computations is a lack of floating-point capabilities. Hardware implementations of mathematical coprocessor cores occupy quite large portion of target devices, which is particularly problematic in case of FPGA chips, which were intended to be used in the project. Thus we decided to perform a conversion between floating-point and fixed-point code. However, instead of internal ImpulseC fixed-point data types and mathematical operations we used SystemC standard, another C-based hardware description language, since it offers more sophisticated fixed-point support.

We conducted several experiments aimed at error-minimization using a simulation on typical benchmark video coming from CCV dataset [15] and audio recordings from 2002 RTBN database [16]. Then, for these input data, we determined the mean square error between floating and fixed point computations and thus determined the most favorable position of the radix point. After obtaining these parameters, we transferred it into our ImpulseC implementation.

Since the transfers between the NoC nodes are also an important factors influencing the computations in hardware structures, we implemented four various optimization techniques in order to minimize them. In all these cases we allowed the user to choose one out of three minimization criteria: the total transfer and the standard deviation in the whole NoC, and maximal flow between two neighboring nodes. Besides the minimization criteria, it is possible to select one of the following minimization algorithms: random change of two various nodes (parameter: number of iterations), simulated annealing (SA, parameters: max. number of iterations, initial temperature, temperature change), genetic algorithm (GA, parameters: max. number of epochs, population size, mutation probability), and hill climbing (HC). For our case of the 4×4 mesh all of these techniques are quite fast. About 10000 random changes are evaluated in few seconds, but the global minimum is rather difficult to be found this way. SA has found the transfer close to the global minimum in about 30 s., whereas GA needs a few minutes using a typical PC computer. HC is capable of finding a local minimum instantly. In our case, by an appropriate node mapping we were capable of decreasing the total transfer from a random seed (81308 bytes per second) to 64288 bytes per second. The final mapping result is depicted in Fig. 3.

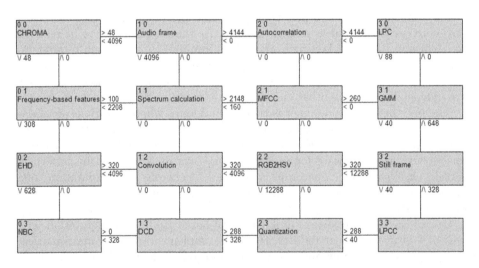

Fig. 3. Multimedia parametrization modules after mapping to the mesh NoC structure

Thanks to some code modifications, aimed at improving its hardware realization, we obtained our final code to be implemented in hardware. These modifications included division of the code into coarse-grain partitions to be implemented in parallel. The transformation is always a trade-off between computational time and resource utilization, hence we analyzed the impact of each module onto the final realization in terms of particular functional blocks, such as adders, multipliers etc., and number of computational stages.

In ImpulseC CoDeveloper package, there exists Stage Master Explorer tool to estimate impact of these modifications into the target chip parameters. This tool is capable of computing two parameters: Rate and Max Unit Delay (MUD), which can be treated as approximations of future hardware performance obtained significantly faster than long-lasting hardware implementation. Taking advantage of these parameters, it is possible to perform a more sophisticated design exploration, analyzing various modifications of the code and their impact on the target hardware. Since our goal is to increase the performance, we use maximal level of possible coarse-grain parallelism by generating multiple instances of each suitable module.

In order to verify our assumptions, we used Xilinx ISE to perform an implementation of the core in Virtex5 FPGA device (XC5VSX50T, Virtex 5 ML506 Evaluation Platform). We got the following device utilization (see Tab. 1).

Table 1. Device usage for individual modules implementation: visual descriptors (a), audio descriptors (b) and classifiers (c)

(a)

Resource	Total	Dom. Color Descriptor	Edge Hist. Descriptor
LUTs	6578	5090	1488
FFs	1777	1104	673

(b)

Resource	Total	FFT	CHROMA	MFCC	FD	AC	LPC	LPCC
LUTs	19391	1381	1899	6089	8860	278	421	463
FFs	7955	645	780	2245	3617	118	281	269

(c)

Resource	Total	NBC	kNN	GMM
LUTs	20266	1519	3354	15393
FFs	4730	324	789	3617

The above assignments mean that about 55 percent of the device has been used, leaving more than 45% for the router and the remaining cores. Thus, in order to implement more than two cores in a single chip, an FPGA chip of larger capacity may be necessary. The more detailed resource requirements and the elementary operators used for the implementation of the presented algorithms are given in Tab. 2. As it can be seen, after splitting the system into three parts: still image descriptors extraction, audio features extraction and classifiers, we have obtained the following hardware resources usage: 8%, 23 % and 24%, respectively.

The simulation performed on benchmark data gave rather encouraging results. The speed-up of the hardware-based computations (FPGA) reached about 155% of software-based approach realized on the PowerPC core embedded in the Evaluation Platform. The maximal frequency of the FPGA chip was less than 250 MHz which is significantly lower than current general-purpose processors. Hence, the electrical power demand of such implementation is far lower than any other

332 P. Forczmański, P. Dziurzański, and T. Mąka

Table 2. Resources and stages requirement for particular modules: visual descriptors (a), audio descriptors (b) and classifiers (c)

(a)

Resource	Total	RGB2HSV	Qunatization	Convolution	DCD Hist	EHD Hist
7-bit adder	2		2			
8-bit adder	33			12		21
10-bit adder	2				2	
14-bit adder	144	2		132		10
32-bit adder	492	21	7	428	14	22
16-bit multiplier	2		2			
32-bit multiplier	15	10				5
32-bit divider	24	10	1	1	1	11
2-bit comparator	6	2	2		2	
32-bit comparator	121	26	32	22	19	22
No. Stages	841	71	46	595	38	91

(b)

Resource	Total	FFT	CHROMA	MFCC	FD	AC	LPC	LPCC
4-bit adder	1		1		ps.			
9-bit adder	8				8			
10-bit adder	53	8	3	24		1	16	1
32-bit adder	223	64	17	23	88	4	17	10
32-bit multiplier	72	14	9	10	32	1	5	1
64-bit multiplier	6			2	3			1
32-bit divider	31	1	4	4	19		2	1
2-bit comparator	7	1	1	1	1	1	1	1
32-bit comparator	119	32	7	22	47	2	6	3
No. Stages	520	120	42	86	198	9	47	18

(c)

Resource	Total	NBC	kNN	GMM
1-bit adder	4	2		2
3-bit adder	5			5
4-bit adder	25	7	14	4
5-bit adder	6			6
6-bit adder	5	2	3	
32-bit adder	77	12	28	37
32-bit multiplier	17	1	6	10
32-bit divider	11	1	6	4
2-bit comparator	1			1
32-bit comparator	55	10	14	31
No. Stages	179	40	54	85

general-purpose CPU. Since we could not find any similar design using NoC presented in the literature, we can not compare it with our results.

5 Summary

In the paper we presented the problem of hardware implementation of selected algorithms used in Content-Based Multimedia Retrieval systems. The analysed modules consist of a number of low-level descriptors for still images, descriptors for audio data and three classifiers. Since the majority of these descriptors are independent each other, it is possible to compute them concurrently and implement into separate cores. Moreover, loops in codes executed by almost every core code can be parallelized at the fine-grain level. Despite these modification increasing the size of the generated hardware, we still managed to fit the analysed algorithms in a rather small-sized FPGA device, leaving almost half of the chip area to necessary NoC routers and computational cores. Increasing the number of sequential computation, we could even manage to fit a slightly larger number of cores at the expense of lower processing speed. It should be stressed that despite growing popularity of GPU usage in various multimedia applications, it is still reasonable to use FPGA architecture, since it offers better energy efficiency. Moreover, it allows us to implement quite sophisticated multi-core system in a flexible way close to full-custom hardware chips.

Acknowledgement. This work was supported by National Science Centre (NCN) within the research project N N516 475540.

References

1. Forczmański, P., Kukharev, G.: Comparative analysis of simple facial features extractors. Journal of Real Time Image Processing 1(4), 239–255 (2007)
2. Forczmański, P., Frejlichowski, D.: Robust Stamps Detection and Classification by Means of General Shape Analysis. In: Bolc, L., Tadeusiewicz, R., Chmielewski, L.J., Wojciechowski, K. (eds.) ICCVG 2010, Part I. LNCS, vol. 6374, pp. 360–367. Springer, Heidelberg (2010)
3. Sniatala, P., Kapela, R., Rudnicki, R., Rybarczyk, A.: Efficient hardware architectures of selected mpeg-7 color descriptors. In: 15th European Signal Processing Conference EUSIPCO 2007, Poznan, pp. 1672–1675 (2007)
4. Xing, B., Fu, P., Sun, Z., Liu, Y., Zhao, J., Chen, M., Li, X.: Hardware Design for Mpeg-7 Compact Color Descriptor Based on Sub-Block. In: 8th International Conference on Signal Processing, Beijing, pp. 16–20 (2006)
5. Skarpathiotis, C., Dimond, K.R.: A Hardware Implementation of a Content Based Image Retrieval Algorithm. In: Becker, J., Platzner, M., Vernalde, S. (eds.) FPL 2004. LNCS, vol. 3203, pp. 1165–1167. Springer, Heidelberg (2004)
6. Noumsi, A., Derrien, S., Quinton, P.: Acceleration of a content-based image retrieval application on RDISK cluster. In: 20th International Parallel and Distributed Processing Symposium, IPDPS, pp. 25–29 (2006)
7. Kestur, S., Davis, J.D., Williams, O.: BLAS Comparison on FPGA, CPU and GPU. In: IEEE Annual Symposium on VLSI (ISVLSI 2010), pp. 288–293 (2010)
8. Fowers, J., Brown, G., Cooke, P., Stitt, G.: A Performance and Energy Comparison of FPGAs, GPUs, and Multicores for Sliding-Window Applications. In: 20th ACM/SIGDA International Symposium on Field-Programmable Gate Arrays (FPGA 2012), pp. 47–56 (2012)
9. Benini, L., de Micheli, G.: Networks on Chips: A New SoC Paradigm. IEEE Computer 35(1), 569–571 (2002)
10. Chojnacki, B., Maka, T., Dziurzanski, P.: Virtual Path Implementation of Multistream Routing in Networks on Chip. In: Malyshkin, V. (ed.) PaCT 2011. LNCS, vol. 6873, pp. 431–436. Springer, Heidelberg (2011)
11. Rabiner, L., Schafer, W.: Theory and Applications of Digital Speech Processing. Prentice-Hall (2010)
12. Mitrovic, D., Zeppelzauer, M., Breiteneder, C.: Features for Content-Based Audio Retrieval. Advances in Computers 78, 71–150 (2010)
13. Maka, T., Dziurzański, P.: Parallel Audio Features Extraction for Sound Indexing and Retrieval Systems. In: 55th International Symposium ELMAR 2013, Zadar, Croatia, September 25-27, pp. 185–189 (2013)
14. Kalogeridou, G., Voros, N.S., Masselos, K.: System Level Design of Complex Hardware Applications Using ImpulseC. In: Proceedings of the 2010 IEEE Annual Symposium on VLSI, ISVLSI 2010, pp. 473–474 (2010)

15. Jiang, Y., Ye, G., Chang, S.-F., Ellis, D., Loui, A.C.: Consumer Video Understanding: A Benchmark Database and An Evaluation of Human and Machine Performance. In: ACM International Conference on Multimedia Retrieval (ICMR), Trento, Italy, pp. 29:1–29:8 (2011)
16. Garofolo, J., Fiscus, J., Le, A.: 2002 Rich Transcription Broadcast News and Conversational Telephone Speech. Linguistic Data Consortium, Philadelphia (2004)

Texture Analysis of Carotid Plaque Ultrasound Images

Krishnaswamy Sumathi[1] and Mahesh Veezhinathan[2]

[1] Department of ECE,
Sri Sairam Engineering College, Chennai, India
ksumathi_0409@yahoo.co.in
[2] Department of BME,
SSN College of Engineering, Chennai, India
maheshv@ssn.edu.in

Abstract. In this work, analysis of carotid plaque ultrasound images have been attempted using statistical method based on Gray Level Co-occurrence matrix (GLCM). The ultrasound imaging of the carotid arteries is a common study performed for diagnosis of carotid artery disease. The first order linear scaling filter was used to enhance the image quality for analysis. Second order statistical texture analysis is performed on the acquired images using GLCM method and a set of 12 features are derived. Principal Component Analysis (PCA) is employed to reduce features used for classifying normal and abnormal images to four which had maximum magnitude in the first principal component. It appears that, during diagnosis this method of texture analysis could be useful to develop an automated system for characterisation and classification of carotid ultrasound images.

Keywords: Atherosclerosis, Cardio vascular diseases, Carotid Plaque, Principal component analysis, Stenosis.

1 Introduction

The world health organisation ranks Cardio Vascular Diseases (CVD) as the third leading cause of death and adult disability in the industrial world [1]. Adults over the age of 65 or older have one or more types of CVD. It is estimated that there could be 20 million deaths due to atherosclerosis that will be associated with coronary heart disease and stroke. Carotid artery disease is a condition where atherosclerosis develops within one or both of the carotid arteries of the body. The carotid arteries are vessels that bring blood which includes oxygen and other nutrients to the brain. When plaque develops on inner wall of these vessels they become narrow and blood flow to brain reduces leading to stroke. Atherosclerosis can affect any of the arteries in the body. The plaque deposit in the coronary (heart) arteries causes heart attack. Asymptomatic plaques are truly asymptomatic as they have never been associated with symptoms in the past [2].

J. Sobecki, V. Boonjing, and S. Chittayasothorn (eds.), *Advanced Approaches to Intelligent Information and Database Systems*, Studies in Computational Intelligence 551,
DOI: 10.1007/978-3-319-05503-9_33, © Springer International Publishing Switzerland 2014

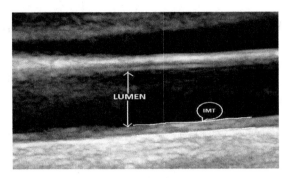

Fig. 1. The anatomical locations of ultrasound image components IMT- Intima-Media Thickness

Ultrasound imaging of the carotid artery is widely used in the assessment of carotid wall status. The anatomical locations of ultrasound image components for atherosclerosis are shown in figure 1. Clinically, Intima Media Thickness (IMT), Media Layer (ML), Intima Layer (IL) is the measure for assessment of atherosclerosis. Loizou *et al* [2] estimated the values for Intima Media Thickness (IMT=0.79mm), Media Layer Thickness (MLT=0.32mm) and Intima Layer Thickness (ILT=0.47mm) for normal subjects. The thickness of Intima layer increases with age from a single cell layer at birth to 250μm at the age of 40 for normal individuals. The results of the Asymptomatic Carotid Atherosclerosis Study (ACAS) have provided the first scientific evidence that in patients with asymptomatic carotid stenosis greater than 60% carotid endarterectomy reduces the risk of stroke from 2% to 1% per year [3]. When the degree of internal artery stenosis is 70% to 90%, the patient must undergo carotid endarterectomy to restore blood flow and reduce the risk of stroke [4, 5]. Further endovascular angioplasty treatment of atherosclerotic carotid stenosis with stenting is an alternative to surgical endarterectomy [5]. It is thus necessary to identify the patients at lower risk.

An attempt has been made to facilitate an early diagnosis of carotid plaque. The acquired ultrasound images were pre-processed for speckle reduction. One of the limitations cited in the literature is the presence of speckle noise which hinders the visual and automatic analysis in ultrasound images [2] and [15]. First order statistical filtering technique is applied to the ultrasound image [11]. The spatial interrelationships and arrangement of the basic elements in an image is referred as texture. The intensity patterns or gray tones are visualised as these spatial interrelationships and arrangements of the image pixels. The Texture provides useful information for characterizing atherosclerotic plaque [14, 15]. Gray Scale Median (GSM) analysis is mainly performed on the plaque and was found to be related with its historical features. Statistical Features (SF): mean, standard deviation, energy, kurtosis and skewness were computed. Gray level co-occurrence matrixes (GLCM): The texture features autocorrelation, contrast, energy, dissimilarities, difference variance, entropy, energy, cluster variance, homogeneity, maximum probability, sum entropy, sum of squares variance and sum variance were extracted from Region of Interest (ROI). Most significant GLCM features were extracted using Principle Component Analysis (PCA). Information measure of correlation, sum entropy, sum of square variance and maximum entropy were four significant texture features used for classification of normal and abnormal images.

2   Methodology

A total of 20 B-mode longitudinal ultrasound images of carotid artery were investigated in this study. The ultrasound images obtained from model GE Logic 7 scanner with multi frequency probe at frequency 5-10 MHz with an axial system multi-frequency probe 16pixel/mm are considered for the analysis of GLCM based statistical texture features .

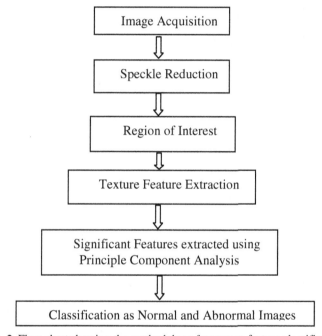

Fig. 2. Flow chart showing the methodology for texture feature classification

Brightness adjustments of ultrasound images were carried out as in the method introduced by C.P.Loizou *et al.* [6, 7] and [10]. Normalization of image improves image compatibility. It has been shown that normalizing carotid artery ultrasound images, improves image quality enabling a better classification [9].

Speckle is a form of multiplicative noise caused when the surface imaged appears rough to the scale of wavelength used. Speckle noise has a significant impact on the boundaries of the arteries. Despeckling is an important operation in the enhancement of ultrasound imaging of the carotid arteries. Filters based on local statistics [6, 7, 8, 11] were successfully used for the pre-processing. The filter using the first order statistics such as variance and mean of the neighbourhood was described with the model as in [6]. This filter improved the class separation between the normal and abnormal images.

The region of interest is selected in the ultrasound carotid images and its texture features were extracted. The GLCM matrices are constructed at a distance of $d = 11$ and for direction of θ given as 0°,45°, 90° and 135°. These matrices are averaged to give a non-directional matrix from which textural features are computed [16]. The statistical texture features extracted from this matrix are autocorrelation, contrast, correlation, cluster prominence, cluster shade, dissimilarity, energy, entropy, maximum probability, sum of squares, sum average, sum and difference variance and difference entropy. Image-processing and pattern-recognition techniques are appropriate tools for automatic screening of stenosis. The automated identification plaque in such images is a challenging task. During the past decade, literature has shown the ability of texture analysis algorithms to extract diagnostically meaningful information from medical images. Texture or structural features represent the spatial distribution of gray values [12]. Medical image texture analysis of different modalities such as ultrasound, magnetic resonance imaging, computed tomography, and light microscopy are found to be useful for automated diagnosis [13]. Texture features have also been used to derive methods for numerical description, objective representation and subsequent classification of cellular protein localization patterns in ultrasound images. An attempt has been made to provide a simple technique for assisting clinical diagnosis.

The GLCM feature extracted are subjected to PCA which is mathematically defined as an orthogonal linear transformation such that the greatest variance of the data lie on the first coordinate called the first principal component, the second greatest variance on the second coordinate and rest of the variance lie on the consecutive coordinates. PCA generally rotates the data around their mean in order to align with the principal components. This moves as much of the variance as possible into the first few dimensions using an orthogonal transformation. The values in the remaining dimensions, therefore, tend to be small and may be dropped with minimal loss of information. The PCA transformation is given by:

$$X^T = XP \tag{1}$$

X represents the data matrix and P the Eigen vectors matrix of X. The first column of Y^T is made up of "scores" of the cases with respect to the "principal" component, the next column has the scores with respect to the "second principal" component and as well as others [17]and [23]. In this analysis the first principal component alone showed more than 45% of the variance and the feature corresponding to maximum magnitude in the first principal component vector are selected. Among these co-occurrence features, Information measure of correlation, sum entropy, sum of square variance and maximum probability are significant features extracted by PCA for this analysis. Sum entropy for abnormal image represents higher values than the normal mages. Entropy is a measure of randomness of intensity image and characterizes the texture non-uniformity. Complex textures tend to have high entropy. Entropy is obtained by summing over single GLCM elements [19-22].

3   Results and Discussion

Figure 2(a) and (b) shows the despeckled normal and abnormal image. Figure(c) shows the plaque in the abnormal image. Statistical features extracted from region of interest showed significant values and Gray Scale Median ($p=0.009$) between normal and abnormal images. The decrease in the GSM values suggests an increase in the hypoechoic structures due to deposition of calcium. Clinically, calcification is the marker of atheroscelerosis which increases the risk of stroke.

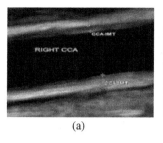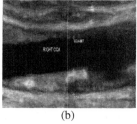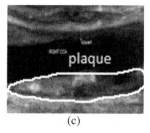

(a)   (b)   (c)

Fig. 3. Typical (a) normal ultrasound carotid image (b) abnormal ultrasound carotid image and (c) showing the plaque in the carotid arteries

Of the twelve features extracted from GLCM, four most significant features are selected using PCA. The selection of the parameters was based on the highest magnitude of Eigen values in the first principal component. The percentage variance of principal components analysis shows above 45% variance. It is observed from the results that the first four components show 90% of the total variance. The variation in Eigen values for the texture features is shown in the table 1. Information measure of correlation, sum entropy, sum of square variance and maximum probability are selected from PCA based on their magnitude of the Eigen values. The magnitude of four significant components is shown in table 2.

Table 1. Variation in Eigen vector component magnitudes with the loadings of principal components

S.No.	Texture features	Magnitude of Eigen values for normal images	Variance for normal images	Magnitude of Eigen values for abnormal images	Variance for abnormal images
1	Autocorrelation	0.05341	48.45307	0.00279	61.19352
2	Energy	0.02290	36.57739	0.03184	18.38136
3	Contrast	0.04203	7.33832	0.00929	13.28694
4	Entropy	0.20663	4.30490	0.04821	3.95198
5	Cluster Shade	0.10189	1.80263	0.01220	2.05858
6	Dissimilarity	0.359252	0.71866	0.09048	0.73653
7	Homogeneity	0.01684	0.64584	0.02464	0.26545
8	Maximum probability	0.43944	0.09216	0.23771	0.12320
9	Sum of squares: Variance	0.413385	0.06704	0.39252	0.00244
10	Sum variance	0.302913	0	0.37541	0
11	Sum entropy	0.331244	0	0.41655	0
12	Information measure of correlation	0.415119	0	0.63155	0

Table 2. Significant features from principle component analysis

Features contributing to PC1	Magnitude	
	Normal	Abnormal
Information measure of correlation	0.4151	0.6315
Sum Entropy	0.3312	0.4165
Sum of square variance	0.4133	0.3925
Maximum probability	0.4394	0.2237

Information measure of correlation is the most discriminatory feature with highest distance value (p=0.0002). In addition to the four significant features sum variance and dissimilarity also showed significant magnitude of Eigen values shown in table 1.

The information measure of correlation has the largest magnitude followed by sum entropy, sum of square variance, maximum probability for normal and abnormal images. The feature information measure of correlation and sum entropy has high variance and hence found to be most significant features. Hence these features have been considered as the distinguishing feature for classification. The magnitudes of most significant GLCM features in the first two principal components for normal and plaque images are shown in the figure 3 (a) and (b) respectively. It is observed that the angle between information measure of correlation and sum entropy is high. It is observed that the orientation and magnitude of information measure of correlation greatly differs between the normal and plaque images. The normalised average values are high for plaque images than normal images.

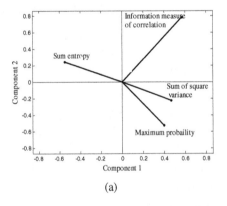
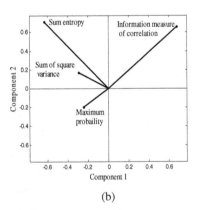

(a)   (b)

Fig. 4. Variation in component magnitudes of selected GLCM features with first four principal components for (a) normal and (b) abnormal images

The figure 4 shows the variations of the normalised average values of GLCM features reduced using PCA analysis for normal and abnormal images. Variations were found between normal and abnormal images for all the four features. The normalised average values are high for information measure of correlation and sum entropy than other features. Due to the presence of plaque there exist variations between normal and abnormal images. The gray level range in plaque images is more than normal images.

The normalised mean and standard deviation for Information measure of correlation function is 0.8776±0.241 for normal and 0.842±0.10 for abnormal images. Sum Entropy values of mean and standard deviation for normal images and abnormal images is 0.82±0.158 and 0.866±0.102.The presence of plaque in the arteries shows increase in the sum entropy feature. The mean and standard deviation for sum of square variance is 0.789±0.167 for normal and 0.683±0.22 for abnormal images. Maximum probability showed mean standard deviation of normal images 0.744±0.19 and 0.64±0.197 for abnormal images. The significant features show lower magnitude of average values for normal images. Figure shows plot comparable between normalized average values and principle components for normal and abnormal images.

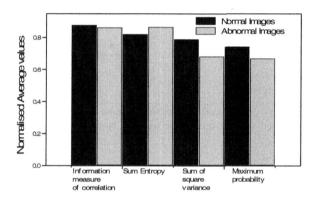

Fig. 5. Comparison of variation in the normalized average values of the GLCM features in normal and abnormal images

The GLCM is analysed using scattergram as shown in the figure 5. All texture features have been distinctly differentiated. The range for mean and standard deviation is a small overlap observed for the information measure of correlation feature. The overlap represents the non uniform morphology exhibited in both normal and abnormal images.

The variation between the normal and abnormal carotid images for information measure of correlation, sum entropy, sum of square variance and maximum probability features are statistically significant as shown in figure 4. The plaque present in the abnormal images results in local variations, and hence values of sum of square variance and maximum probability are more scattered as shown in the figure (c) & (d). The presence of this deviation is due to large variation in the gray level.

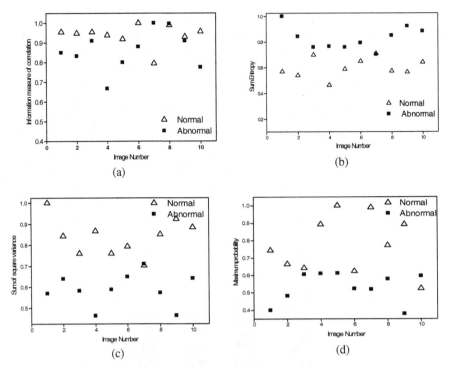

Fig. 6. Variations in (a) Normalized information measure of correlation (b) Normalized sum entropy (c) Normalized sum of square variance (d) Normalized maximum probability for different images

4   Conclusion

In this work an attempt has been made to characterise ultrasound carotid images using GLCM texture features as normal and abnormal images. GLCM texture measures computed for autocorrelation, contrast, energy, dissimilarities, difference variance, entropy, energy, cluster variance, homogeneity, maximum probability, sum entropy, sum of squares variance and sum variance. The extracted feature values are subjected to PCA. PCA is able to extract four significant features such as information measure of correlation, sum entropy, sum of square variance and maximum probability. It appears that statistical features based on GLCM analysis demonstrates that normalized average values of sum entropy feature are higher for abnormal images than normal images. It appears that statistical analysis based on GLCM could be used for classification in the future diagnosis of the disease. These values contribute towards the implementation of most effective strategy to minimise the risk of stroke, avoid unnecessary surgery and offering a better service to the patient. The texture analysis performed in this study reveals that the extraction of multiple features from high resolution ultrasound images of atherosclerotic carotid images are used for classification and identification of individuals with symptomatic and asymptomatic carotid stenosis at the risk of stroke.

Acknowledgement. The authors sincerely thank Dr.C.Emmanuel, Director, Academics and Research, Global Hospitals and Health city, Chennai and Dr.Ganesan Visvanathan, Head of department of Radiology and Imaging, Global Hospitals and Health city, Chennai for helpful discussion and support provided to carry out this research.

References

1. American Heart Association: Heart Disease and Stroke Statistics, update, Dallas, Texas (2007)
2. Loizou, C.P., Murray, V., Pattichis, M.S., Pantziaris, M., Pattichis, C.S.: Multiscale Amplitude-Modulation Frequency-Modulation (AM–FM)Texture Analysis of Ultrasound Images of the Intima and Media Layers of the Carotid Artery (2011)
3. Nicolaides, A., Sabetai, M., Kakkos, S.M.: The Asymptomatic Carotid Stenosis and Risk of Stroke (ACSRS) study. Aims and results of quality control. Int. Angiol. 22(3), 263–272 (2003)
4. Geroulakos, G., Domjan, J., Nicolaides, A.: Ultrasonic carotid artery plaque structure and the risk of cerebral infraction on computed tomography. J. Vasc. Surg. 20(2), 263–266 (1994)
5. Christodoulou, C.I., Pattichis, C.S., Kyriacou, E., Nicolaodes, A.: Image Retrieval and Classification of Carotid Plaque Ultrasound Images. The Open Cardiovascular Imaging Journal 2, 18–28 (2010)
6. Loizou, C.P.: Constantinos: Comparitive Evaluation of despecke filtering in ultrasound imaging of the carotid artery. IEEE Transaction on Ultrasonics, Ferroelectronics and Frequency Control 52(10) (October 2005)
7. Loizou, C.P., Pattichis, C.S., Pantiziaris, M., Tyllis, T., Nicolaides, A.: Snake based segmentation of the common carotid artery intima media. Med. Bio. Eng. Comput. 45, 35–49 (2007), doi:10.1007/s11517-006-0140-3
8. Lee, J.S.: Redefined filtering of image noise using local stastistics. Computer Graph Image Processing 15, 380–389 (1981)
9. Nicolaides, A.N., Kakkos, S.M., Griffin, M.: Effect of image normalization on carotid plaque classification and the risk of ipsilateral hemispheric ischemic events: results from the asymptomatic carotid stenosis and risk of stroke study. Vascular 13(4), 211–221 (2005)
10. Loizou, C.P., Pattichis, C.S., Nicolaides, A.N., Pantziaris, M.: Manual and automated media and Intima thickness measurements of the common carotid artery. IEEE Transactions on Ultr. Fer. Freq. Contr. 56(5), 983–994 (2009)
11. Christodoulou, C.I., Kyriacou, E., Pattichis, M.S., Pattichis, C.S., Nicolaides, A.: A Comparative Study of Morphological and other Texture Features for the Characterization of Atherosclerotic Carotid Plaques. In: Petkov, N., Westenberg, M.A. (eds.) CAIP 2003. LNCS, vol. 2756, pp. 503–511. Springer, Heidelberg (2003)
12. Kim, J.K., Park, H.W.: Statistical Texture features for Detection of Microcalcifications in Digitised Mammograms. IEEE Transactions on Medical Imaging 18(3) (March 1999)

13. Haralick, R.M., Shanmugam, K., Dinstein, I.: Texture features for image classification. IEEE Trans. Syst. Man Cybern. 3, 610–621 (1973)
14. Wagner, R.F., Smith, S.W., Sandrik, J.M., Lobes, H.: Statistics of speckle in ultrasound B scans. IEEE Trans. Sonics Ultrason. 30, 156–163 (1983)
15. Loizou, C.P., Pattichis, E.C.S., Pantziaris, E.M., Tyllis, T., Nicolaides, E.A.: Quality evaluation of ultrasound imaging in the carotid artery based on normalization and speckle reduction filtering. Med. Biol. Eng. Computation 44, 414–426 (2006)
16. Bian, N., Eramian, M.G., Pierson, R.A.: Evaluation of texture features for analysis of ovarian follicular development. Med. Image Comput. Comput. Assist. Interv. 9(pt. 2), 93–100 (2006)
17. Lopes, R., Betrouni, N.: Fractal and multifractal analysis: A review. Med. Image Anal. 13(4), 634–649 (2009)
18. Zeng, Y., Zhang, J., van Genderen, J.L., Zhang, Y.: Image fusion for land cover change detection. Int. J. Image Data Fusion 1(2), 193–215 (2010)
19. Ion, A.L.: Methods for knowledge discovery in images. Inf. Technol. Con. 38(1) (2009)
20. Ozdemir, I., Norton, D.A., Ozkan, U.Y., Mert, A., Senturk, O.: Estimation of Tree Size Diversity Using Object Oriented Texture Analysis and Aster Imagery. Sensors 8(8), 4709–4724 (2008)
21. Hassan, H.H., Goussev, S.: Texture analysis of high resolution aeromagnetic data to identify geological features in the Horn river basin, NE British Columbia, Recovery, CSPG CSEG CWLS Convention (2011)
22. Dong, K., Feng, Y., Jacobs, K.M., Lu, J.Q., Brock, R.S., Yang, L.V., Bertrand, F.E., Farwell, M.A., Hu, X.-H.: Label-free classification of cultured cells through diffraction imaging, Biomed. Optics Express 2(6), 1717–1726 (2011)
23. Priya, E., Srinivasan, S., Ramakrishnan, S.: Differentiation of digital TB images using texture analysis and RBF classifier. Biomedical Sciences Instrumentation 48, 516–523 (2012)

Visualization-Based Tracking System Using Mobile Device

Phuoc Vinh Tran, Trung Vinh Tran[*], and Hong Thi Nguyen

University of Information Technology,
Lintrung, Thuduc, Hochiminh City, Vietnam
`phuoc.gis@uit.edu.vn`,
`{hongnguyen1611,tvtran0825}@gmail.com`

Abstract. Services monitoring freight vehicle on route have been responding an increasing number of demands of goods owners. This chapter proposes a visualization-based tracking system using mobile device to enable users to keep track of the status of their freights ubiquitously in real time by viewing comprehensive graphs on mobile devices. The system comprises a moving device implemented on vehicle, a mobile device serving as owner's controller, data transmission line, and map service. Data of location and conditions of vehicle from moving device are transmitted to mobile device to display as graphs. Based on space-time cube, we developed a cube to visualize location and freight weight, where location changes over time and goods weight at locations and over time. A software of diverse display modes was implemented to convert collected data into visual graphs to facilitate users' comprehension on the location of vehicle and the status of their goods in real time.

Keywords: mobile device, visualization, space-time cube, tracking system, visualization-based tracking system.

1 Introduction

Innovations in information and communication technology (ICT) have resulted in sophisticated information systems referred to as telematics systems to support freight transportation since the mid-80s [1,2]. Telematics systems have been adopted to improve tracking and monitoring the vehicles and freights [3]. These systems provide freight owners with valuable information regarding locations and conditions of both the vehicles and freights in a real-time manner [2]. Such information helps the owners to effectively handle daily operation of distribution centers, freight forwarding, long-

---

[*] Phuoc Vinh Tran is Associate Professor in Informatics, Vice Rector of the University as well as Dean of the Department of Information Science and Engineering at the University of Information Technology (UIT) in Vietnam. His interests include real-time spatial information system, multivariate data visualization, real-time geo-visualization, GIS for administration, access control in government GISystems, real-time GIS for disaster and climate change.

J. Sobecki, V. Boonjing, and S. Chittayasothorn (eds.), *Advanced Approaches to Intelligent Information and Database Systems*, Studies in Computational Intelligence 551,
DOI: 10.1007/978-3-319-05503-9_34, © Springer International Publishing Switzerland 2014

haul freight transport, loading and unloading planning, and route scheduling [2]. The benefit of telematics systems is, therefore, the avoidance of transportation risks including accident, theft and change in quality of goods that are of high value, perishable, time-dependent, or hazardous [4,5].

A telematics system often composes of a vehicle system and a communication system. Data about the vehicles and freights are transmitted from the vehicle systems over a telecommunication network (e.g. a cellular network) to the communication systems where the data are processed by a computational application (e.g. a geographic information system -- GIS) [6,3]. Thanks to its capability of spatially and temporally integrating, managing, and analyzing a large amount of data collected from the vehicle systems, GIS for transportation (GIS-T) has been adopted and implemented in the communication systems for a variety of transportation purposes including administration, planning, design, and operation at both public and private organizations since 1960s [7,8]. Traditionally, these GIS-T applications use computational analysis approach for data analysis [6]. However, this approach may be ineffective in cases that require human knowledge for pattern recognition, imagination, association, and analytical reasoning due to the multitude of spatial data [9]. To go over this limitation, the geovisualization approach has been suggested. This geovisualization approach provides an interactive and dynamic visual environment, in which humans (i.e. freight owners) and computers can work in synergy to analyze data about the vehicles and freights in a real-time manner.

In this chapter, we proposed a visualization-based tracking system. Unlike conventional tracking systems, this proposed system used mobile devices (e.g. smartphones) for geovisualization applications. This system, therefore, not only respected knowledge of freight owners in exploring and analyzing vehicle and freight data but also helped them to keep track of their vehicle and freights ubiquitously in a real-time manner.

The rest of the paper is organized as follows. The next section presents the overall system architecture. In the third section, we briefly discuss our experiences when testing the system. A broader discussion is presented in the fourth session. This paper is ended in the fifth section where we not only summarize our study but also propose our future expansion for this system.

2   System Architecture

The overall architecture of the proposed visualization-based tracking system is presented in Fig. 1. Technically, the system was designed to have four components: a set of moving devices, a telecommunication network, a map service, and a set of mobile devices (Fig. 1). While the moving devices were implemented on vehicles, the mobile devices (e.g. smartphones) were used by freight owners. Data about the vehicle positions (i.e. latitude and longitude) and conditions (e.g. fuel and weight) generated by the moving devices were sent to the mobile devices through a telecommunication network (e.g. the General Packet Radio Service – GPRS – network). The position and condition data were collected by the GPS (Global

Positioning System) receivers and sensors. The frequency of these collection processes was determined by the timer. The map service provided a set of base maps (e.g. administrative boundaries and physiography) forming a spatio-temporal context for the geovisualization processes occurring at the mobile devices.

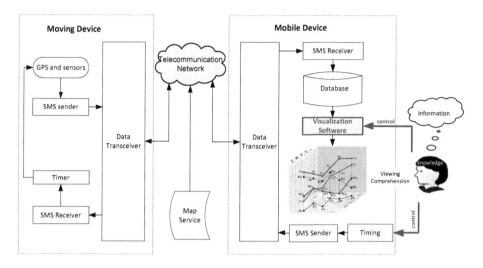

Fig. 1. Overall architecture of the proposed visualization-based tracking system using mobile device

2.1   Database

In this system, the relational database management system SQLite was used to manage the data. This most widely deployed SQL database engine supported the SQL-92 standard for query and also provided several libraries for Android development [10]. In this study, a table with seven fields was created as presented in Table 1. Data in this system included the vehicle positions and conditions observed at various time points.

Table 1. Attribute table of the database

Field	Data type	Description
ID_Device	Text	Device identification string
Name_Device	Text	Device name
Time	Text	Time of observation (YYYY-MM-DD-hh-mm-ss)
Longitude	Real	Longitude of a vehicle location
Latitude	Real	Latitude of a vehicle location
Fuel	Real	Percentage of vehicle fuel (0-100%) at a location
Weight	Real	Percentage of vehicle weight (0-100%) at a location

It should be noted that at a specific time point, the vehicle position was unique. However, as the time goes on, the vehicle location might change as the vehicle moved. Additionally, as far as the vehicle became moving, its conditions might change as well due to internal (e.g. vehicle engine condition) or external (e.g. weather condition) factors.

2.2   Visualization Software

2.2.1   Visualization Frameworks

The visualization software for the mobile devices in this study was developed to utilize three visualization frameworks: (1) the space-time cube [11], (2) the space-feature cube [12], and the simplified-space-time-feature cube [12]. The space-time cube (Fig. 2a) represented one or multiple space-time paths connecting multiple space-time points representing positions of a vehicle at various time points [13]. The base of the cube was used to represent locations whereas its height represented the time. Therefore, the function of the space-time cube was to visualize vehicle positions by time and vice versa.

However, the space-time cube did not support the visual exploration and analysis of vehicle conditions by time because the condition information was not presented in the cube. Thus, an alternative approach to the space-time cube could be to link the space-time cube with multiple documents (e.g. charts and/or tables) [13]. However, this approach was not selected for this proposed system because viewing multiple documents at the same time on a mobile device with small screen would be impossible. Rather, for the purpose of visualizing vehicle conditions, the space-feature cube (Fig. 2b) was used.

Similarly to the space-time cube, the base of the space-feature cube was used to represent locations. However, the height of the space-feature cube represented thematic attributes (e.g. vehicle weight) [12]. The issue with the space-feature cube was that it was not helpful to visualize durations of vehicle states (e.g. run or stop). For instance, according to Fig. 2b, the vehicle stopped at position P_1 and its weight decreased from W_a to W_b and then increased from W_b to W_c. Deciding whether this stop was a regular loading/unloading or an irregular stop due to a risk (e.g. accident or theft) would require information about the stop duration. To go over this issue, the simplified-space-time-feature cube was introduced.

The simplified-space-time-feature cube (Fig. 2c) was similar to the space-feature cube in which its height represented a thematic attribute. However, the Y-axis of the simplified-space-time-feature cube represented the time. Vehicle positions were shown on the X-axis and were simplified as representative points whose coordinates were not real as they were in the space-time cube. For instance, point P_0 at the origin of the X-axis in Fig. 2c was the departing location. P_1 was determined based on its network distance from point P_0. The simplified-space-time-feature cube helped to reveal the stopping time at P_1 of the vehicle in the above example (Fig. 2c).

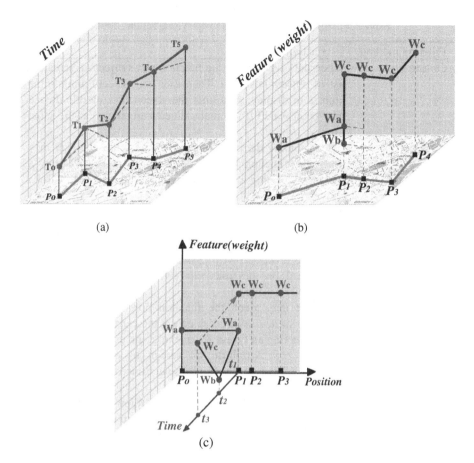

Fig. 2. Visualization frameworks. (a) The space-time cube, (b) the space-feature cube, (c) the simplified-space-time-feature cube. Please see text for detailed discussion.

Tran and Nguyen [12] suggested that the Time axis should have only been visualized when the vehicle was in stop state. For instance, the Time axis in Fig. 2(c) was visualized at P1 because the vehicle stopped at location P_1 between time t_1 and t_3. This flexibility in visualizing the Time axis ensured that the simplified-space-time-feature cube provided a simple view for the freight owners to better track of their vehicle/goods positions but at the same time ensured that the owners were able to identify significant changes in their vehicle/goods conditions when the vehicles stopped.

2.2.2 Software Development

The visualization software was developed on the Android platform using a variety of tools including Java Development Kit and Android Software Development Kit. Additionally, Google Maps Android API (Application Program Interface) and OpenGL API were also used to handle GIS data. In addition to the available functions of a mobile device such as zooming, moving, rotating, our software provided several filtering methods (Table 2).

Table 2. Filtering methods provided by the visualization software

Filtering method	User activity
Object	When the user touches the function box **Object** on screen, a list of objects is shown. Then user may select one or some objects to display. Note that the screen of a mobile device may be crowded if too many objects are shown at the same time.
Study area	After zooming the map to a convenient scale, the user may touch the function box **Study Area** on screen to draw a rectangle defining a study area within which the data will be shown.
Current area	The user may touch the function box **Current Area** on screen to see the selected objects.
Study time	After zooming the map to a convenient scale, the user may touch the function box of **Study Time** on screen and mark an interval on the time axis for which the data will be visualized on screen.
Current time	The user may touch the function box **Current Area** on screen to see the selected objects at the current time.
Attribute	After selecting an object for tracking, the user may touch the function box **Attribute** on screen to open the list of attributes. Then, the user may select one or more attributes to be visualized on screen.
View all	The user may touch the function box **View All** on screen to visualize the attributes of all objects. Note that the screen of a mobile device may be crowded in this case due to too many objects.

3   System Implementation

The proposed visualization-based tracking system described in the previous section was implemented for testing purpose. The goal of the test was to assess the performance of the system architecture as well as the visualization software running on popular smartphones. Two testing approaches were conducted. In the first testing approach, the system was installed on a truck running for a distance of about 300 km from Rạch Giá (Kiên Giang province, Vietnam) to Ho Chi Minh City (Vietnam). In this test, two smartphones (Acer E210 and HTC Desire HD) were used as moving devices. With Wi-Fi and GPS embedded, these smartphones were able to record and send their positions to the mobile device, which was also a smartphone (Sony Xperia Acro S) through the SMS/GPRS network. Due to the impossibility of installing sensors on the testing truck, data about vehicle conditions (e.g. fuel and weight) were taken from a simulation (Table 3).

Table 3. Data table of the testing trip

Time (t)	Longitude (x)	Latitude (y)	Fuel (b)	Weight (w)	Note
15:30:26	**106.68456**	**10.75634**	100.00	100	Vehicle in running mode
15:32:56	106.68349	10.75594	99.85	100	
15:34:27	**106.68150**	**10.75512**	**99.71**	**100**	Vehicle in stopped mode
15:47:57	**106.68150**	**10.75512**	**99.71**	70	Weight reduces
16:05:27	**106.68150**	**10.75512**	**99.71**	85	
16:45:56	**106.68150**	**10.75512**	**99.26**	**100**	Weight increases
16:47:32	106.67569	10.75397	99.11	100	Vehicle in running mode again
16:49:02	106.67483	10.75396	98.95	100	
16:51:35	106.67316	10.75346	98.80	100	
16:53:18	106.67095	10.75353	98.60	100	

Legend: Time(T),Weight(W),Fuel(F),Speed(S),Position(P)

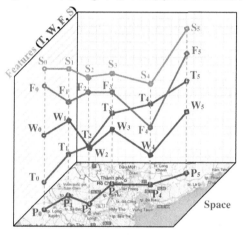

Fig. 3. The simplified-space-time-feature cube demonstrating the variations in positions and conditions of the testing vehicle

Data in Table 3 were saved in the SQLite database and were visualized by the Visualization software (Fig. 3 and 4). Unlike the first testing approach conducted for one vehicle, in the second testing approach, simulated data of more than ten vehicles were sent from the moving devices (i.e. Acer E210 and HTC Desire HD smartphones) to the mobile device (i.e. Sony Xperia Acro S smartphone) for processing. The tests demonstrated that all moving devices were working properly whereas the mobile device became overloaded when processing data from more than ten vehicles, especially when the number of data records for each vehicle was large (i.e. more than a hundred).

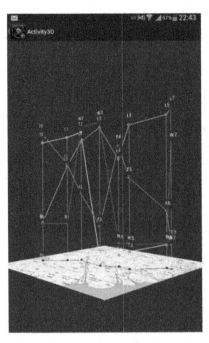

Fig. 4. The simplified-space-time-feature cube demonstrating the variations in positions and conditions of the testing vehicle on a smart phone

4   Discussions

The proposed visualization-based tracking system is an easy-to-use and effective tool for the freight owners to keep track of their vehicles and freights. The strength of the system relies on (1) the use of the geovisualization approach to explore and analyze vehicle data and (2) the use of mobile devices (e.g. smartphones) to run the visualization software. The visualization approach provides a visual, dynamic, and interactive environment for the freight owners to analyze vehicle data. This is especially important for owners (e.g. agrifood owners) who do not have enough time and knowledge to understand, input, and run computational analysis tools. In addition, these owners often do not have fixed office but rather work closely with farmers at their farms. Thus, the owners need mobile devices (rather than desktop computers) to perform such activities as managing goods loading/unloading or scheduling routes. The drawback of using mobile devices for visualization software is its capability of processing and visualizing massive data set. However, the owners (e.g. agrifood owners) at whom this system targets often do not need to track more than ten vehicles at the same time.

In this system, Google Maps is used as a map service to provide base maps for the tracking. The benefit of using Google Maps along with open-source libraries (e.g. Java Development Kit and Android Software Development Kit) is low-cost implementation. However, the use of Google Maps may result in static and low quantity as well as quality of the base maps. For instance, at the time of writing this paper (January 2014) the newest aerial photo of Rach Gia (Kien Giang province, Vietnam) is in 2007 whereas the photo of Ho Chi Minh City is in 2013. This temporal mismatch of the base data may affect the identification as well as the understanding of land cover/use impacts on the transportation risks (e.g. accident) perhaps arising during the transportation periods [14].

5   Conclusion

In this study, we proposed a visualization-based tracking system using mobile devices to help freight owners effectively keep track of their vehicles and freights as well as identify potential risks perhaps arising during transportations ubiquitously in a real-time manner. The system composed of a set of moving devices setting on the vehicles/freights, a telecommunication network, a map service, and a set of mobile devices used by freight owners. Vehicle data including positions and conditions were recorded at the moving devices and transmitted to the mobile devices through the telecommunication network. A visualization software was developed for the mobile device to visually track vehicles/freights. We introduced the use of a combination of three different visualization approaches (space-time cube, space-feature cube, and simplified-space-feature cube) to improve the system capability of visual exploration and analyses of the vehicle positions and conditions by time.

One of our next steps is to upgrade the visualization software to include visual analytics. In fact, it has been recommended that computational analysis methods (e.g. generalization) need to be used in combination with visualization when massive data sets need to be explored and analyzed [9,15,16]. In addition, we also want to expand the map service to use not only Google Maps but also other local services so that the base data are more recent and detailed.

References

1. Giannopoulos, G.A.: The application of information and communication technologies in transport. European Journal of Operational Research 152, 302–320 (2004)
2. Marentakis, C.A.: Telematics for Efficient Transportation and Distribution of Agrifood Products. In: Bourlakis, M., Vlachos, I., Zeimpekis, V. (eds.) Intelligent Agrifood Chains and Networks, pp. 87–108. Wiley-Blackwell (2011)
3. Santa, J., Zamora-Izquierdo, M.A., Jara, A.J., Gómez-Skarmeta, A.F.: Telematic platform for integral management of agricultural/perishable goods in terrestrial logistics. Computers and Electronics in Agriculture 80, 31–40 (2012), doi:http://dx.doi.org/10.1016/j.compag.2011.10.010
4. Siror, J.K., Sheng, H., Dong, W., Wu, J.: Application of RFID Technology to Curb Diversion of Transit Goods in Kenya. In: Fifth International Joint Conference on INC, IMS and IDC, NCM 2009, August 25-27, pp. 1532–1539 (2009), doi:10.1109/ncm.2009.84

5. Ruiz-Garcia, L., Barreiro, P., Rodríguez-Bermejo, J., Robla, J.I.: Review. Monitoring the intermodal, refrigerated transport of fruit using sensor networks. Spanish Journal of Agricultural Research 5(2), 142–156 (2007)

6. Goel, A.: Fleet Telematics: Real-time management and planning of commercial vehicle operations. Operations Research/Computer Science Interfaces, vol. 40. Springer US (2008)

7. Kim, T., Choi, K.: GIS for Transportation. In: Kresse, W., Danko, D.M. (eds.) Springer Handbook of Geographic Information, pp. 503–521. Springer, Heidelberg (2012), doi:10.1007/978-3-540-72680-7_26

8. Goodchild, M.: GIS and transportation: Status and challenges. GeoInformatica 4(2), 127–139 (2000), doi:10.1023/a:1009867905167

9. Andrienko, N., Andrienko, G.: Designing visual analytics methods for massive collections of movement data. Cartographica: The International Journal for Geographic Information and Geovisualization 42(2), 117–138 (2007), doi:10.3138/carto.42.2.117

10. SQLite, About SQLite (2013), http://www.sqlite.org/about.html (accessed March 1, 2013)

11. Hägerstraand, T.: What about people in regional science? Papers in Regional Science 24(1), 7–24 (1970), doi:10.1111/j.1435-5597.1970.tb01464.x

12. Tran, P.V., Nguyen, H.T.: Visualization cube for tracking moving object. Paper presented at the Computer Science and Information Technology. Information and Electronics Engineering, Bangkok, Thailand (2011)

13. Kraak, M.J.: The space-time cube revisited from a geovisualization perspective, Durban, South Africa, August 10-16 (2003)

14. Kim, K., Brunner, I.M., Yamashita, E.Y.: Influence of land use, population, employment, and economic activity on accidents. Transportation Research Record: Journal of the Transportation Research Board 1953, 56–64 (2006)

15. Kraak, M.J.: Geovisualization and Visual Analytics. Cartographica: The International Journal for Geographic Information and Geovisualization 42(2), 115–116 (2007), doi:10.3138/carto.42.2.115

16. Andrienko, N., Andrienko, G.: Spatial generalization and aggregation of massive movement data. IEEE Transactions on Visualization and Computer Graphics 17(2), 205–219 (2011)

The Improvement of Video Streaming Security in Communication with Multiple Modes Ciphering for Handheld Devices

Yi-Nan Lin and Kuo-Tsang Huang

Department of Electronic Engineering, Ming Chi University of Technology, Taiwan
Department of Electrical Engineering, Chang Gung University, Taiwan
jnlin@mail.mcut.edu.tw, d9221006@gmail.com

Abstract. Based on the popularity of handheld computing devices such as Smartphones and notepads, people take photos or videos for a main purpose from personal memories transforming to social communication. In digital communications, the confidentiality of personal information is protected by ciphering. The most straightforward video encryption method is to encrypt the entire stream using standard encryption methods such as DES or AES. However, in block ciphering, there exists an attack with the weakness of plaintext-ciphertext pairs (a plaintext-ciphertext pair problem). Therefore several modes of operation are proposed to solve such above problem. Most of these techniques can be only executed one selected operational mode during a communication session. We propose and implement the multiple modes ciphering to improve video streaming security in communication. Our android implementation shows that an overhead of multiple modes ciphering is just two percent decreased throughput competing with straightforward mode ciphering.

Keywords: Ciphering, Confidentiality, Multiple Modes.

1 Introduction

Due to the increasing popularity of social networking services such as Twitter, Flickr and YouTube, many researchers and industries have been focused on the convergence of mobile networked multimedia systems and social web technology. The convergence of multimedia communications and the constantly increasing capability of web technologies to deliver digital media such as video, images, and audio will allow for more media rich content to be delivered over the Web. With the proliferation of video data on the web and the widespread use of multimedia contents via mobile devices such as Smartphones and so on, there has been a tremendous growth in video data that needs to be considered with bandwidth, storage capabilities, and processor speed in networked multimedia applications during the last few years.

The growing requirements for high-speed, high level secure communications forces the system designers to propose the hardware implementation of cryptographic algorithms. A block cipher by itself allows encryption only of a single

J. Sobecki, V. Boonjing, and S. Chittayasothorn (eds.), *Advanced Approaches to Intelligent Information and Database Systems*, Studies in Computational Intelligence 551,
DOI: 10.1007/978-3-319-05503-9_35, © Springer International Publishing Switzerland 2014

data of the ciphers block length. Block encryption may be vulnerable to cipher-text searching, replay, insertion, and deletion; because it encrypts each block independently. Unfortunately, the non-feedback conventional block ciphers have plaintext-ciphertext pair problem with the disadvantage of limit block region scramble. A disadvantage of this method is that identical plaintext blocks are encrypted into identical ciphertext blocks. Non-feedback Conventional block ciphers do not hide data patterns well. A crack can use it to do known-plaintext attack. In some senses, it doesn't provide serious message confidentiality. A non-feedback conventional block cipher is not recommended for use in cryptographic protocols at all. A striking example of the degree to which non-feedback electronic codebook (ECB) mode can leave plaintext data patterns in the ciphertext can be seen when the electronic codebook mode is used to encrypt a bitmap image which uses large areas of uniform colour. The overall image may still be discerned as the pattern of identically-coloured pixels in the original remains in the encrypted version while the colour of each individual pixel is encrypted. Block cipher modes of encryption beside the electronic codebook mode have been suggested to remedy these drawbacks.

In the following images, a pixel-map version of the image on the left was encrypted with electronic codebook (ECB) mode to create the centre image, versus a non-ECB mode for the right image. The image on the right is how the image might appear encrypted with any of the other more secure modes indistinguishable from random noise. In Fig. 1, the random appearance of the image on the right does not ensure that the image has been securely encrypted; many kinds of insecure encryption have been developed which would produce output just as random-looking.

Original Encrypted   using ECB mode   Modes other than ECB result
                                                    in pseudo-randomness

Fig. 1. A striking example when different modes are used to encrypt a bitmap image [1]

Block cipher confidentiality modes of operation are the procedure of enabling the repeated use of a block cipher under a single key. Confidentiality modes of encryption beside Electronic codebook (ECB) have been suggested to remedy these drawbacks. However, one disadvantage still exists with these techniques when they are applied to practice encryption: any changing mode in encrypting session (between encrypting blocks) requires the re-configuration of crypto module and has no mechanism guaranteeing a smooth transition in such encrypting single session.

The improvement of video streaming security in this paper is based on an idea which is making multiple modes ciphering of mre secure communication. The rest of this paper is organized as follows. Section 2 contains acknowledge of ciphers and Video Encryption Algorithm (VEA). Section 3 describes how to make multiple modes ciphering with a secured mode sequence. Section 4 contains the practice of handheld devices implementations. Finally Section 5 is the conclusions.

2   Related Works

Data confidentiality is one of the security services in cryptography. The major concept in information security today is to continue to improve encryption algorithms. There are two major types of encryption algorithms for cryptography, symmetric-key algorithms and public-key algorithms. Symmetric-key algorithms also referred to as conventional encryption algorithms or single-key encryption algorithms are a class of algorithms that use the same cryptographic keys for both encryption of plaintext and decryption of ciphertext. It remains by far the most widely used of the two types of encryption algorithms.

Symmetric-key ciphers are a class of ciphers for cryptography that use trivially related cryptographic secret keys for both encryption of plaintext and decryption of ciphertext. Symmetric-key cryptography is to be contrasted with asymmetric-key cryptography, and symmetric-key cryptography was the only type of encryption in use prior to the development of asymmetric-key cryptography [2], [3]. The Advanced Encryption Standard (AES) [4] algorithm approved by NIST in December 2001 uses 128-bit blocks.

The standard modes of operation described in the literature [5], such as non-feedback electronic codebook (ECB) mode, cipher block chaining (CBC) mode, output feedback (OFB) mode, and cipher feedback (CFB) mode provide confidentiality. How to choose an appropriate operation mode? The different mode has each one's characters. For example, both of CFB and OFB can be design operating without padding with bit-based size keystream output; both of CBC and CFB can self sync to avoid channel noise error propagation; and both of CFB and OFB encryption and decryption applications need an encryption module only to reach both usages. In addition, only the forward cipher function of the block cipher algorithm is used in both encryption and decryption operations, without the need for the inverse cipher function.

Video data have special characteristics, such as its coding structure, large amount of data, and real-time constraints. A Video stream is quite different from traditional textual data because it has special data structure and it is compressed. To provide the security requirements of confidentiality, video security often uses software encryption to ensure the safety of computational complexity

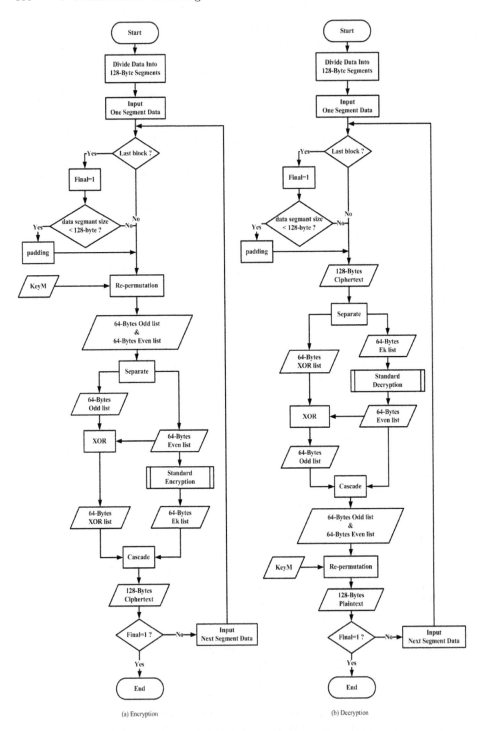

Fig. 2. Flow diagram of encryption and decryption for VEA

in the media. Notice that the common thing between compression and encryption is that both try to remove redundant information [6].

There exist several encryption algorithms such as Naive Algorithm, Selective Algorithm, and others, to secure video content. The naive method is to encrypt the entire video content using standard encryption methods such as DES, AES and so on. Each of them has its strength and weakness in terms of security level, speed, and resulting stream size metrics. For example of MPEG video content, the basic selective algorithm is based on the MPEG IPB frames structure. Encrypting only I-frames does not provide a satisfactory secure level, because great portions of the video are visible partly dependent on inter-frame correlation and mainly from unencrypted I-blocks in the P-frames and B-frames [7].

One ZigZag-Permutation Algorithm [8] where the encryption is an integral part of the MPEG compression process uses a random permutation list to map the individual 8x8 block to a 1x64 vector in zig-zag. ZigZag-Permutation Algorithm consists of three steps. Step 1, generate a permutation list with cardinality of 64. Step 2, do splitting procedure. Step 3, apply the random permutation list to the split block.

In 1998, [6] described, evaluated, and compared five representative MPEG encryption algorithms with respect to not only their encryption speed but also security level and stream size. It indicates that there are trade-offs among performance comparison metrics and the choice of encryption algorithm will depend on security requirements of the considered multimedia applications. Study of MPEG properties and its statistical behavior leads to the Video Encryption Algorithm (VEA). In Fig. 2, VEA is mainly interested in dealing with MPEG stream in byte-by-byte fashion for the following reasons: it is easier to handle data byte-wise; a single byte is meaningless in video stream because, normally, video content is encoded in several bytes (This is different from text information where one byte or a character has its own meaning.). Therefore, VEA in dividing the MPEG bit stream into a byte stream, each unit is an integer between 0 and 255.

VEA [6] in Fig. 2 creates a ciphering-keystream by function E to encrypt itself. Note that the seed of ciphering-keystream could also be a 256-bit random 0-1 sequence (128 0s and 128 1s). Since video file sizes are typically several Mbytes, we can afford to have such a long seed. It is easy to show that, if ciphering-keystream has no repeated pattern, then the secrecy depends on function E because ciphering-keystream is a one-time pad [2], [3] which is well-known to be perfectly secure.

3   Multiple Modes Ciphering

We describe the idea of using multiple confidentiality modes with a secured mode sequence to increase security level of video streaming in this Section. Traditional block ciphers have a plaintext-ciphertext pair problem with the disadvantage of limit block region scramble. In the literatures [5], [10], standard operational modes for data confidentiality are specified for using with any approved block cipher, such as AES algorithm [4]. The operational modes are Electronic Codebook (ECB), Cipher Block Chaining (CBC), Cipher Feedback (CFB), Output Feedback (OFB), and Counter (CTR). These standard operational modes are used to provide the data confidentiality function. An initialization vector (IV) or initial counter value must be required and given for CBC, CFB, OFB, and CTR modes performing the encryption operation in the beginning of any communication session and the decryption operation as well.

Multiple Modes Ciphering means that these cipher modes of operation could dynamically switch among them. In our previous research [11], we describe a problem and one hardware design solution for a real-time environment. In this paper , we focus on the implementation of Android APP programming in handheld devices practice. We design that the mode change depends on a control sequence such as an example in Table 1 and Fig. 3. The control sequence is synchronous between encryption and decryption parts and it is pre-shared as a secret (secure information). However, the additional cost of a target control sequence can be flexibly reduced according to the practical requirements of security.

Table 1. One example of a pre-shared mode sequence

Block #i	1	2	3	4	5	6	7	...
Block *mode*	CFB	CBC	OFB	CTR	ECB	CTR	CFB	...

Our idea can extend to contain and spread to other confidentiality modes which are not included in the standards [5], [10] yet. But we simplify our contribution with all those standard confidentiality modes here.

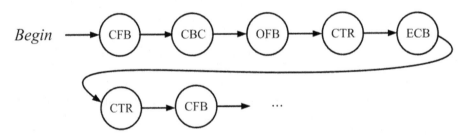

Fig. 3. The crypto task flow diagram of mode sequence in Table 1

The improvement of video streaming security is based on a secure secret. This secret, mode sequence, could be created by a synchronous pseudo-random number generator.

4   The Implementations of Multiple Modes Ciphering

We program a multiple modes AES-128 ciphering APP with eclipse JUNO 64-bit version IDE platform [12]. There are compare results with different handheld devices in Fig 4, according to three devices for running this Android APP. The processing performance and specifications of each handheld device for testing are described in Table 2. We use two multimedia files to be the testing samples in Fig. 5 and Fig. 6. The characters of our multimedia test files shows in Table 3(a) and Table 3(b).

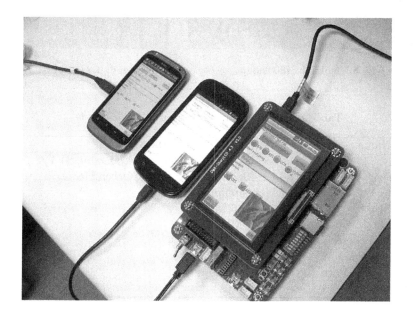

Fig. 4. Our handheld devices in simulations

Table 2. The specifications of our handheld devices

Specifications	DMA-6410L platform	HTC Desire S	Google Nexus S i9023
Processor	S3C6410XH-66	Snapdragon MSM8255	Exynos 3110
(instruction set)	ARMv6	ARMv7	ARMv7
(core)	RM11	Cortex-A8	Cortex-A8
(manufacture)	Samsung	Qualcomm	Samsung
(frequency)	667MHz	1GHz	1GHz
ROM	128MB	1.1GB	16GB
RAM	256MB	768MB	512MB
Android	1.6 upto 2.1	2.3	2.3 upto 4.1.2
Battery	none (DC 5V)	1450mAh	1500mAh

Fig. 5. elephants_1800k.3gp   **Fig. 6.** 4.2.04.tiff

Table 3. (a) The characters of our VIDEO sample

File Name	**elephants_1800k.3gp**
URL	*http://download.wavetlan.com/SVV/ Media/HTTP/H264/Other_Media/ elephants_1800k.3gp*
File Size	*60.6* MB
Length	*4* Min. *48* Sec.
Video Format	AVC-Baseline
Audio Format	AAC-LC
Overall Bit Rate	*1762* kbps
Frames/Sec.	*24* fps

Table 3. (b) The characters of of our IMAGE sample

File Name	**4.2.04.tiff**
URL	*http://en.wikipedia.org/wiki/File:Lenna.png*
File Size	*786,572* bytes
File Type	TIFF
Image Size	*512 x 512* pixels
ompression	Uncompressed
Photometric Interpretation	RGB
Bits Per Sample	*8, 8, 8*

Table 4. (a)Simulation results with multiple modes ciphering

Samples	DMA-6410L platform	HTC Desire S	Google Nexus S i9023
elephants_1800k.3gp	193.627 sec.	22.906 sec.	9.090 sec.
4.2.04.tiff	2514.206 ms.	297.118 ms.	117.932 ms.
Processing Time (Sec./MB)	**~3.199**	**~0.378**	**~0.149**

(b)Simulation results with straightforward, ECB mode, ciphering

Samples	DMA-6410L platform	HTC Desire S	Google Nexus S i9023
elephants_1800k.3gp	189.799 sec.	22.543 sec.	8.902 sec.
4.2.04.tiff	2463.275 ms.	293.932 ms.	115.939 ms.
Processing Time (Sec./MB)	**~3.132**	**~0.372**	**~0.147**

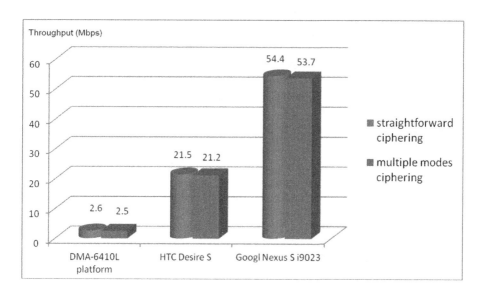

Fig. 7. The throughput performance with results in Table 4

We test one hundred times with each file and then calculate the average value of processing time results. The processing results of multiple modes ciphering are in Table 4(a)(b) and Fig. 7. It shows that the multiple modes ciphering throughput is between 2.5 to 54.4 Mbps on current general class mobile devices.

5   Conclusions

According to most of the content in mobile communication are products from social applications or multimedia applications. People share their own programs frequently, especially a large data such as video streaming. In this article, we propose an idea to improve video streaming security in communication with multiple modes ciphering for current handheld devices.

We also present the simulation of seamless changing confidentiality modes of ciphering operation and simplify to use standard modes for simulations. In software video encryption, our android implementation shows that an overhead of multiple modes ciphering is 1.3 to 2 percent and just a few decreased throughput competing with straightforward ciphering.

References

1. A striking example,
   http://en.wikipedia.org/wiki/Block_cipher_modes_of_operation
2. Simmons, G.J.: Contemporary cryptology: The science of information integrity. Wiley (1999)
3. Stallings, W.: Cryptography and network security: Principles and practices, 3rd edn. Pearson Education (2003)
4. Daemen, J., Rijmen, V.: The design of Rijndael: AES - the advanced encryption standard. Springer, Heidelberg (2002)
5. National Institute of Standards and Technology (NIST), NIST. Gov - Computer Security Division - Computer Security Resource Center, Recommendation of block cipher security methods and Techniques, NIST SP800-38 (2001)
6. Qiao, L., Nahrstedt, K.: Comparsion of MPEG encryption algorithms. Computers and Graphics 22(4) (1998)
7. Agi, I., Gong, L.: An empirical study of MPEG video transmissions. In: Proceedings of the Internet Society Symposium on Network and Distributed System Security, pp. 137–144 (1996)
8. Tang, L.: Methods for encrypting and decrypting MPEG video data efficiently. In: Proceedings of Fourth ACM Multimedia, vol. 96, pp. 219–230 (1996)
9. Wu, L., Weaver, C., Austin, T.: CryptoManiac: a fast flexible architecture for secure communication. In: Proceedings of ISCA 2001 (2001); SIGARCH Comput. Archit. News 29(2), 110–119 (2001)
10. International Organization for Standardization (ISO), Information Technology-Security Techniques-Modes of Operation for an n-bit Block Cipher, ISO/IEC 10116 (2006)
11. Huang, K.-T., Lin, Y.-N., Chiu, J.-H.: Real-time mode hopping of block cipher algorithms for mobile streaming. International Journal of Wireless & Mobile Networks 5(2), 127–142 (2013)
12. Android Developers, http://developer.android.com/

Author Index

Printed in the United States
By Bookmasters